AUTISM:
NEURAL BASIS AND TREATMENT POSSIBILITIES

The Novartis Foundation is an international scientific and educational charity (UK Registered Charity No. 313574). Known until September 1997 as the Ciba Foundation, it was established in 1947 by the CIBA company of Basle, which merged with Sandoz in 1996, to form Novartis. The Foundation operates independently in London under English trust law. It was formally opened on 22 June 1949.

The Foundation promotes the study and general knowledge of science and in particular encourages international co-operation in scientific research. To this end, it organizes internationally acclaimed meetings (typically eight symposia and allied open meetings and 15–20 discussion meetings each year) and publishes eight books per year featuring the presented papers and discussions from the symposia. Although primarily an operational rather than a grant-making foundation, it awards bursaries to young scientists to attend the symposia and afterwards work with one of the other participants.

The Foundation's headquarters at 41 Portland Place, London W1B 1BN, provide library facilities, open to graduates in science and allied disciplines. Media relations are fostered by regular press conferences and by articles prepared by the Foundation's Science Writer in Residence. The Foundation offers accommodation and meeting facilities to visiting scientists and their societies.

Information on all Foundation activities can be found at http://www.novartisfound.org.uk

Novartis Foundation Symposium 251

AUTISM:
NEURAL BASIS AND TREATMENT POSSIBILITIES

2003

Published in 2003 by John Wiley & Sons Ltd,
The Atrium, Southern Gate,
Chichester PO19 8SQ, UK

National 01243 779777
International (+44) 1243 779777
e-mail (for orders and customer service enquiries): cs-books@wiley.co.uk
Visit our Home Page on http://www.wileyeurope.com
or http://www.wiley.com

Other Wiley Editorial Offices

John Wiley & Sons Inc., 111 River Street, Hoboken, NJ 07030, USA

Jossey-Bass, 989 Market Street, San Francisco, CA 94103-1741, USA

Wiley-VCH Verlag GmbH, Boschstr. 12, D-69469 Weinheim, Germany

John Wiley & Sons Australia Ltd, 33 Park Road, Milton, Queensland 4064, Australia

John Wiley & Sons (Asia) Pte Ltd, 2 Clementi Loop #02-01, Jin Xing Distripark, Singapore 129809

John Wiley & Sons Canada Ltd, 22 Worcester Road, Etobicoke, Ontario, Canada M9W 1L1

Wiley also publishes its books in a variety of electronic formats. Some content that appears in print may not be available in electronic books.

Novartis Foundation Symposium 251
ix+310 pages, 15 figures, 12 tables

Library of Congress Cataloging-in-Publication Data

Autism : neural basis and treatment possibilities [editors, Gregory Bock and Jamie Goode].
p. cm. – (Novartis Foundation symposium ; 251)
Symposium on autism: neural basis and treatment possibilities, held at the Novartis Foundation, London, 18–20 June 2002.
Includes bibliographical references and indexes.
ISBN 0-470-85099-X (alk. paper)
1. Autism–Congresses. 2. Autism in children–Congresses. 3. Neurophysiology–Congresses. I. Bock, Gregory. II. Goode, Jamie. III. Series
RJ506.A9A9228 2003
618.92′8982–dc21 2003049714

British Library Cataloguing in Publication Data

A catalogue record for this book is available from the British Library

ISBN 0 470 85099 X

Typeset in $10\frac{1}{2}$ on $12\frac{1}{2}$ pt Garamond by Dobbie Typesetting Limited, Tavistock, Devon.
Printed and bound in Great Britain by Biddles Ltd, Guildford and King's Lynn.
This book is printed on acid-free paper responsibly manufactured from sustainable forestry, in which at least two trees are planted for each one used for paper production.

Contents

Symposium on Autism: neural basis and treatment possibilities, held at the Novartis Foundation, London, 18–20 June 2002

This symposium is based on a proposal made by Professor Sir Michael Rutter

Editors: Gregory Bock (Organizer) and Jamie Goode

Participants

David G. Amaral UC Davis Center for Neuroscience, 1544 Newton Court, Davis, CA 95616, USA

Anthony J. Bailey Department of Psychiatry, University of Oxford, Park Hospital, Old Road, Headington, Oxford OX3 7LQ, UK

Gillian Baird Newcomen Centre, UMDS, Guy's Hospital, St Thomas Street, London SE1 9RT, UK

Margaret L. Bauman Children's Neurology Service, Massachusetts General Hospital, 55 Fruit Street, Boston, MA 02114, USA

Dorothy Bishop Department of Experimental Psychology, University of Oxford, South Parks Road, Oxford OX1 3UD, UK

Patrick Bolton Department of Psychiatry, University of Cambridge, Level E4, Addenbrooke's Hospital, Hills Road, Cambridge CB2 2QQ, UK

Jan Buitelaar Department of Psychiatry and Academic Center for Child and Adolescent Psychiatry, University Medical Center St Radboud (Internal Post 333), P.O. Box 9101, 6500 HB Nijmegen, The Netherlands

Tony Charman Behavioural and Brain Sciences Unit, Institute of Child Health, 30 Guilford Street, London WC1N 1EH, UK

Geraldine Dawson UW Autism Center, Center on Human Development and Disability, Box 357920, Seattle, WA 98195, USA

Susan Folstein NEMC Department of Psychiatry, 750 Washington Street, Box 1007 Boston, MA 02111, USA

Eric Fombonne McGill University Division of Psychiatry, Montréal Children's Hospital, 4018 Ste Catherine Ouest, Montreal, Québec, Canada H3Z 1P2

Chris Frith Wellcome Department of Imaging Neuroscience, Institute of Neurology, 12 Queen Square, London WC1N 3BG, UK

Uta Frith Institute of Cognitive Neuroscience & Department of Psychology, University College London, Alexandra House, 17 Queen Square, London WC1N 3AR, UK

Francesca Happé Social, Genetic and Developmental Psychiatry Research Centre, Institute of Psychiatry, 111 Denmark Hill, London SE5 8AF, UK

Chris Hollis Division of Psychiatry, E Floor/South Block, Queen's Medical Centre, Nottingham NG7 2UH, UK

Patricia Howlin Department of Psychology, St George's Hospital Medical School, Cranmer Terrace, London SW17 0RE, UK

W. Ian Lipkin Center for Immunopathogenesis and Infectious Diseases, Mailman School of Public Health, 722 West 168th St, Columbia University, New York, NY 10032, USA

Catherine Lord University of Michigan, Autism and Communication Disorders Center (UMACC), 1111 Catherine Street, Ann Arbor, MI 48109, USA

Anthony Monaco The Wellcome Trust Centre for Human Genetics, Roosevelt Drive, University of Oxford, Oxford OX3 7BN, UK

Margaret Pericak-Vance Center for Human Genetics, Department of Medicine, Duke University Medical Center, Box 3445, Durham, NC 27710, USA

Nicole Rinehart Department of Psychological Medicine, Monash Medical Centre, 246 Clayton Road, Victoria 3168, Australia

Sally Rogers MIND Institute, University of California Davis, 4860 Y Street, Room 3020, Sacramento, CA 95817, USA

Sir Michael Rutter (*Chair*) Box 080 SGDP Research Centre, Institute of Psychiatry, De Crespigny Park, Denmark Hill, London SE5 8AF, UK

Robert Schultz Yale University Child Study Center, 230 South Frontage Road, New Haven, CT 06520, USA

Marian D. Sigman Department of Psychology, University of California Los Angeles, Franz Hall 2344 C, Box 951563, Los Angeles, CA 90095-1563, USA

David Skuse Behavioural and Brain Sciences Unit, Institute of Child Health, 30 Guilford Street, London WC1N 1EH, UK

Introduction: autism — the challenges ahead

Michael Rutter

Social, Genetic and Developmental Psychiatry Research Centre, Institute of Psychiatry, De Crespigny Park, Denmark Hill, London SE5 8AF, UK

There have been many important advances in research into the nature of autism and, as a result, our concepts of autism have undergone a radical change (Rutter 1999). At one time, the prevailing view was that autism was an unusually early variety of schizophrenia that had been caused, in large part, by so-called refrigerator parenting. It became clear that that was a wholly mistaken concept and that, instead, autism constitutes a neurodevelopmental disorder with a rather distinctive pattern of cognitive deficits, and that it is strongly genetically influenced.

Nevertheless, we are a long way from understanding the basic pathophysiology, and numerous puzzles and paradoxes remain. The aim of this symposium is to grapple with these issues, tackling the challenges from a range of different perspectives in the hope that a coming together of minds, and of different research strategies, may point the way ahead. My task is to set the scene by outlining some of these challenges in order to provoke us all to abandon the safety of our own research territory, and of the findings that are well established, in order to focus on the difficulties that are inherent in our favoured theories.

We need to begin with implications of the huge rise in diagnosed autism (Baird et al 2000, Chakrabarti & Fombonne 2001, Fombonne 1999). To a substantial extent, this rise is a consequence of a major broadening of the concept of autism together with better ascertainment. However, is that all? When like is compared with like, has there been a real rise in the rate of autism? If that should prove to be the case, what is the environmental factor that has brought this about (the rise is unlikely to have been genetically determined)? There have been claims that the rise is due to the use of the combined measles/mumps/rubella (MMR) vaccine but that does not seem very likely. The rise began before the introduction of MMR and it continued to rise, without any plateauing, after MMR was used with the vast majority of the population (Dales et al 2001, Farrington et al 2001, Taylor et al 1999). But, if that is not the cause, what is?

The prevailing consensus at the moment is that autism spectrum disorders constitute a continuum extending from mild autism to severe handicap. That could prove to be the case but, if so, why is it that individuals with the so-called broader phenotype do not have associated mental retardation and do not seem to have an increase in the rate of epilepsy, both being very characteristic correlates of autism (Rutter 2000)? The question has to be addressed if only because the limited genetic evidence from twin studies indicates that the broader phenotype seems to share the same genetic liability (Le Couteur et al 1996). Could there be some kind of two-hit mechanism? If so, what is it that provides impetus for the shift from the broader phenotype to major handicap?

Asperger syndrome appears to involve exactly the same qualitative deficits as those associated with autism, but, unlike autism, this has not been associated with any delay in early language development and abnormalities are usually not clearly manifest until after infancy. Of course, that is not to suggest that language development is necessarily fully normal (indeed there are good reasons to suppose that it is not) but the existence of the syndrome provides a challenge to those who have viewed the language deficits as basic. It also provides a challenge to those who have argued that autism is almost always manifest from at least the age of 18 months, if not considerably earlier (Osterling et al 2002). Of course, it may be that a careful analysis of the social and communicative behaviour of individuals with Asperger syndrome would show early abnormalities, but what is clear is that the abnormalities are usually not recognized by either parents or professionals until quite a lot later (Gilchrist et al 2001). If Asperger syndrome is synonymous with mild autism, what does this mean?

In most cases, autism involves no developmental regression or loss of skills. However, numerous studies have shown that in about a quarter of cases, there is a temporary loss of language skills usually in the second half of the second year (Kurita 1985, Rogers & DiLalla 1990). This is often accompanied by a change in social interaction and a loss of pretend play, but it is not usually accompanied by a loss of motor, or other, skills. So far, evidence suggests that there is nothing distinctive about autism that is accompanied by regression. Interestingly, regression seems to be as common in autism when it occurs in two or more members of the same family (Parr et al 2003), with the implication that regression is neither more nor less common when it is likely that there is a strong genetic liability. What, therefore, does the regression mean?

Over the years, evidence has accumulated that the clinical picture of autism in early childhood is seen in several atypical circumstances. Thus, for example, it was described in children with congenital rubella, the follow-up indicating that, although the children tended to remain severely handicapped, the autistic features lessened (Chess 1977). Autistic-like syndromes have also been described in congenitally blind children (Hobson et al 1999), and in children who have suffered severe institutional deprivation (Rutter et al 1999). Careful analysis suggested that the picture is in some respects slightly atypical and, at least in the case of the institution-reared children, the autistic features tend to diminish as the children grow older. What do these findings tell us about the nature of autism or the cause of the syndrome? Some may be tempted just to dismiss the descriptions as representing phenocopies but there is still the need to account for the emergence of the picture strongly resembling autism.

There is then the further question of the overlap with semantic–pragmatic language disorders (Bishop 2000). It is clear that the two cannot be regarded as entirely synonymous because by no means all children with semantic–pragmatic language disorder show the features of autism (Bishop & Norbury 2002). The follow-up of the sample of boys with severe developmental disorder of receptive language, first seen in early childhood has brought out two further findings that need taking into account. First, although the children with language disorder did not appear at all autistic when young, at least half of them showed substantial social impairment early and mid-adult life (Howlin et al 2000). Second, the adults who had shown this severe developmental disorder of receptive language, were found to have impairments in 'theory of mind' skills at follow-up (Clegg 2002). It is not known, of course, whether the impaired 'theory of mind' skills had been present in early childhood but it seems likely that they must have been. If so, why were they not showing autistic features when young?

One further epidemiological finding requires highlighting. Autism is very much more common in males than females. The ratio is usually given as about 3 or 4:1, but evidence from recent epidemiological studies suggest that the male preponderance is very much greater in the case of autism that is not accompanied by severe mental retardation (Baird et al 2000). As we shall hear, hypotheses have been put forward to account for this sex ratio in autism. However, it is not self evident that the explanation will be found in a feature that is specific to autism. It is noteworthy that a similar marked male preponderance is found in most neurodevelopmental disorders such as dyslexia, attention deficit disorder with hyperactivity, and developmental language disorders (Rutter et al 2003). Is this just coincidence or is there some common factor that is responsible for the male preponderance across this range of disorders? They are all associated with cognitive deficits of one kind or another that are evident in the preschool period. This is quite different from what is seen with female preponderant psychopathological disorders such as depression or eating disorders, which typically have an onset in adolescence and are not accompanied by any marked cognitive deficits. Does this provide a clue as to a causal explanation? Do we need to consider epigenetic mechanisms and, if so, what might they be? Is it likely that prenatal differences in sex hormone pattern have been influential?

Another well established clinical finding concerns the tendency for autism to be associated with increased head size (Lord & Bailey 2002). Probably, this arises after birth and possibly, too, is associated with an increased head size in other members of the family. If the emergence of increased head size after birth is confirmed, what does this imply with respect to the neural processes that are responsible?

As the pioneering studies of Hermelin and O'Connor (1970) demonstrated, it has long been evident that autism is associated with an unusual pattern of cognitive deficits. During the 1980s and 1990s, attention particularly focused on what came to be called 'theory of mind' deficits—meaning an impairment in mentalizing skills that enabled children to use context to assess what another person was likely to be thinking. However, impairments in executive planning, the use of central coherence, and in facial processing have also been found (Hobson 1993, Lord & Bailey 2002, Medical Research Council 2001). It may certainly be accepted that deficits in social cognition constitute an intrinsic part of autism. Nevertheless, questions remain. What are the interconnections, if any, among these various deficits? If theory of mind skills are so crucial, why is autism manifest such a long time before theory of mind skills can be clearly demonstrated? Of course, the answer could lie in cognitive precursors of theory of mind but, if so, what is the explanatory power of theory of mind as such? Even infants are highly social, and so should the explanation be sought in some aspect of social relationships, rather than cognition, as Hobson has suggested? How might these highly specific cognitive deficits account for the language delay and mental retardation that are so commonly associated with autism (Rutter & Bailey 1993)? What accounts for the savant skills or special cognitive talents that occur in a substantial minority of individuals with autism (Hermelin 2001)? Conversely, if these are closely associated with the specific cognitive deficits, why are such talents not found in most individuals with autism? How might the cognitive deficits account for the repetitive stereotyped behaviours that are so characteristic of autism? I have spent quite some time outlining the epidemiological and clinical background because these are the findings that require explanation.

Let me turn now to the genetic findings. The findings suggest that genetic factors account for over 90% of the population variance in the underlying liability (Folstein & Rosen-Sheidley 2001, Rutter 2000). In view of the new evidence indicating that the prevalence of autism is considerably higher than used to be believed, there must be some caution about the precise heritability, because it will be affected to some extent by the assumptions made about the general population base rate. Nevertheless, even if the true rate of autism is as high as 0.6%, the rate in siblings would still be at least 10 times that. The marked fall off rate between monozygotic and dizygotic pairs, together with the fall off in the broader phenotype from first degree to second degree relatives, suggests that it is likely that some three to 12 genes are involved in the susceptibility to autism, and that there is a synergistic interaction among the susceptibility genes (Pickles et al 1995). But, what are the effects of each of these genes? Do they provide a vulnerability to autism as such or, rather, do they involve susceptibilities for individual components of autism (Bradford et al 2001, Folstein et al 1999)? If they do operate on different components, why is not the rate of each component very much higher than the rate of the syndrome as a whole?

Of course, there are no epidemiological studies that provide precise estimates of each component but such evidence as there is provides no indication that the rates are high. Also, one might expect that individual members of families with a proband showing autism might have only single elements, because they are likely to have only a few of the susceptibility genes. Findings suggest that, although that is sometimes the case, familial loading is mainly for a combined pattern that is similar to autism in quality, although much milder in degree. The history of medical genetics indicates that it must be expected that autism will prove to be genetically heterogeneous. To some extent, we know that it is heterogeneous because of the associations with tuberous sclerosis and with the fragile X anomaly (Lord & Bailey 2002, Medical Research Council 2001). Nevertheless, it is not yet quite clear why either of these conditions predisposes to autism.

If autism is genetically heterogeneous, we have to ask whether the heterogeneity is indexed by clinical variability. Of course, it need not be. The findings on concordant monozygotic pairs show that there is huge clinical variability in the manifestations of autism and of the associated cognitive impairment, even when one may assume that the genetic liability is the same (Le Couteur et al 1996). It is also known that even single gene conditions such as Rett syndrome or tuberous sclerosis show surprisingly wide clinical expression (Sharbazian & Zogbi 2001). What is not known, however, is what causes such variable expression. When variable expression is not properly understood even with single gene disorders, elucidation is likely to prove even more challenging with a multifactorial disorder such as autism.

In sorting out genetic heterogeneity, there must be consideration of the possibility of either multiple mutations of the same gene, as found in Rett syndrome or multiple different genes, as is the case with tuberous sclerosis. As already noted, although the heritability of autism is very high, it does appear to be a multifactorial disorder in which environmental factors also play a role in the overall susceptibility. What are those environmental susceptibility factors? Of course, these may not necessarily involve specific environmental hazards. Thus, they could reflect developmental perturbations of one kind or another (Rutter 2002). The increase in the rate of minor congenital anomalies is perhaps consistent with this suggestion. Recently, it has been argued that the rate of twinning in autism is much increased (Greenberg et al 2001) but it seems likely that this is an artefact of ascertainment. No substantial increase in twinning was found in the British twin studies of autism (Bailey et al 1995). Nevertheless, the possibility that developmental perturbations might play a role in aetiology is worth further exploration.

Neuropathological studies have been consistent in showing abnormalities but the findings are inconsistent on just what these are (Bailey et al 1996, Lord & Bailey 2002, Medical Research Council 2001). Some reports have emphasized abnormalities in the cerebellum; some have drawn attention to abnormalities in the cerebral cortex and some have focused on neurochemical features. How does this picture fit together? To what extent are the findings a consequence of the fact that most of the brains examined have come from individuals who are severely retarded as well as autistic and most of whom have had epilepsy? How do the human findings fit in with what has been shown with animal models? Are the neuropathological findings informative about the area of the brain that is affected in autism or, rather, do the findings reflect variations in the time point at which development went awry? How do the neuropathological findings fit in with the brain localization findings that have derived from functional imaging studies? In what way, if any, are the findings helpful in understanding why the epilepsy associated with autism so frequently has an onset in late adolescence and at early adult life, rather than the more common onset in early childhood? Are the neuropathological findings informative about the origins of the increased head size in autism? Do they help with respect to the phenomenon of regression?

It has long been known that blood serotonin levels tend to be raised in autism (Cook & Leventhal 1996). However, levels are similarly raised in many other neuropsychiatric disorders and, at least so far, this finding has not helped in understanding the basis of autism. It is clear that reduction of serotonin levels by drugs has not helped. There is a mixed bag of other positive findings in the field of neurochemistry but few have been replicated and they do not seem to add up to any coherent story (Bailey et al 1996, Medical Research Council 2001). On the other hand, it has to be said that the quality of this field of research has not been as high as one might have wished. Is there a potential for doing more and, if there is, what are the strategies that ought to be employed? Similar queries arise with respect to the immune system. Although the claims in relation to MMR do not seem to be well based, there are some pointers indicating that it is too early to rule out the possibility of some form of immune disorder as the basis for at least some cases of autism. Is this possibility a research priority and, if it is, how should it be pursued?

In the field of psychiatry as a whole there are reasonably good pointers that neurotransmitter abnormalities are likely to play some role in disorders as diverse as schizophrenia, depression, and attention deficit disorder with hyperactivity. With each of these conditions, too, there are drugs that have been shown to have quite marked beneficial effects in many, although not all, individuals with the conditions in question. Attention has been drawn to neuropeptide abnormalities in autism, but these seem not to differentiate autism from mental retardation (Nelson et al 2001). Given the expectation that autism is likely to prove to be some kind of systems disorder, it is surprising that there is so little evidence of either neurotransmitter abnormalities or major benefits from pharmacological interventions. Has research been looking in the wrong place, or are there lessons to be drawn from the largely negative findings? Where do we go from here?

Finally, I need to turn to the benefits associated with psychological interventions. There is no doubt that developmentally modulated, behavioural interventions can bring worthwhile short-term and long-term benefits in autism (Howlin & Rutter 1987). But, how much do they achieve? There have been recent strong claims that early intervention can make a difference (National Research Council 2001) but what is the evidence that this is so? Why are the benefits of intervention specifically focusing on psychological deficits, such as 'theory of mind' that are supposed to underlie autism, so disappointing (Hadwin et al 1996, Ozonoff & Miller 1995)? If the early interventions do make such a major difference, what are the implications for our understanding of the neural basis of autism? What is the evidence that early interventions can alter the neural substrate?

I am hugely impressed by the immense amount that has been achieved through systematic, thoughtful, innovative research into autism. Views have been transformed as a result of that research. That constitutes a considerable achievement. I am equally impressed, however, by the major questions that remain and by the puzzles involved in putting together diverse research findings. I hope that, by the end of this symposium, we will at least have narrowed down this list of questions. Also, I am hopeful that where the questions cannot as yet be properly answered, we will have identified at least the outlines of the research programme that will be needed in order to provide the answers. Those are the challenges that I am counting on all of you to meet over the next few days.

References

Bailey A, Le Couteur A, Gottesman I et al 1995 Autism as a strongly genetic disorder: evidence from a British twin study. Psychol Med 25:63–77

Bailey A, Phillips W, Rutter M 1996 Autism: towards an integration of clinical, genetic, neuropsychological, and neurobiological perspectives. J Child Psychol Psychiatry Annu Res Rev 37:89–126

Baird G, Charman T, Baron-Cohen S et al 2000 A screening instrument for autism at 18 months of age: A 6-year follow-up study. J Am Acad Child Adolesc Psychiatry 39:694–702

Bishop DVM 2000 Pragmatic language impairment: a correlate of SLI, a distinct subgroup, or part of the autistic continuum? In: Bishop DVM, Leonard LB (eds) Speech and language impairments in children: causes, characteristics, intervention and outcome. Psychology Press, Hove, p 99–113

Bishop DVM, Norbury CF 2002 Exploring the borderlands of autistic disorder and specific language impairment: a study using standardized diagnostic instruments. J Child Psychol Psychiatry 43:917–929

Bradford Y, Haines J, Hutcheson H et al 2001 Incorporating language phenotypes strengthens evidence of linkage to autism. Am J Med Genet 105:539–547

Chakrabarti S, Fombonne E 2001 Pervasive developmental disorders in preschool children. J Am Med Assoc 285:3093–3099

Chess S 1977 Follow-up report on autism in congenital rubella. J Autism Childhood Schizophr 7:69–81

Clegg J 2002 Developmental language disorders: a longitudinal study of cognitive, social and psychiatric functioning. PhD Thesis, University of Nottingham

Cook EH Jr, Leventhal BL 1996 The serotonin system in autism. Curr Opin Pediatr 8:348–354

Dales L, Hammer SJ, Smith NJ 2001 Time trends in autism and MMR immunization coverage in California. J Am Med Assoc 285:1183–1185

Farrington CP, Miller E, Taylor B 2001 MMR and autism: further evidence against a causal association. Vaccine 19:3632–3635

Folstein SE, Rosen-Sheidley B 2001 Genetics of autism: complex aetiology for a heterogeneous disorder. Nat Rev 2:943–955

Folstein SE, Santangelo SL, Gilman SE et al 1999 Predictors of cognitive test patterns in autism families. J Child Psychol Psychiatry 40:1117–1128

Fombonne E 1999 The epidemiology of autism: a review. Psychol Med 29:769–786

Gilchrist A, Green J, Cox A et al 2001 Development and current functioning in adolescents with Asperger syndrome: a comparative study. J Child Psychol Psychiatry 42:227–240

Greenberg DA, Hodge SE, Sowinski J et al 2001 Excess of twins among affected sibling pairs with autism: implications for the etiology of autism. Am J Hum Genet 69: 1062–1067

Hadwin J, Baron-Cohen S, Howlin P et al 1996 Can we teach children with autism to understand emotions, belief or pretence? Dev Psychopathol 8:345–365.

Hermelin B 2001 Bright splinters of the mind: a personal story of research with autistic savants. Jessica Kingsley, London

Hermelin B, O'Connor N 1970 Psychological experiments with autistic children. Pergamon, Oxford & New York

Hobson RP 1993 Autism and the development of mind. Lawrence Erlbaum Associates, Hillsdale, NJ

Hobson RP, Lee A, Brown R 1999 Autism and congenital blindness. J Autism Dev Disord 29:45–56

Howlin P, Mawhood L, Rutter M 2000 Autism and developmental receptive language disorder—a comparative follow-up in early adult life. II: Social, behavioural, and psychiatric outcomes. J Child Psychol Psychiatry 41:561–578

Howlin P, Rutter M 1987 Treatment of autistic children. Wiley, Chichester

Kurita H 1985 Infantile autism with speech loss before the age of thirty months. J Am Acad Child Psychiatry 24:191–196

Le Couteur A, Bailey AJ, Goode S et al 1996 A broader phenotype of autism: the clinical spectrum in twins. J Child Psychol Psychiatry 37:785–801

Lord C, Bailey A 2002 Autism spectrum disorders. In: Rutter M, Taylor E (eds) Child and adolescent psychiatry. Blackwell Scientific, Oxford, p 664–681

Medical Research Council 2001 MRC review of autism research: epidemiology and causes. MRC, London

National Research Council 2001 Educating children with autism. Committee on Educational Interventions for Children with Autism. Division of Behavioral and Social Sciences and Education. National Academy Press, Washington, DC

Nelson KB, Grether JK, Croen LA et al 2001 Neuropeptides and neurotrophins in neonatal blood of children with autism or mental retardation. Ann Neurol 49:597–606

Osterling JA, Dawson G, Munson JA 2002 Early recognition of 1-year-old infants with autism spectrum disorder versus mental retardation. Dev Psychopathol 14:239–251

Ozonoff S, Miller JN 1995 Teaching theory of mind: a new approach to social skills training for individuals with autism. J Autism Dev Disord 25:415–433

Parr J, Baird G, Fombonne E et al 2003 Autistic regression in a sample of multiplex families. In preparation

Pickles A, Bolton P, Macdonald H et al 1995 Latent-class analysis of recurrence risks for complex phenotypes with selection and measurement error: a twin and family history study of autism. Am J Hum Genet 57:717–726

Rogers S, DiLalla D 1990 Age of symptom onset in young children with pervasive developmental disorders. J Am Acad Child Adolesc Psychiatry 29:863–872

Rutter M 1999 The Emanuel Miller Memorial Lecture 1998: Autism: two-way interplay between research and clinical work. J Child Psychol Psychiatry 40:169–188

Rutter M 2000 Genetic studies of autism: from the 1970s into the Millennium. J Abnorm Child Psychol 28:3–14

Rutter M 2002 Nature, nurture, and development: from evangelism through science toward policy and practice. Child Dev 73:1–21

Rutter M, Bailey A 1993 Thinking and relationships: mind and brain (some reflections on theory of mind and autism). In: Baron-Cohen S, Tager-Flusberg H, Cohen DJ (eds) Understanding other minds: perspectives from autism. Oxford University Press, Oxford, p 481–504

Rutter M, Andersen-Wood L, Beckett C et al 1999 Quasi-autistic patterns following severe early global privation. J Child Psychol Psychiatry 40:537–549

Rutter M, Caspi A, Moffitt T 2003 Using sex differences in psychopathology to study causal mechanisms: unifying issues and research strategies. J Child Psychol Psychiatry, in press

Sharbazian MD, Zogbi HY 2001 Molecular genetics of Rett syndrome and clinical spectrum of MECP2 mutations. Curr Opin Neurobiol 14:171–176

Taylor B, Miller E, Farrington CP et al 1999 Autism and measles, mumps, and rubella vaccine: no epidemiological evidence for a causal association. Lancet 353:2026–2029

Epidemiology and early identification of autism: research challenges and opportunities

Tony Charman

Behavioural and Brain Sciences Unit, Institute of Child Health, 30 Guilford Street, London WC1N 1EH, UK

Abstract. Recent studies suggest that the prevalence of autism spectrum disorders may be as high as 60 per 10 000, considerably greater than the long-accepted figure of 5 per 10 000 for classic autism. Increased recognition, the broadening of the diagnostic concept and methodological differences across studies may account for most or all of the apparent increase in prevalence, although this cannot be quantified. In addition to the implications for families and services, these conceptual changes will affect the scientific study of autism. At present, case definition is reliant on the behavioural and developmental picture alone. Because the behavioural phenotype of autism and the broader autism spectrum disorders includes individuals with different ultimate aetiologies, even when biological or genetic markers are found they will not be present in all individuals with the phenotype. The fact that autism is not a unitary 'disorder' presents a significant challenge to genetic, biological, neurological and psychological research. Progress has recently been made in the earlier identification of autism both through screening programmes and by increased understanding and enhanced surveillance. This offers an opportunity to better understand the early developmental course of autism and may provide additional clues to the underlying pathology.

2003 Autism: neural basis and treatment possibilities. Wiley, Chichester (Novartis Foundation Symposium 251) p 10–25

Best estimate of prevalence up until 1999 and three recent studies

Since the first epidemiological study of autism was published by Lotter (1966) over 30 epidemiological surveys have been published (Fombonne 2002, Wing & Potter 2002). The majority of studies published up until the end of the 1990s found prevalence rates for childhood autism close to the figure of 0.4/1000 obtained by Lotter. Prevalence rates of 2.0/1000 for the broader autism spectrum disorder (ASD) were found in several studies (e.g. Wing & Gould 1979). The 'best-estimate' prevalence figures derived from two meta-analyses published in 1999

TABLE 1 Comparison of the studies by Baird et al (2000), Chakrabarti & Fombonne (2001) and Bertrand et al (2001)

	Baird et al	*Chakrabarti & Fombonne*	*Bertrand et al*
Base population size	16235[a]	15500	8896[b]
Age	7 years	2.5–6.5 years	3–10 years
Proportion of direct assessments	46%	95%	71%
Prevalence autism[c]	3.08/1000	1.68/1000	4.0/1000
Prevalence other PDDs[d]	2.71/1000	4.58/1000	2.7/1000
Prevalence all ASDs	5.79/1000	6.26/1000	6.7/1000
Boys: Girls all ASDs	88%:12%	79%:21%	73%:27%
IQ >70/<70 all ASDs	78%:22%[e]	74%:26%[f]	51%:49%[g]

[a] 16 235 children from 40 818 screened with the Checklist for Autism in Toddlers
[b] Estimated from 7117 1990 census by school-role factor
[c] ICD-10 childhood autism in Baird et al, DSM-IV Autistic disorder in Chakrabarti & Fombonne and Bertrand et al
[d] Asperger syndrome categorized as autism in Baird et al, as other PDD in Chakrabarti & Fombonne and Bertrand et al
[e] IQ data available on $n = 36$ cases
[f] IQ data available on $n = 91$ cases
[g] IQ data available on $n = 42$ cases

were 1.0/1000 for autism and 2.0/1000 for the broader ASD (Fombonne 1999, Gillberg & Wing 1999).

However, three studies published in the past two years found prevalence rates between 1.7/1000 and 4.1/1000 for autism and rates between 5.8/1000 and 6.7/1000 for all ASDs (Baird et al 2000, Bertrand et al 2001, Chakrabarti & Fombonne 2001). The main findings from these studies are summarized in Table 1: base population, age, proportion of direct and indirect assessments, prevalence of autism and other pervasive developmental disorders (PDDs), sex ratio and IQ distribution. There is considerable overlap between the studies in terms of base population size and age, and all shared the methodological features of multiple ascertainment methods and use of standard research diagnostic instruments (e.g. Autism Diagnostic Interview, ADI-R, Lord et al 1994). The prevalence figures for all ASDs across the three studies are similar. However, there is a difference is in the relative proportion of cases given autism vs. other PDD diagnosis. In the Chakrabarti & Fombonne study this proportion is 26:71(0.37:1.0), compared to 50:44 (1.14:1.0) in the Baird et al study and 36:24 (1.5:1.0) in the Bertrand et al study. This suggests that the diagnostic criteria for classic vs. atypical forms of autism were applied differently across the studies. The sex ratio and IQ ratio also vary across the studies.

Possible reasons for an apparent increase in prevalence

Possible explanations for higher prevalence figures for ASD in recent studies include: artefacts that have produced a false increase in prevalence estimates; factors that indicate that the current rate is correct but that do not indicate a true increase; and factors that indicate that the current rate is correct and indicate a real increase.

Artefacts that might have produced a false increase include an over-expansion of the diagnostic category of ASD such that children receive a 'false positive' diagnosis and a particular sampling bias in the studies that have found higher prevalence estimates (although no clear 'candidate bias' is apparent in the three recent studies). The first point is difficult to prove either way as no 'litmus test' such as a specific biological or genetic marker is currently available to determine true 'caseness'. Thus, case definition is reliant on the behavioural and developmental picture alone. Further, the behavioural phenotype of autism and the broader ASDs includes individuals with different ultimate aetiologies, so even when a marker is found it would not be present in all individuals with the phenotype. These considerations go beyond mere niceties of nosology and classification. They are critical to an understanding of whether the increase in prevalence is real or apparent and present a considerable challenge to scientific investigation of the genetic, biological, neurological and psychological causes of autism.

Wing & Potter (2002) have estimated that only one third to one half of children meeting ICD-10 (WHO 1993) criteria for childhood autism would meet Kanner's criteria. Thus, the diagnostic boundaries of even the core presentation of autism have broadened over the decades. Another significant factor is the increasing recognition of a broader spectrum of autistic disorders. With sub-threshold severity or combination of symptoms or atypical onset an individual can meet diagnostic criteria for atypical autism, PDD or Asperger syndrome. Another way in which the diagnostic concept of ASD has been broadened (both in application and conceptualization) is the increasing recognition that an ASD can co-exist with other disorders. These include Down syndrome, cerebral palsy, Tourette syndrome, Turner syndrome, tuberous sclerosis and individuals with hearing and visual impairment (see Gillberg & Coleman 2000, for a review). Another factor is the increasing recognition that individuals with average IQ may have an ASD.

One critical methodological consideration in epidemiological studies relevant to the apparent increase is the effectiveness of case-finding. The majority of studies have relied on a two-stage procedure where an initial screening phase is followed by a more intensive case ascertainment and diagnostic phase. Although estimates of the specificity of initial screens can be calculated (see Fombonne 1999), estimates of the sensitivity have rarely been ascertained (see Baird et al 2000, for an exception). Another critical factor is the population initially surveyed. Some prevalence studies have only included individuals within the special educational system, by definition excluding cases of ASD of average IQ within mainstream education. Cases are also missed when prevalence studies only include cases already identified and diagnosed by clinical services (i.e. they do not attempt to search for as yet unidentified cases in the population). The size of the population sampled has been shown to systematically relate to the prevalence rates found, with higher rates being found in smaller samples, presumably due to more intensive ascertainment and comprehensive coverage, although at a cost to the width of the confidence intervals. Multi-phase detection mechanisms that target a whole population of medium size are likely to provide the most accurate prevalence estimates.

In the absence of a clear demonstration that the increase in prevalence found in recent studies has been real and not apparent, speculation on putative reasons for a real increase are premature. However, confidently answering whether the increase is real or apparent is difficult (if not impossible), as we are unable to estimate quantitatively the impact of the diagnostic and methodological factors summarized above. This does not mean that attempts to answer the question of a true versus an apparent increase should not be made. Rather, our ability to do so retrospectively on the extant empirical base is limited.

The putative factor that has received most attention is the suggestion that regressive autism was increasing and was associated with bowel symptoms, and furthermore that this increase in cases was associated with the MMR (measles, mumps and rubella) vaccination (Wakefield et al 1998). Subsequent research has demonstrated no temporal association between the introduction of the MMR vaccination and increases in the number of registered cases of autism on health and education service databases (Dales et al 2001, Fombonne 2001, Kaye et al 2001, Taylor et al 1999). Further, research evidence provided no plausible causal association between MMR, bowel symptoms and autism (Halsey et al 2001), nor evidence for an increase in autism in association with a regressive course or bowel symptoms (Fombonne & Chakrabarti 2001, Taylor et al 2001). Reviews by the USA Institute of Medicine, the American Academy of Pediatrics and the UK Medical Research Council have concluded that no proven association existed between the MMR vaccination and prevalence rate of autism (Halsey & Hymen 2001, Medical Research Council 2001, Stratton et al 2001).

Directions for future research on prevalence and implications for scientific investigation of autism

Understanding that ASD is a phenomenological and not an aetiological classification has important implications for future attempts to establish prevalence. As discussed above, the behavioural phenotype of autism might be arrived at by a number of different pathogenic routes and thus the autistic spectrum includes individuals with different ultimate aetiologies. Even when biological or genetic markers are found they may not be present in all individuals with the phenotype. Thus, case definition for the whole spectrum of disorders will continue to be reliant on the behavioural and developmental picture alone. Further, the presentation within individuals and within a population cohort changes with development, reflecting our understanding of ASD as a developmental disorder. This places a heavy load on our ability to measure and establish reliable thresholds for characteristic autistic behaviours in order to decide 'caseness'. Considerable progress has been made in this direction in the past 15 years, most notably in the development of the ADI-R (Lord et al 1994) and the Autism Diagnostic Observation Schedule (ADOS-G; Lord et al 2000). However, even these instruments have important limitations. For example, they are better at identifying classic autism than less severe or atypical presentations of the phenotype and may work better at identifying the phenotype at its most prototypic age point (4–5 years, in the case of the ADI-R) than at earlier and later points in development. This reflects the clinical experience that the diagnosis of atypical autism, PDD and Asperger syndrome is less reliable than that of childhood autism (Mahoney et al 1998). The reliability with which a lower threshold can be set for the boarder spectrum of autistic disorders will be a significant challenge for future epidemiological studies. The conceptual issue of whether ASD should be considered a 'lifetime diagnosis' is also relevant. Once an individual has met criteria at one point in their development should they be considered 'a case' (for the purposes of scientific investigation) throughout their lifespan, whatever the improvement in their symptoms over time? We know much less about the diagnosis and prevalence of ASD in adulthood than childhood and this is a priority for future research. The end product of future prevalence studies may be a range of prevalence figures, depending on what threshold criteria are met in terms of symptoms or what age of presentation is being considered.

The challenge to scientific investigation of the genetic, biological, neurological and psychological causes of autism is also considerable. For example, the primary methodology in many areas of scientific investigation is to identify a phenotypically homogenous group of individuals. Typically this consists of cases meeting diagnostic criteria for ICD-10 (WHO 1993) childhood autism or DSM-IV (APA 1994) autistic disorder, buttressed by the application of ADI-R and ADOS algorithms. However, we do not yet have evidence that even the most classic cases have a unitary aetiology. As a consequence, we may be studying the neuropsychology or neuroanatomy of genetically or neurologically distinct subgroups. This may go some way towards explaining the pattern of inconsistent findings that has plagued many areas of investigation into ASD, for example in terms of identifying neurochemical or neuroanatomical abnormalities. One hope for the future is that once clear aetiological causes are identified for at least a subgroup of individuals who have ASD, scientific studies can include aetiologically, as opposed to phenotypically, homogeneous samples. This would also apply to epidemiological studies and in the future it may be possible to determine more precise prevalence figures for 'regressive autism' or 'serotonin autism'.

Alternative methodological approaches in future prevalence studies may provide more data on as yet unanswered scientific questions regarding the nature, course and aetiology of ASD. Fombonne (2002) suggests that a symptom as well as a syndromic approach needs to be adopted. This is critical since autism is not a unitary disease entity or disorder but an end phenotype of a number of complex, distinct and overlapping aetiologies at several levels of causation including genes and brain development. Further, the nosology of the current classification systems may well change as new scientific evidence about the nature of ASDs emerges (Lord & Risi 1998).

Another strategy would be in-depth prospective study of a population cohort from birth in order to marry prevalence data with onset and disease course data. The relative rarity of ASDs would make this prohibitively expensive to do on a population wide basis (e.g. 5000 children might only yield 30 cases of ASD and perhaps only 15 or fewer of autism). One suggestion is that younger siblings of already-diagnosed children make a suitable high-risk sample to study prospectively (London 2001). This is attractive from the point of view of tracking early development of autism (Baron-Cohen et al 1992). However, the disease course of individuals from families who have more than one child with ASD may differ from those in which the genetic loading for autism is lower and where other factors may have a greater determination on the development of ASD.

Future epidemiological research should be conducted hand-in-hand with other relevant branches of science so that as answers emerge about ASD this new knowledge can be applied post-hoc to epidemiological datasets. The nesting of biological and genetic research designs within future epidemiological studies will provide information to better answer questions about heterogeneity of presentation and aetiology. Other relevant areas for such nested epidemiological designs include familial medical and behavioural information and neuropsychology. However, the field needs to be open for unexpected and as yet unexplained findings, such as the recent discovery of elevated neuropeptides and neurotrophins in neonatal blood of children subsequently diagnosed with autism (Nelson et al 2001). Such research may provide data that in future will enable us to answer questions regarding the prevalence of ASD that have not yet been framed.

Opportunities from screening and early identification

One positive development has come from attempts to prospectively identify cases of autism using screening instruments (Baird et al 2001). These have been applied both to general populations (Checklist for Autism in Toddlers [CHAT], Baird et al 2000) and to referred populations (Modified-CHAT, Robins et al 2001, CHAT, Scambler et al 2001). These studies have demonstrated that it is possible to identify some cases of autism by the age of 18 months. However, in the only general population screening study completed to date, the CHAT screen had a high positive predictive value but its sensitivity was moderate at best and cannot support a recommendation for total population screening at a single time point (Baird et al 2000). There is some evidence that screening for ASD in referred children, where a concern about development has already been identified, may result in better sensitivity (Charman et al 2001, Robins et al 2001). Whether or not the sensitivity and specificity of future screens are adequate to recommend universal screening for ASD, a by-product of these initial studies has been a better understanding of the presentation of autism in the second and third years of life (Charman & Baird 2002). At the same time, increased awareness and better detection of cases through routine health surveillance and practice has enabled scientific study of younger cohorts of children than previously (e.g. Lord 1995, Stone et al 1999).

Findings from such studies are beginning to uncover novel information about the early course of development in autism that may have implications for our understanding of the disorder at the level of behaviour, psychology and neurology. For example, at least 2 studies have found that repetitive and stereotyped behaviours were identified less consistently in the second and third years of life (Cox et al 1999, Stone et al 1999), compared to older samples of 4 and 5 year old children with autism. It may be that in at least a subgroup of children with ASD repetitive, restricted and stereotyped abnormalities only begin to emerge in children with autism *after* infancy, later than the social and communication deficits are apparent. Consistent with this, two recent studies have not found executive function deficits in 3 year olds with autism relative to controls, in contrast to studies with school-age children with ASD (Dawson et al 2002, Griffith et al 1999). In contrast, autism-specific impairments in early social communication behaviour (e.g. joint attention) were found in both studies (Dawson et al 2002, Griffith et al 1999). In the psychological realm, Sigman et al (1999) have demonstrated that joint attention ability measured at 3 years of age was associated with gains in language and pro-social peer behaviour over a 9-year period. This suggests that joint attention may be a pivotal marker of the underlying psychopathology, as well as an important target for early intervention. Courchesne et al (2001) have presented intriguing findings on changes in brain volume during early development in a sample of children with autism aged between 2 and 16 years. Whilst birth records showed the sample had normal brain volume as indexed by head circumference, magnetic resonance imaging data indicated larger brain volume in children between the ages of 2 to 4 years. However, older children and adolescents did not have large brain volume. Courchesne et al (2001) suggested that abnormal regulation of brain growth may result in early overgrowth followed by abnormally slowed growth.

It is likely that in all or almost all cases of ASD the underlying organic cause is present at birth. However, these examples illustrate that the ability to study cases from late infancy through the early childhood years is adding incrementally to our knowledge of the course and pathogenic mechanisms of the disorder at the behavioural, psychological and neurological levels. If it is possible for such studies of early development to be embedded within an epidemiological and familial framework, they hold out the promise of important gains in our knowledge of ASD and our in ability to develop and test different types of early intervention.

Acknowledgements

I am grateful to Gillian Baird, Emily Simonoff and Andrew Pickles for discussions on this topic. The author was supported by a grant (060633) from The Wellcome Trust.

References

American Psychiatric Association 1994 Diagnostic and statistical manual of mental disorders 4th edn. (DSM-IV). American Psychiatric Association, Washington, DC

Baird G, Charman T, Baron-Cohen S et al A 2000 A screening instrument for autism at 18 months of age: a six-year follow-up study. J Am Acad Child Adolesc Psychiatry 39:694–702

Baird G, Charman T, Cox A et al 2001 Current topic: Screening and surveillance for autism and pervasive developmental disorders. Arch Dis Child 84:468–475

Baron-Cohen S, Allen J, Gillberg C 1992 Can autism be detected at 18 months? The needle, the haystack, and the CHAT. Br J Psychiatry 161:839–843

Bertrand J, Mars A, Boyle C, Bove F, Yeargin-Allsopp M, Decoufle P 2001 Prevalence of autism in a United States population: the Brick Township, New Jersey, investigation. Pediatrics 108:1155–1161

Chakrabarti S, Fombonne E 2001 Pervasive developmental disorders in preschool children. J Am Med Assoc 285:3093–3099

Charman T, Baird G 2002 Practitioner review: diagnosis of autism spectrum disorder in 2- and 3-year-old children. J Child Psychol Psychiatry 43:289–305

Charman T, Baron-Cohen S, Baird G et al 2001 Commentary: The Modified Checklist for Autism in toddlers: an initial study investigating the early detection of autism and pervasive developmental disorders. J Autism Dev Disord 31:145–148

Courchesne E, Karns CM, Davis HR 2001 Unusual brain growth patterns in early life in patients with autistic disorder: an MRI study. Neurology 57:245–254

Cox A, Klein K, Charman T et al 1999 Autism spectrum disorders at 20 and 42 months of age: stability of clinical and ADI-R diagnosis. J Child Psychol Psychiatry 40:719–732

Dales L, Hammer SJ, Smith NJ 2001 Time trends in autism and MMR immunization coverage in California. JAMA 285:1183–1185

Dawson G, Munson J, Estes A et al 2002 Neurocognitive function and joint attention ability in young children with autism spectrum disorder versus developmental delay. Child Dev 73:345–358

Fombonne E 1999 The epidemiology of autism. Psychol Med 29:769–786

Fombonne E 2001 Is there an epidemic of autism? Pediatrics 107:411–412

Fombonne E 2002 Epidemiological trends in rates of autism. Mol Psychiatry 2:S4–S6(suppl 7)

Fombonne E, Chakrabarti S 2001 No evidence for a new variant of measles-mumps-rubella-induced autism. Pediatrics 108:E58

Gillberg C, Wing L 1999 Autism: not an extremely rare disorder. Acta Psychiatr Scand 99: 399–406

Gillberg C, Coleman M 2000 The biology of the autistic syndromes, 3rd edn. MacKeith Press, London

Griffith EM, Pennington BF, Wehner EA, Rogers SJ 1999 Executive functions in young children with autism. Child Dev 70:817–832

Halsey NA, Hyman SL 2001 Measles-mumps-rubella vaccine and autism spectrum disorders: report from the New Challenges in Childhood Immunizations Conference, Oak Brook, Illinois, June 2000, Pediatrics 107:E84

Kaye JA, del Mar Melero-Montes M, Jick H 2001 Mumps, measles, and rubella vaccine and the incidence of autism recorded by general practitioners: a time trend analysis. Br Med J 322: 460–463

London E 2001 The 'Baby Sibs' project. *http://www.naar.org/grants/babysibs.pdf*

Lord C 1995 Follow-up of two-year-olds referred for possible autism. J Child Psychol Psychiatry 36:1365–1382

Lord C, Risi S 1998 Frameworks and methods in diagnosing autism spectrum disorders. Ment Retard Dev Dis Res Rev 4:90–96

Lord C, Rutter M, Le Couteur A 1994 Autism Diagnostic Interview-Revised: a revised version of a diagnostic interview for caregivers of individuals with possible pervasive developmental disorders. J Autism Dev Disord 24:659–686

Lord C, Risi S, Lambrecht L et al 2000 The autism diagnostic observation schedule-generic: a standard measure of social and communication deficits associated with the spectrum of autism. J Autism Dev Disord 30:205–223

Lotter V 1966 Epidemiology of autistic conditions in young children. Soc Psychiatry 1:124–137

Mahoney WJ, Szatmari P, MacLean JE et al 1998 Reliability and accuracy of differentiating pervasive developmental disorder subtypes. J Am Acad Child Adolesc Psychiatry 37:278–285

Medical Research Council 2001 Review of autism research: epidemiology and causes. Medical Research Council, London

Nelson KB, Grether JK, Croen LA et al 2001 Neuropeptides and neurotrophins in neonatal blood of children with autism or mental retardation. Ann Neurology 49:597–606

Robins DL, Fein D, Barton ML, Green JA 2001 The Modified Checklist for Autism in Toddlers: an initial study investigating the early detection of autism and pervasive developmental disorders. J Autism Dev Disord 31:131–144

Scambler D, Rogers SJ, Wehner EA 2001 Can the checklist for autism in toddlers differentiate young children with autism from those with developmental delays? J Am Acad Child Adolesc Psychiatry 40:1457–1463

Sigman M, Ruskin E, Arbeile S et al 1999 Continuity and change in the social competence of children with autism, Down syndrome, and developmental delays. Monogr Soc Res Child Dev 64:1–114

Stone WL, Lee EB, Ashford, L et al 1999 Can autism be diagnosed accurately in children under three years? J Child Psychol Psychiatry 40:219–226

Stratton K, Gable A, Shetty P on behalf of the Immunization Review Committee 2001 Measles-mumps-rubella vaccine and autism. Institute of Medicine, National Academy Press, Washington DC

Taylor B, Miller E, Farrington CP et al 1999 Autism and measles, mumps, and rubella vaccine: no epidemiological evidence for a casual association. Lancet 353:2026–2029

Taylor B, Miller E, Lingam R, Andrews N, Simmons A, Stowe J 2001 Measles, mumps, and rubella vaccination and bowel problems or developmental regression in children with autism: population study. Br Med J 324:393–396

Wakefield AJ, Murch SH, Anthony A et al 1998 Ileal-lymphoid-nodular hyperplasia, non-specific colitis, and pervasive developmental disorder in children. Lancet 351:637–641

Wing L, Gould J 1979 Severe impairments of social interaction and associated abnormalities in children: epidemiology and classification. J Autism Dev Disord 9:11–29

Wing L, Potter D 2002 The epidemiology of autistic spectrum disorders: is the prevalence rising? Ment Retard Dev Disabil Res Rev 8:151–161

World Health Organisation 1993 Mental disorders: a glossary and guide to their classification in accordance with the 10th revision of the International Classification of Diseases: research diagnostic criteria (ICD-10). WHO, Geneva

DISCUSSION

Fombonne: The three recent studies that you focused on all dealt with pervasive developmental disorder (PDD) as case definition for the survey. This represents a big shift in the method of these epidemiological studies. In the past, all studies focused solely on autistic disorders narrowly defined and left out large groups of children who failed to strictly meet diagnostic criteria for autism. These children were referred to with various labels (e.g. atypical autism, triad of impairments) and in some studies it was nevertheless apparent that they formed a sizeable group of children with serious developmental impairments. These three recent epidemiological studies have really set out to find cases looking at the broad spectrum of conditions. One consequence of this is that you cannot compare these prevalence rates to the previous ones. It would be misleading to say that there is a big increase because you are not comparing like with like. The other important point is that there was also a striking convergence of estimates in these three surveys. In addition, we have now completed a new survey in the Stafford area in the UK based on birth cohorts, born in 1996, 1997 and 1998 (our first survey included children born from 1992 to 1995). We have a rate of 59 per 10 000, which replicates in the same area and with rigorously identical methods our earlier findings. Therefore, there are now four convergent studies looking at the broad spectrum of conditions with consistent prevalence estimates (around 60/10 000) for the prevalence of PDD as a whole. What is puzzling is that if you look at subtypes, the estimates for specific subtypes of PDD are all over the place, as if the capacity of the investigators to draw the line between autism, PDD and Asperger's in this age group is extremely limited and very unreliable. This issue will need to be addressed in future studies. There are also issues about very mild PDD presentations that were incorporated in some studies. It would be important if we had follow-up data on these kids five years later. Some children with an early diagnosis of PDD might have made subsequent improvements and some diagnoses might have to be reconsidered during school-age years.

Lipkin: How are we going to sort out whether there has or has not been an increase in the incidence of autism? You suggested a birth cohort approach based on studies of adults, but were less than enthusiastic.

Charman: I think it would be an arduous study that would be likely to prove inconclusive. The likelihood of establishing whether a rise in prevalence has occurred on the historical extant database is very low. I don't think we have the ability to do this. I've seen epidemiologists trying to analyse various factors and studies to look at ascertainment and the use of various diagnostic instruments, and to quantitatively estimate whether one can relate the prevalence figure that different studies have in relation to aspects of their sample or methodology. The information we have and the way in which studies have been conducted is so varied that I don't think that quantitatively we are able to extrapolate from different studies to see whether methodological factors can account for all or some of the apparent increase.

Fombonne: There is evidence that method factors do account for most of the variations. There have been four studies published in the UK in the last two or three years (Chakrabarti & Fombonne 2001, Taylor et al 1999, Powell et al 2000, Baird et al 2000), and the rates vary from 10–62 per 10 000. These are studies conducted in the same country, with the same age groups, over a similar period, and we therefore expect to find similar estimates. Yet, there is a fourfold variation in prevalence estimates. Clearly, the way you set up your method for case identification in surveys allows for huge variation in the estimates. You could say the same thing for four US studies published in recent years: there was a 13-fold variation in estimates at the same point in time for the same country. This really confirms that method factors in prevalence studies can account for huge variation in rates. To test for secular changes in the incidence of autism or PDD, we need prospective studies or registries to monitor the accumulation of cases over time in defined populations. These sorts of studies have been done in cancer, for example.

Monaco: I was quite struck by the decrease in mental retardation. We need an explanation for this since most mental retardation has genetic causes and is heterogeneous. In addition we know that mental retardation is caused by new mutations in many cases.

Charman: That might have been misleading. Those aren't prevalence studies; they are the diagnoses that are registered on the California Department of

Developmental Services database. It is really what the school system is agreeing to call these children's condition.

Monaco: I agree that the shift of mental retardation into autism is a plausible hypothesis, because mental retardation cannot decrease unless our genome is better at repairing mutations than it was before!

Charman: I agree. I don't think that this is an indication that the prevalence of mental retardation in California has changed at all. It is just that the school system is now choosing to call some of those children 'autistic' who would previously have been called 'mentally retarded'.

Bolton: Another explanation for the potential change of prevalence is that more cases with subtle impairments and normal intelligence are being identified. There is some support for this notion in that there appears to be a higher proportion of cases with normal intelligence identified in the recent epidemiological studies. Although it is a plausible argument, I am not sure that it can account for all the apparent change in prevalence over time.

Charman: Again, I probably wasn't clear enough. It wasn't supposed to be necessarily accounting for the majority of any putative increase. It was just a demonstration of how differences in the way in which children with the same presentation are described can affect the prevalence rates. You are right: one of the major factors that has changed is that individuals with IQ in the normal range are now given autism spectrum diagnosis. One good example is the Wing & Gould (1979) study from Camberwell: none of those children had IQs in the normal range, because they all came from the special school system. By definition, that study was only looking at a subgroup of the population.

Monaco: What is the actual increase in severe autism over the last 15 years? You then have to dissect out the normal IQ milder phenotype if this area does not apply. Is the severe form of autism increasing?

Bailey: Its recognition has increased.

Fombonne: The safe answer is to say that the rate of narrowly defined autism was around 4–5/10 000 in old studies. In most recent studies, the minimum estimate of narrowly defined autism is around 10 per 10 000. This twofold increase most probably reflects improved recognition and identification of autism over the last 20 years.

Rutter: The methodological issues that apply to autism with severe mental retardation and autism associated with normal non-verbal intelligence are rather different. In the case of autism with severe retardation the main change over time has been a reduction in the tendency to exclude the diagnosis of autism if there is some other medical condition. That may have led to small increase in the prevalence of autism as diagnosed. In the case of autism in individuals with a normal IQ, there are the major effects of both better ascertainment and a considerable broadening of the diagnostic concept. Both have clearly led to an increase in the rate of diagnosed autism. However, it remains uncertain whether, in combination, these account for all of the rise in the rate of diagnosed autism. Is the evidence good enough for us to rule out a real rise in autism? That is the problem.

Dawson: One possible strategy that may be helpful is to focus on the alternative hypothesis. If there is a true increase, what might it be related to? There are methodologically sound epidemiological approaches that would tell us whether in a particular geographical location children are exposed to these kinds of events. Do we find increases in prevalence there? It is not answering the same question, but I do think we have more hope for studying the other side of the question of what the factors might be. If we find that all the hypothesized factors we can imagine are negative, this is an important piece of circumstantial evidence.

Fombonne: You would need a strong hypothesis to start with. All those proposed so far have been unsupported by the evidence.

Dawson: Look at cancer biology, where researchers took a very broad look at almost everything. We haven't really done large scale descriptive epidemiological studies. Broad approaches might at least lead to some directions and hypotheses.

Rutter: I agree that epidemiology comes into its own when there is a postulated specific causal influence to examine. That is where the MMR hypothesis was potentially useful. In my introduction I suggested that, at least as I read the epidemiological evidence, there really is no support for MMR being a cause of the rise in autism. Whether it is responsible for a small number of individual cases is an entirely separate issue.

Fombonne: I agree. There are six good epidemiological population-based studies that have looked for associations between MMR and autism. None of them have suggested that there is any connection. How can we take this further with evidence like that?

Dawson: I'm not suggesting that specifically MMR is what needs to be studied. I am saying that there are lots of things that one could potentially examine.

Hollis: Tony Charman, you raised the important issue that there isn't just one prevalence rate, and that prevalence relates to the age of the population being sampled. We think of classical, Kanner-type, autism as a very early onset disorder. However, if we think about a broader autistic spectrum, then these disorders, with more subtle impairments, may not be detected until later in childhood. In some ways, this is similar to the onset pattern seen in another neurodevelopmental disorder, schizophrenia. We know that while there are early abnormalities in neurodevelopment, the clinical manifestation of schizophrenia very rarely occurs before adolescence. The prevalence of a chronic disorder such as schizophrenia or autism will increase with age in the population. Hence, prevalence studies must refer both the severity definition of the disorder (e.g. 'narrow' or 'broad' autism) and to the age characteristics of the sample. In autism, a sample of older children will have a higher prevalence than a younger one as new cases on the 'broad' autistic spectrum are detected throughout childhood.

Amaral: It is disturbing that we can't determine whether there has been a real increase in autism or not. My concern is that there are so many potential hypotheses out there that we will get side-tracked. At our group at UC Davis we are now looking at the links between autism, and mercury and PCBs. As we have started looking at the relationship between these two, all of a sudden our data have been dominated by phenyls that are a component of computer keyboards and cases. They are one of the most neurotoxic substances found, and they are vapourized from keyboards. We started wondering whether this might be another potential environmental factor. How many other thousands of toxic chemicals might there be in the environment? We could take an eternity looking at all these hypotheses if we don't have some better notion that there really is a real contribution.

Charman: We are in a bind in some ways. The increased motivation for wanting to identify potential environmental factors is related to the notion that there is an increased prevalence of autism. It has always been known that not all of the outcome and variability within the population of individuals with autism is specifically due to genetic factors. At some level, developmental perturbations and other environment factors have always been part of the cause and manifestation of autism in at least some individuals. Separating the question out about potential environmental factors, or interactions between genes and the environment, from the question of whether there has been a real or only an apparent increase is probably helpful in terms of picking some of the environmental factors most likely to have any contributing cause to autism. Linking those two things might be unhelpful (indeed misleading) in terms of what are the best environmental factors to investigate.

Buitelaar: Another strategy would be to look at whether other disorders are also increasing in prevalence. It is unlikely that if an environmental factor is involved that it will just affect autism. For example, if in the same cohort there is an increase in learning disabilities or ADHD, it may be indicative of the involvement of environmental factors.

Baird: We were thinking about all these issues in our current prevalence study. One approach we decided to take is to go back to Lotter's papers and look at the criteria on which he made his judgements. It has proved to be interesting and also extremely difficult. This is partly because in his publications it is quite difficult to be clear about exactly which criteria he used. From the examples he gave, it is quite clear that now we would all have diagnosed the ones he excluded as autistic.

Bolton: In my view, we are in the early stages of epidemiological research into the prevalence of *autism spectrum disorders*. Our attention ought to be focused on trying to improve our methods of case identification and case diagnosis. We ought to be looking at establishing good epidemiological data on prevalence in well-defined birth cohorts at specific age points. Until we do this, we don't really have a firm enough basis on which to start testing more hypotheses concerning putative environmental risk factors.

Rutter: How do you suggest we should move to this better definition?

Bolton: It is a question, to begin with, of trialing methods for screening for autism spectrum disorders within a population. We know very little about the properties of screening instruments for identifying possible cases. We have this uncertainty about what sort of thresholds we should use for a broader definition of the autistic spectrum. There is a subsequent need to develop and test ways of combining data from parent reports and observations of the child. Once our screening and diagnostic procedures are well understood and validated, we need to study prevalence in birth cohorts rather than population-based samples which might be biased because families move into the area in order to access services. Moreover, we need to be determining age specific prevalence rates at different time points in development and longitudinal studies with repeat measures: this is an unfolding developmental disorder so some children may no longer meet case criteria later in development, whereas others might later meet criteria.

Rutter: Jan Buitelaar, does your study in Utrecht meet up to the challenges that Patrick Bolton has laid out?

Buitelaar: Yes, if you give me another 10 years! In principle, it will. We have just completed a population-based screening of 30 000 children at 14 months of age, using four key screening items that had to be completed by a trained doctor of the well-baby clinic. Screen-positive children have been rescreened during a home visit, and definitely screen-positive children have undergone extensive clinical evaluations. We are in the process of analysing our data of the screening items versus clinical diagnoses around 18–20 months and around 36 months. In addition, we have started to send a longer questionnaire of 60 items on early social and communicative/language behaviours to the parents of 10 000 children at age 14 months, 18 months and around age 28 months and 36 months. Around 3 years of age, cases with PDD (broad spectrum) will be identified. This will enable us to explore which set of items is most predictive for later PDD across various ages. This study includes early measures of anxiety, inattention and hyperactivity as potential predictors of PDD.

Hollis: In terms of investigating aetiological factors with epidemiology, we need to describe the incidence of the disorder and not its prevalence.

Bolton: Incidence is a difficult concept to apply when studying developmental disorders because in many instances the disorder does not have a clear onset in the way that say myocardial infarction does.

Lord: I think you can see the effect of the case definitions in the differences between the two British studies (Baird et al 2000, Chakrabarti & Fombonne 2001) and the Bertrand study (Bertrand et al 2001). The Bertrand study used the

ADOS as the major instrument, and they had some parent information, and they were less likely to recruit mild cases because of the US medical system and the methods they asked for referrals. We do not know how well the ADI discriminates autism from PDD-NOS, but we do know that the ADOS doesn't make this discrimination well on an individual basis (Lord et al 2000). The ADOS tends to be over-inclusive for autism for children with clinical reports of atypical autism or PDD-NOS, and also misses some of the highest-functioning subjects. This is what Bertrand et al got: a higher prevalence of autism than the other studies and a lower prevalence of ASDs than British studies.

Dawson: I wanted to comment about something that was mentioned earlier about early development. There was a comment about executive function in young children — specifically, that children with autism at age 3 don't show executive function impairments but do later. I think that is a misinterpretation of the data. I have heard it so many times I want to clarify this issue. What we find at age 3 is that children with autism don't differ from mental-age matched children with developmental delay without autism in their executive function profile (Dawson et al 2002). Children with autism clearly have executive function impairments at age 3. You just can't say that their executive function impairment is more severe than a child with developmental delay. Whether or not some autism-specific executive function signature arises with development later is a different question.

References

Baird G, Charman T, Baron-Cohen S et al A 2000 A screening instrument for autism at 18 months of age: a 6-year follow-up study. J Am Acad Child Adolesc Psychiatry 39:694–702

Bertrand J, Mars A, Boyle C, Bove F, Yeargin-Allsopp M, Decoufle P 2001 Prevalence of autism in a United States population: the Brick Township, New Jersey, investigation. Pediatrics 108:1155–1161

Chakrabarti S, Fombonne E 2001 Pervasive developmental disorders in preschool children. J Am Med Assoc 285:3093–3099

Dawson G, Munson J, Estes A et al 2002 Neurocognitive function and joint attention ability in young children with autism spectrum disorder. Child Dev 73:345–358

Lord C, Risi S, Lambrecht L et al 2000 The autism diagnostic observation schedule-generic: a standard measure of social and communication deficits associated with the spectrum of autism. J Autism Dev Disord 30:205–223

Powell J, Edwards A, Edwards M, Pandit BS, Sungum-Paliwal SR, Whitehouse W 2000 Changes in the incidence of childhood autism and other autistic spectrum disorders in preschool children from two areas of the West Midlands, UK. Dev Med Child Neurol 42:624–628

Taylor B, Miller E, Farrington CP et al 1999 Autism and measles, mumps, and rubella vaccine: no epidemological evidence for a causal association. The Lancet 353:2026–2029

Wing L, Gould J 1979 Severe impairments of social interaction and associated abnormalities in children: epidemiology and classification. J Autism Dev Disord 9:11–29

Implications of the broader phenotype for concepts of autism

Anthony Bailey and Jeremy Parr*

*Department of Psychiatry, University of Oxford, Park Hospital, Old Road, Headington, Oxford OX3 7LQ and *Department of Child and Adolescent Psychiatry, Institute of Psychiatry, De Crespigny Park, London SE5 8AF, UK*

Abstract. Autism, like many new diseases, was initially characterized by its most severe phenotypic manifestation and the ability to explain these distinctive features has been the benchmark against which explanatory models have subsequently been judged. Our understanding of the significance of milder phenotypes in other relatives has shifted from presumed environmental aetiological factors to variable manifestations of a complex disease process. In this paper we outline how the challenge of explaining the full range of phenotypic expression inevitably leads to more complex models of disease process than previously supposed. The implications of milder phenotypes for genetic, neurobiological and cognitive models of autism will be considered in relationship to several key features of complex diseases: complexity, hierarchy, emergence and coherence.

2003 Autism: neural basis and treatment possibilities. Wiley, Chichester (Novartis Foundation Symposium 251) p 26–47

This Novartis Foundation symposium occurs at an opportune time in autism research. Developments in rapid-throughput genotyping, structural and functional neuroimaging and an increased awareness of the utility of postmortem studies offer the possibility of significant advances in our understanding of the brain basis of autism and related disorders. Nevertheless, if these approaches, and the complimentary strategies of neuropsychological and behavioural phenotyping, are to be most effective, then future research questions require the optimal conceptual framework. The focus of this chapter is not on the components of milder phenotypic expression and their measurement (for a review see Bailey et al 1998a), but rather on how incorporation of the full spectrum of phenotypic expression, particularly of milder or 'broader' phenotypes, into our conceptualization of autism inevitably leads to a complex disease model.

Since autism (Kanner 1943) and Asperger's syndrome (Asperger 1944) were first described research has focused predominantly upon the individual components of these severe disorders. Nevertheless, both Kanner (1943) and Asperger (1944) noted mild behavioural characteristics in some parents that appeared related to their children's difficulties. Why were these behaviours neglected for so long? A major factor was Eisenberg & Kanner's (1956) claim that these parental traits were environmental causes of autism, a suggestion that inevitably caused distress to many families. Consequently the focus of early studies was upon refuting this hypothesis and parents were also described as unremarkable in terms of broad personality attributes (see Cox et al 1975). That Asperger's account was not translated into English until the early 1980s is also relevant, as both the convergent evidence for milder phenotypes and the suggestion of a genetic mechanism underlying related behavioural difficulties in fathers and sons (Asperger 1944) were largely overlooked for nearly 40 years. Lastly, psychological and biological researchers have been preoccupied with seeking explanations for the most peculiar and distinctive aspects of autistic behaviour and even since milder phenotypes have been generally recognised, their status has usually been that of 'lesser variants', rather than phenomena requiring an explanation in their own right.

In the absence of a research agenda seeking to incorporate both mild and severe phenotypes, psychological and biological models for autism have concentrated on defining the parts of the disorder. Psychological models have been particularly influential in driving this research, and although some early studies (see Hermelin & O'Connor 1970) focused on general high-level cognitive dysfunctions (and others have imputed deficits in processes such as memory, attention and executive function), it has been the concept of a central deficit in a narrow cognitive function that has dominated the field (see Morton & Frith 1995 for an account of this approach). One of the first specific cognitive accounts proposed that language difficulties were central to autism (Rutter 1978), but attention subsequently shifted to impairments in social understanding and this type of model has reached its apogee in the elaborated modular account of autism as a disorder of mind blindness (Baron-Cohen 1995). Three features of these narrow cognitive conceptualizations are worth noting. Firstly, it is usually argued that other cognitive abnormalities represent nothing more than secondary consequences of the primary deficit. Secondly, there has been a particular focus on the specificity of primary deficits, especially with respect to preserved skills in individuals with autism, but also with respect to other disorders. Thirdly, these accounts are single mechanism explanations of a complex behavioural phenotype.

Neurobiological models of autism have led to much less theoretically driven research. Historically, localizing anatomical accounts have been preferred, mainly because of the need to explain symptom specificity and the early assumption that mental handicap was simply a consequence of whatever process had damaged the critical systems. The two dominant anatomical accounts have been the hippocampal/amygdala and the cerebellar. Hippocampal abnormalities were first suggested on the basis of pneumoencephalographic findings, but it is reports of increased neuronal packing density in the hippocampus, some subnuclei of the amygdala and related structures that have particularly focused interest on this region (Kemper & Bauman 1998). The cerebellar account of autism is based on postmortem observations of decreased Purkinje cell number and also neuroimaging findings (Courchesne et al 1994). These localizing accounts suggest that symptomatology is a consequence of abnormal function in these structures plus less well specified long-range effects on the development and function of other brain regions. Neurochemical explications of autism have usually been less well elaborated, but have also taken as their starting point severe symptomatology; for instance the observation of reduced sensitivity to pain led to the suggestion of increased endogenous brain opioids as a causal factor for autism (Panksepp 1979).

We refer to these and similar cognitive and biological explanations of autism as 'simple' models; not 'simple' with respect to the complexity of the invoked mechanism, which is often elaborate, but 'simple' because of their focus on unitary mechanisms and the limited scope of their explanatory power. Thus while these models may provide an adequate account of one narrow cognitive deficit or localised abnormality or neurophysiological process, they frequently ignore other characteristic behavioural or biological abnormalities (Bailey et al 1996). Most strikingly, given that these models are predicated on features seen in severely affected individuals, they offer no satisfactory account for the strong association between autism and general mental retardation, or for the association with epilepsy and electroencephalogram (EEG) abnormalities (Rutter & Bailey 1993). The problem is usually sidestepped, often by invoking the notion of 'pure autism' (Minshew et al 1997); the implication being that the associated mental handicap represents nothing more than 'noise'.

The main empirical data supporting the conceptualization of autism as a complex disorder derives from genetic investigations, the pivotal findings coming from Folstein & Rutter's (1977) study of same-sex twins. The results indicated that the liability to develop autism was largely a consequence of complex genetic influences, but that the resultant behavioural phenotype could extend to milder cognitive difficulties (largely language based) and possibly also to social difficulties. A follow-up and re-diagnosis of the original sample and examination of a new sample of twins (Bailey et al 1995, Le Couteur et al 1996) further refined these conclusions. It was now evident that the genetic liability extended to social and/or communication difficulties in co-twins in whom the diagnosis of autism had never been entertained. That the mild autism related phenotypes were not simply a consequence of identical twinning was confirmed by a parallel family history study of relatives of singletons with autism (Bolton et al 1994), which used identical measures. The findings indicated a similar pattern of social and/or communication difficulties in singleton siblings, but went further in showing that these milder phenotypes also affected parents and more distant relatives, albeit less frequently and less severely. As with autism proper, male relatives were more frequently and severely affected than females. Strikingly the autism-related difficulties occurred in relatives of normal intelligence, although several studies have found that affected individuals show a significant decrement in verbal IQ compared to unaffected relatives (Fombonne et al 1997, Folstein et al 1999). Although there is still much uncertainty about the boundaries and components of milder phenotypes, there is substantial agreement that the phenomenon represents variable expression of a genetic susceptibility to autism and possibly other pervasive developmental disorders (Bailey et al 1998a). With most epidemiological studies also finding that only a small minority of cases of autism are associated with recognized medical disorders or severe obstetric hazards, it now seems that most cases represent severe expression of a specific, strongly genetic disease process.

Variable phenotypic expression does not necessarily imply complex mechanisms, raising the question of whether the cognitive and biological findings for autism can somehow still be accommodated in a 'simple' disease framework? With respect to explanatory models based on single cognitive deficits the answer is clearly no. The general arguments were first made by Wing & Wing (1971) and subsequently by Goodman (1989), who highlighted that autism and Asperger's syndrome are characterized by similar social impairments but differ in the extent to which language is affected, the implication being that neither social nor language abnormalities can be construed as secondary phenomena. Of course this argument is predicated on the similarity of the social deficits in both disorders. The fractionation of social and language phenotypes in non-autistic identical co-twins, as well as other relatives, provides much stronger support for the premise of multiple cognitive deficits. Developmental psychologists have responded to these findings by conceding the need for more than one primary cognitive abnormality; for instance a combination of weak central coherence and theory of mind (ToM) impairment (Frith & Happé 1994) has been suggested to account for the autistic phenotype, an accommodation still within the framework of a 'simple' model. Nevertheless this additive approach still faces explanatory challenges at the two extremes of phenotypic expression. Firstly the range of autism-related social, language and repetitive/rigid behaviours seen in milder phenotypes, sometimes in relative isolation, suggests that the number of separate deficits that may have to be invoked may be so large that the notion of a narrow impairment ceases to have heuristic value. Secondly, the various narrow cognitive models have been constructed to explain severe autistic behaviours. Usually it is not evident how the postulated deficits can be minimised in order to account for milder abnormalities. The problem is not always insurmountable. For instance, Happé (1999) has suggested that relatives may inherit a cognitive style that biases them towards reduced central coherence. Nevertheless, with respect to milder social difficulties, it is less obvious how many of the relatively subtle behaviours (and possibly also personality traits) seen in some relatives can easily be accounted for by deficits in ToM. Of course theoretical ingenuity might prevail, but the persisting problem for explanatory accounts based on a small number of specific deficits is their failure to explain the association with general intellectual impairment in severely affected individuals.

How do the various types of biological model fare when judged against the yardstick of milder phenotypic expression? Clearly any explanatory mechanism based largely upon features found only in severely affected individuals struggles to account for mild expression; thus neurochemical explanations of the kind typified by the opiate hypothesis appear inadequate. With respect to localized anatomical accounts, variants of the hippocampal/amygdala hypothesis that emphasise social deficits are badly stretched to account for language delay in relatives in the absence of significant social difficulties. As regards cerebellar involvement it is hard to draw firm conclusions, as multiple mechanisms have been suggested to underlie the claimed effects on cognitive function (Courchesne et al 1994); consequently it is difficult to generate falsifiable hypotheses.

In the face of such challenges to 'simple' models it is appropriate to consider alternatively embedding our knowledge about autism and milder phenotypes in the context of a complex disease model (Bailey et al 1996). Such models have been developed to aid the study of complex, common multifactorial diseases such as coronary heart disease, diabetes and asthma (Sing & Riley 1993). These diseases arise on the basis of a complex genetic predisposition in interaction with environmental factors; the intervening biochemical and physiological processes are also complex; symptoms may only arise at an advanced stage of disease progression and the disorders are non-deterministic i.e. the phenotype can not be predicted on the basis of knowledge about the presence or absence of risk factors. These types of model focus on understanding the biology of the whole organism, and contrast with approaches that assume that by describing individual deterministic systems, the whole will eventually be understood by characterizing the parts.

Sing & Riley (1993) outline several key features of a complex disease, some of which are especially relevant to an understanding of autism and related disorders. Firstly, in contrast to Mendelian diseases the genetic architecture of these disorders is *complex*. As noted by Folstein & Rutter (1977), the concordance findings in twins are not compatible with single gene inheritance and statistical modelling, utilizing data on milder phenotypes in co-twins and the relatives of singletons, suggests that interactions between three or four susceptibility loci are likely to be implicated, although the possibility of as many as 10 susceptibility genes can not be excluded (Pickles et al 1995). It is evident that these susceptibility loci act through non-deterministic mechanisms; thus even in identical twin pairs concordant for autism there are often remarkable differences in the extent to which both twins are affected, with little evidence for concordance either for behavioural features or intellectual level (Le Couteur et al 1996). It is unclear whether the substantial phenotypic variation between individuals sharing the same susceptibility loci is attributable to differences in exposure to environmental risk factors or is entirely a consequence of stochastic processes. Currently, postmortem studies point to a prenatal onset of neurodevelopmental abnormalities, suggesting that if any environmental factors are implicated that they are likely to operate prenatally. Identifying susceptibility genes, the proteins they encode and their subsequent biochemical roles may provide important clues as to the type of environmental factors that should be sought.

Secondly, in contrast to the usual shorthand of genes 'for' autism, complex models are explicit that the genetic and environmental causes of disease act in a *hierarchical* fashion, through a network of intermediate biological traits. In behavioural disorders many of these biological traits also influence the development of cognitive processes in the next level of the hierarchy. To date most autism research has focused on the upper levels of this hierarchy but, as detailed by Barnby & Monaco (2003, this volume), identifying susceptibility genes for autism has recently become an area of intense study. One goal of refining models of disease is to generate new research strategies and if milder phenotypes are manifestations of genetic risk, then measuring these behavioural phenotypes can help identify susceptibility genes. There are two contrasting approaches to this task. Folstein et al (1999) have suggested that susceptibility loci must be contributed by both parents, that individual components of the behavioural phenotype are inherited separately and that these phenotypes are manifested independently in non-autistic family members. In our terminology this is a 'simple' non-hierarchical model of gene action, with genes conceptualized as having a very close link with cognitive/behavioural deficits. By contrast, Pickles et al (1995) found that their data on autism and milder phenotypes were best fitted by a model in which interactions between loci contributed to a single latent trait or measure of autism, rather than one locus predisposing to one component of the phenotype. In a complex disease model this latent trait could be conceptualized as indexing some aspect of the intervening biological level of the disease hierarchy. Bradford et al (2001) have characterized language abnormalities in parents in multiplex families and found that including parents with a history of language problems in linkage analysis increased the evidence of linkage to the susceptibility region on chromosome 7. Alternatively, if the biological consequences of susceptibility alleles simultaneously affect multiple cognitive/ behavioural processes, then alleles will be identified most efficiently by measuring simultaneously these separate processes. This is the logic underlying multivariate approaches to the identification of susceptibility loci. Presently most researchers are characterizing milder phenotypes at the behavioural level but there has been some interest in developing specific cognitive tasks for relatives based on existing narrow cognitive models (see Bailey et al 1998b).

A third feature of complex diseases is that the symptoms of disease are not tightly defined by the genetic and environmental inputs; rather the disease is an *emergent* property of interrelated complex biological traits. Equally, an understanding of the disease does not equate with a description of the state of the organismic hierarchy at only *one* point along a spectrum of disease severity. This 'bottom up' concept of disease contrasts with the 'top down' thinking that has dominated autism research, in which the goal has been to identify mechanisms underlying the most severe features and then to conceptualize milder phenotypes within the same framework. Contrasting mild and severe expression illustrates the types of phenomena which we can consider emergent: the language problems seen in some relatives do not include features such as echoing or regression; the social difficulties do not seem to be typified by joint attentional impairments in infancy; and the rigid and repetitive behaviours do not include the unusual preoccupations, sensory interests and motor mannerisms seen in autism proper. Clearly, as disease expression becomes more severe, the number of impaired mechanisms increases. What are the implications of emergence for our understanding of autism? Firstly, it is probable that some relatives are affected at the level of brain structure and/or function, but do not show obvious cognitive or behavioural impairments. Identifying the processes associated with increasingly severe phenotypic expression requires linking genetic susceptibility with measures of appropriate endophenotypes across the full range of phenotypic expression. Secondly, it might be useful to start 'thinking outside the box' with respect to the cognitive mechanisms underlying milder phenotypes, rather than assuming that these are all simply lesser variants of described abnormalities. Thirdly, the concept is useful for understanding the association of severe phenotypes with EEG abnormalities, epilepsy and mental handicap. These phenomena are generally non-localized and are often assumed to represent some sort of 'second hit' process; i.e. the underlying mechanisms are considered unrelated to those causing autism specific deficits. If these properties are emergent, however, then as milder phenotypes are not associated with mental handicap or epilepsy, the biochemical mechanisms seem unlikely to affect the fundamental properties of neurons. Rather handicap and epilepsy presumably emerge because more severe expression involves increasingly abnormal interactions between neurons. Indeed over the last few years it has been appreciated that autism *is* associated with other non-localized neurodevelopmental abnormalities, including increased head circumference (first noted by Kanner in 1943), increased brain volume and weight (Piven et al 1995, Bailey et al 1998a) and developmental cortical abnormalities (Bailey et al 1998a).

Finally, we consider the property of *coherence*, the concept that initiation and development of disease is determined by the same coherent network of interrelated biological traits that define normal biochemical and physiological processes (Sing & Reilly 1993). Coherence of the network implies that variation in each trait and co-variation between traits is constrained in order to maintain homeostatic responses. As yet we know very little about the biochemistry of autism, but the concept of compensatory mechanisms is useful at a higher level of analysis. These ideas have not figured prominently in thinking about autism, perhaps because in the search for specific deficits it has been convenient to assume that unremarkable behavioural performance is synonymous with normal underlying mechanisms. One exception has been the recognition of the role of language abilities in ToM skills in autism (Happé 1994, Kazak et al 1997), the findings suggesting that individuals with autism who pass ToM tasks may be using language-based strategies.

Of course functional imaging provides a direct method of identifying the use of alternative brain pathways during cognitive tasks and studies of face processing (Schultz et al 2000, Pierce et al 2001) and ToM (Happé et al 1996) have found abnormal cortical localization of task-associated activity; the inference being that subjects are passing these tasks using somewhat different brain mechanisms. Of most significance, however, is the finding of abnormal patterns of cerebral motor activation in able individuals with autism (Muller et al 2001), as this is an example of a functional domain usually thought to be unimpaired in autism.

What are the implications of these findings for our understanding of autism? Firstly, as apparently intact (or indeed perhaps also superior) behavioural performance may be mediated by unusual mechanisms, the distinctions between specifically impaired and intact behaviours may sometimes be more apparent than real. Secondly, the findings raise the possibility that some deficits emerge only if potential compensatory brain pathways are also dysfunctional. Thirdly, autistic spectrum disorders in able individuals appear to be associated with quite widespread abnormal cortical organisation, raising the possibility that the neurobiological basis of supposedly specific deficits and mental handicap may be more similar than previously supposed.

In summary, psychological accounts of autism have usually focused upon explaining only the most severe and apparently specific features of the syndrome with almost no regard to phenotypic variability; an approach that has encouraged localising biological explanations. The recognition that autism is usually a specific genetic disease necessitates a conceptual model in which variability across the entire spectrum of expression is analysed within a hierarchical framework of genetic and environmental risk factors, intervening biological processes, cognitive (and emotional) deficits and abnormal behaviours. Within this framework the challenge is to describe the hierarchy at successive stages of disease progression and to identify the mechanisms underlying the emergence of increasingly severe symptoms, including mental handicap and epilepsy. The problem is complex as recent studies suggest that neurodevelopmental abnormalities may be more pervasive than previously supposed and that in future our explanatory models must also incorporate the role of compensatory mechanisms. We are optimistic, however, that the tools of accurate behavioural and neuropsychological phenotyping, functional and structural neuroimaging, postmortem studies and molecular biology will rapidly provide insights into the underlying disease mechanisms ultimately leading to improved preventative and treatment strategies.

Acknowledgement

This work is supported by the UK Medical Research Council.

References

Asperger H 1944 Die 'Autistischen Psychopathen' im Kindesalter Archiv für Psychiatrie und Nervenkrankhelten 117:76–136

Bailey A, Le Couteur A, Gottesman I et al 1995 Autism as a strongly genetic disorder: evidence from a British twin study. Psychol Med 25:63–77

Bailey A, Phillips W, Rutter M 1996 Autism: towards an integration of clinical, genetic, neuropsychological, and neurobiological perspectives. J Child Psychol Psychiatry 37:89–126

Bailey A, Luthert P, Dean A et al 1998a A clinicopathological study of autism. Brain 121:889–905

Bailey A, Palferman S, Heavey L, Le Couteur A 1998b Autism: the phenotype in relatives. J Autism Dev Disord 28:369–392

Barnby G, Monaco AP 2003 Strategies for autism candidate gene analysis. In: Autism: neural basis and treatment possibilities. Wiley, Chichester (Novartis Found Symp 251) p 48–69.

Baron-Cohen S, Cosmides L, Tooby J 1995 Mindblindness: an essay on autism and Theory of Mind. MIT Press, Cambridge, MA

Bolton P, Macdonald H, Pickles A et al 1994 A case–control family history study of autism. J Child Psychol Psychiatry 35:877–900

Bradford Y, Haines J, Hutcheson H et al 2001 Incorporating language phenotypes strengthens evidence of linkage to autism. Am J Med Genet 105:539–547

Courchesne E, Townsend J, Saitoh O 1994 The brain in infantile autism: posterior fossa structures are abnormal. Neurology 44:214–223

Cox A, Rutter M, Newman S, Bartak L 1975 A comparative study of infantile autism and specific developmental receptive language disorder: II Parental characteristics. Br J Psychiatry 126:146–159

Eisenberg L, Kanner L 1956 Early infantile autism 1943–1955. Am J Orthopsychiatry 26: 556–566

Folstein S, Rutter M 1977 Infantile autism: a genetic study of 21 twin pairs. J Child Psychol Psychiatry 18:297–321

Folstein SE, Santangelo SL, Gilman SE et al 1999 Predictors of cognitive test patterns in autism families. J Child Psychol Psychiatry 40:1117–1128

Fombonne E, Bolton P, Prior J, Jordan H, Rutter M 1997 A family study of autism: cognitive patterns and levels in parents and siblings. J Child Psychol Psychiatry 38:667–683
Frith U, Happé F 1994 Autism: beyond "theory of mind". Cognition 50:115–132
Goodman R 1989 Infantile autism: a syndrome of multiple primary deficits? J Autism Dev Disord 19:409–424
Happé FG 1994 Annotation: current psychological theories of autism: the "theory of mind" account and rival theories. J Child Psychol Psychiatry 35:215–229
Happé F 1999 Autism: cognitive deficit or cognitive style? Trends Cogn Sci 3:216–222
Happé F, Ehlers S, Fletcher P et al 1996 'Theory of mind' in the brain. Evidence from a PET scan study of Asperger syndrome. Neuroreport 8:197–201
Hermelin B, O'Connor N 1970 Psychological experiments with autistic children. Pergamon Press, London
Kanner L 1943 Autistic disturbances of affective contact. Nervous Child 2:217–250
Kazak S, Collis GM, Lewis V 1997 Can young people with autism refer to knowledge states? Evidence from their understanding of "know" and "guess". J Child Psychol Psychiatry 38:1001–1009
Kemper TL, Bauman M 1998 Neuropathology of infantile autism. J Neuropath Exp Neurol 57:645–652
Le Couteur A, Bailey A, Goode S et al 1996 A broader phenotype of autism: the clinical spectrum in twins. J Child Psychol Psychiatry 37:785–801
Minshew NJ, Goldstein G, Siegel DJ 1997 Neuropsychologic functioning in autism: profile of a complex information processing disorder. J Int Neuropsychol Soc 3:303–316
Morton J, Frith U 1995 Causal modelling: a structural approach to developmental psychopathology. In: Cichetti D, Cohen DJ (eds) Manual of developmental psychopathology. John Wiley & Sons Inc, New York, p 357–390
Muller RA, Pierce K, Ambrose JB, Allen G, Courchesne E 2001 Atypical patterns of cerebral motor activation in autism: a functional magnetic resonance study. Biol Psychiatry 49: 665–676
Panksepp J 1979 A neurochemical theory of autism. Trends Neurosci 2:174–177
Pickles A, Bolton P, Macdonald H et al 1995 Latent-class analysis of recurrence risks for complex phenotypes with selection and measurement error: a twin and family history study of autism. Am J Hum Genet 57:717–726
Pierce K, Muller R, Ambrose J, Allen G, Courchesne E 2001 Face processing occurs outside the fusiform 'face area' in autism: evidence from functional MRI. Brain 124: 2059–2073
Piven J, Arndt S, Bailey J, Havercamp S, Andreasen NC, Palmer P 1995 An MRI study of brain size in autism. Am J Psychiatry 152:1145–1149
Rutter M 1978 Language disorder and infantile autism. In: Rutter M, Schopler E (eds) Autism: a reappraisal of concepts and treatment. Plenum Press, New York, p 85–104
Rutter M, Bailey A 1993 Thinking and relationships: mind and brain. In: Baron-Cohen S, Tager-Flusberg H, Cohen D (eds) Understanding other minds: perspectives from autism. Oxford University Press, Oxford p 481–504
Schultz RT, Gauthier I, Klin A et al 2000 Abnormal ventral temporal cortical activity during face discrimination among individual with autism and Asperger syndrome. Arch Gen Psychiatry 57:331–340
Sing C, Reilly S 1993 Genetics of common diseases that aggregate but do not segregate in families. In: Sing CF, Harris CL (eds) Genetics of cellular individual family and population variability. Oxford University Press, New York, p 140–161
Wing L, Wing JK 1971 Multiple impairments in early childhood autism. J Autism Child Schizophr 1:256–266

DISCUSSION

Bishop: One of the models you put forward was that individual components of the phenotype may have different genes associated with them. How do you account for monozygotic (MZ) twins where there are such different expressions of symptom pattern?

Bailey: I didn't say that individual genes cause individual components of the phenotype; indeed the whole thrust of my paper was in the opposite direction!

Bishop: You didn't make that claim, but that's what Susan Folstein seemed to be saying.

Bailey: There are two contrasting views. There is very little empirical evidence to distinguish between them in terms of identifying genes. One view is that there is a tight link between susceptibility alleles and individual components of the phenotype. At the simplest level, this idea is that there is a gene for social deficits, a gene for language abnormalities and a gene for repetitive behaviours. The argument that I was putting forward is the converse: we are probably dealing with a disorder in which there are epistatic interactions between genes that are acting together to cause abnormal brain development which is very variable in its phenotypic expression.

Bishop: Could you not rule out the first model by evidence from MZ twins with very different phenotypes?

Bailey: I don't strictly think so, simply because one has to take into account the strategies that people are using to achieve normal behavioural performance.

Bishop: Let me push you a little bit further. If you take Isabelle Rapin's (1996) big survey of autism and specific language impairment (SLI), she has a pair of MZ twins where one met the criteria for SLI and the other met criteria for autism. Is this just a difference in strategy? They are genetically identical.

Bailey: My own view is very similar to yours, which is that I think it is unlikely that individual genes contribute in the way that is being suggested. We are faced with the puzzle of Asperger's syndrome. Here we have a disorder in which affected individuals can show quite abnormal use of language in an absence of a history of language delay and the semantic and syntactic problems that are sometimes seen in individuals with autism. It is striking that there have been no brain imaging studies of language function in Asperger's syndrome. It is widely assumed that their normal intelligence and apparently normal language skills represent intact functioning of the relevant brain systems; but this is an assumption. One has to be somewhat cautious about equating normal behavioural performance with an intact biological mechanism.

Folstein: Dorothy Bishop, Mike Rutter and I would say that after all these years we still don't understand the twin data. They don't really go with the family data. I don't understand why the twins can be so different.

Bishop: There is variable expression for other single-gene disorders which could be due to completely random factors.

Folstein: That is the view that Mike has been putting forward. In the first twin study (and this wasn't replicated in the others), there seemed to be a clear relationship between what happened prenatally and postnatally, and severity in the MZ twins. The idea of a second hit is one possibility. There are many different kinds of potential second hits.

Lipkin: Do we know anything about the positions of twins in the womb and what sorts of effects positioning might have on brain development? Positioning could influence nutrition and exposure to cytokines, growth factors, toxins or infection.

Bailey: One of the obstacles for all the twin studies has been a lack of good data on pregnancy factors. We don't even know the placentation status of most of the twins in these different series. It is striking in both of the UK twin studies that there were some cases in which there were clear biological differences between the twins — either a birth weight difference or rhesus compatibility. It was always the most severely affected individual who had the worst outcome. But the obstetric data in general show clearly that significant obstetric hazards have a very weak association with autism. There is no association with very low birth weight or extreme prematurity. The sort of intrauterine factors that we might be looking for might be very subtle indeed. We can't assume they are intrauterine in the way we have traditionally conceptualized such factors. That is why although the strategy that Geraldine Dawson suggested in the previous discussion is theoretically a useful one, without having some clues from identifying susceptibility alleles and the associated biochemical pathways, and perhaps generating animal models, it is going to be rather difficult to know where to look for relevant environmental factors. They may be completely non-noxious factors, if one doesn't happen to carry the susceptibility allele.

Monaco: I want to come back to Dorothy Bishop's point and shed some light on the problem from work that is going on in dyslexia. Groups have done quantitative trait loci linkage analysis in dyslexia with a range of different measures, using univariate analysis. For example, they may find linkage on chromosome 15 to one measure, but on chromosome 6 to a different measure, and then make predictions that genes controlling these separate measures are at these different loci. A number of papers were published finding linkage to these same regions but with different measures. One way Angela Marlow, Lon Cardon and our group have gone about this goes beyond analysis of the univariate case with multivariate linkage analysis to see how these measures actually co-vary. Very interestingly, we found, on chromosomes where we do have evidence for linkage with some measures and not others in univariate analysis, in the multivariate model they all contribute to the linkage. For autism, this might relate to different Autism Diagnostic Interview (ADI) domains and similar analysis. If the ADI domain scores are highly correlated, the stochastic process of a univariate measure providing linkage at a locus is going to happen in some studies and not others. But if you perform multivariate analysis, you may uncover how they co-vary and contribute to the linkage. I agree that it is difficult to try to make those arguments about genes controlling individual measures in a clinical picture, or quantitative measures in something like dyslexia. It probably just requires better statistical methods to look at the co-variation of these traits. How correlated are the measures taken in the different domains of the ADI?

Pericak-Vance: Was that multivariate analysis based on simulation data or looking at the actual data?

Monaco: It's done using the actual genotype data. It is the first time that it has been done on the whole genome.

Pericak-Vance: Have you published that yet?

Monaco: The manuscript is in preparation.

Rogers: Tony Bailey, it seemed like you were arguing for a unity between mental retardation and autism, and not to think of them as separate aspects of the phenotype. Is that true?

Bailey: What I'm suggesting is that the neurobiological mechanisms that can give rise to general intellectual impairment may be more closely related to those causing specific deficits than we previously supposed. Historically the overarching model has been a narrow localizing abnormality affecting a critical region. What the postmortem and functional imaging data push us towards is the notion that there may be quite widespread involvement of the cerebral cortex: there may be some species-specific behaviours for which it is not easily possible to rewire or use compensatory strategies. If this is a continuum, then at some point in expression one will reach a point where connectivity is sufficiently impaired to cause general intellectual impairments. The interest for me is in the absolute neglect of Asperger's syndrome. We really have no clue as to whether intact or superior intelligence in Asperger's syndrome is arising on the basis of normal or somewhat unusual cortical connectivity.

Rogers: Isn't that argument a little bit circular? We define Asperger's partly by IQ level.

Bailey: I'm not arguing for a definition of the disorder. Indeed all the available evidence suggests it is very hard to separate out Asperger's from autism empirically. The follow-up studies indicate that if individuals from both groups are matched for intelligence and language, they are indistinguishable in their outcome in adult life. Historically, we have tended to group together individuals who seem to be more phenotypically similar to each other. But the available genetic evidence suggests that although pervasive developmental disorders (PDDs) may be genetically heterogeneous, the heterogeneity doesn't seem to easily map on to the historical diagnostic groupings.

Rogers: One of the things that has always made me want to think of mental retardation aspects of the phenotype separately from the autistic symptoms comes from Susan Folstein and Mike Rutter's twin paper. In this, it seemed that neither intelligence nor severity of symptoms correlated between the affected twin and the co-twin. If I interpret the findings in the paper correctly, 90% of the co-twins had either autism or evidence of the broader phenotype. They were affected, it seemed, in some way by the autistic part of the phenotype. But I didn't read in that paper the same kind of relationship with cognitive impairment. Even if it wasn't correlated, I didn't read that the co-twins had a higher level of cognitive impairment. Is that true? That paper seemed to really differentiate the two.

Rutter: In terms of the follow-up of the twins first studied by Susan Folstein and myself, what was striking were the major differences within concordant MZ pairs in both IQ level and severity of social and communicative abnormalities (Le Couteur et al 1996).

Bishop: I'd like to throw another spanner in the works. If you are going to say that we need to consider a broader phenotype which might include language impairment, where do you stop? Should we start including ADHD and dyspraxic problems? Where do we draw the line?

Bailey: The short answer is that we don't yet know. Therefore, the way to progress is by gradually inching out from those characteristics that we are most confident about. Of course, if brain development is going badly awry, then in a small number of individuals other apparently unrelated disorders will also arise, but one would expect them to co-occur with the core features. Strategically, the way to proceed is to start with those features that we are most certain about and to use those to identify susceptibility alleles. We can then look at the problem from the other way round: identify individuals in the general population with susceptibility alleles and characterize the range of the behavioural phenotype.

Bishop: Our hope of finding genes is dependent on our conceptualization of the phenotype.

Bailey: And therefore the most efficient way of doing that is to take a definition on which there is reasonably good agreement. So one can either draw the threshold at PDD, or one can presume that in families selected for having two individuals affected by PDD and where there are also other relatives who show language or social abnormalities, that these relatives also carry some or all of the susceptibility alleles. One is somewhat less confident making that leap when assessing families of singletons because we are then assuming a genetic aetiology. Nevertheless it is still a necessary step as it is possible that there has been assortative mating in the parental generation in multiplex families which will reduce the power to identify susceptibility alleles.

Skuse: Thinking outside the box, I wondered whether there is a case to be made for looking at liability to the three major dimensions of autistic behaviour within a general population sample, and seeing whether these tend to go together, or whether in fact they are separable and occur independently of one another. For example, is there evidence that the pragmatic disorder of language, which is characteristic of high-functioning autistic individuals, is found in the general population of children irrespective of their diagnosis? I feel we could be making a big mistake if we assume these abnormalities of language processing are specifically associated with autistic disorders. We should separate a liability to cognitive deficit from liability to disorder as such. Is there a discontinuity in the distribution of a cognitive ability such as pragmatic language skills, in the sense for example of bimodality in the general population, or are we just looking at a normal distribution, with the autistics at an extreme? Does pragmatic disorder, for example, occur together with one or other features of the autistic phenotype, for example, over-sensitivity to certain sensory stimuli, or clumsiness? I don't think we know the answer to these questions yet, but we certainly have pretty good evidence from our own research that pragmatic deficits in language processing are far more widely distributed than in the autistic or PDD populations (Gilmour et al 2003). I'm not sure this answers the question of whether the same alleles are predisposing to language disorder, social disorder or repetitive behaviours, but one might get a clue from a novel approach to phenotype definition.

Rutter: Will the Utrecht study or the Checklist for Autism in Toddlers (CHAT) study provide that kind of information?

Charman: The CHAT study won't because we only targeted one domain of impairment: early social communication behaviours.

Buitelaar: In the study I mentioned earlier, we have the opportunity to look at the distribution in the general population of the separate dimensions of social interaction skills/deficits, communication skills/deficits and other problems. We will also be able to look at whether these dimensions are relatively independent from each other in the general population; or, alternatively, whether deficits in social interaction, communication, language, etc. co-occur. We may also be able to pull apart those components to see whether there is differential prediction across time for PDD at age 4–5 years. It may be that 14 months is too early to separate those components, so we might have to re-do this study at an older age.

Hollis: One of the problems when we start to move into studying the general population is that social difficulties are both common and diagnostically non-specific. So behavioural measures alone are likely to be fairly unhelpful in screening for a rare disorder such as autism. In the field of schizophrenia research (which faces similar issues of non-specific premorbid behaviour impairments) there is quite a lot of interest in refining early detection by combining behavioural measures with biological markers of abnormal brain activity, such as event related potentials (ERPs) and functional magnetic resonance imaging (fMRI).

Happé: We are going to try to look within the TEDS twin study at some limited number of items that we can examine in terms of social communicative versus non-social traits that might relate to autism. Tony Bailey, do we still have a pressing need to explain the association between autism and mental retardation if we take into account the apparently much higher number of individuals with Asperger's syndrome? If we look at the whole spectrum, how strong is the association between the autism spectrum and mental retardation?

Bailey: At the severe end it is very strong. One also has to think about this from a practical perspective: What causes morbidity and mortality in this population? A major cause of lifelong morbidity is intellectual impairment and a significant cause of mortality is epilepsy. Whilst we can become preoccupied with the fine details of impairments in narrow cognitive functions, if we are really interested in reducing impairment, then one of the priorities once susceptibility alleles are identified will be identifying those factors that cause more severe phenotypic expression. There is certainly no evidence that mild phenotypes represent a genetically distinct disorder from severe phenotypes and given that we are not interested in altering the gene pool, what we have to consider is influencing those interacting factors.

Happé: My question is not about clinical need, although that is obviously very important, as would be the association of autism with blindness or deafness, for example — anything that makes the picture more complex. Instead, it was just in terms of numbers: if you imagine that, hypothetically, in the future we found many more people that we wanted to put on the autism spectrum at high IQs, there would come a point at which autism wasn't associated with mental retardation above the base rate. I am not saying that we have approached that point, but I'm wondering how far along we have got.

Rutter: We should switch to Tony Charman, because his study is the one with the highest proportion of individuals with autism whose IQs were in the normal range. Tony, if we take Francesca Happé's question and apply it to your data, what odds ratio do you end up with in your total population for the rate of mental retardation in the autism group as compared with the general population.

Charman: I'm slightly hesitant here, because one of the limitations of our study on the existing data is that we have only directly assessed half the cases, in contrast to Chakrabarti & Fombonne's (2001) study where they assessed 95%. But the figures aren't so dissimilar. If you take the rate of one-quarter of individuals with an IQ under 70, we know from the normal distribution of IQs in the population that this would be 2.5% of the general population. There is still a difference of a factor of 10.

Happé: How old are they when you are assessing them?

Charman: In our study they are a variety of ages, from 3–7.

Happé: Seven is still too early to pick up most of the Asperger's cases.

Charman: That could be the case, but it is still a tenfold difference. The association of autism with mental retardation and with epilepsy is telling you something quite significant about the brain pathology.

Bailey: I'm not quite sure why you are pressing this line.

Happé: I am just curious: given the numbers at the high ability end of the spectrum, I'm just wondering whether we are getting anywhere near a point where there wouldn't be a strong association between autism and mental retardation. I'm quite happy to accept that there is such an association. Then there is the challenge of having to explain this.

Rutter: If the rate of autism is 60 per 10 000, and you assume a tenfold increase of mental handicap in that group, you are going to have to go up to a prevalence of 600 per 10 000 in order to wipe that out, assuming that all new cases are in the normal range. Anything is possible, but that seems pretty implausible.

U. Frith: I don't agree at all. I am glad we are discussing this association, but I do think there might be some progress made about separating this. If there is a general agreement that there are similarities between these normally intelligent people whether they have Asperger's syndrome or not, and others without this normal intelligence, then the intelligence question is something separate. Logically, this must be the case.

Rutter: How do you deal, then, with the fact that in the study Susan Folstein and I did, in one of the pairs of MZ twins with autism, there was more than a 50 IQ points difference between them?

U. Frith: Isn't that precisely a case in point for considering IQ as a separate issue? In answer to the question about the actual correlation between IQ and severity of autistic features, we just heard that in general, the spread of cognitive impairment in the more impaired proband is just as wide as in the less impaired co-twin. The degree of the difference between the co-twins is another matter.

Folstein: I'd like to bring people back to something we all know, and that is that IQ is not really a unitary concept in autism. Even the ones with normal intelligence have a very peculiar scatter of subtest scores. In terms of sorting out the genes, it is much more helpful to talk not about IQ *per se* but particular aspects such as executive function. If you look at it this way, abnormalities are not universal, but close to it.

Dawson: I was going to make a similar point. We need to keep in mind that mental retardation is defined by how individuals perform on an IQ test. It will be important to think about information processing measures that can differentiate between children that we think of as having real mental retardation as opposed to poor performance ability on an IQ test that is associated with autism. The social motivational, language and attentional impairments of autism play a role here. If we could think of a way to separate this out, it would be helpful. A lot of this might come from studying the field of mental retardation and some of the measures there that seem to be capturing this phenomenon. If we did this, in a genetics study we might be able to separate this out a little. These scores move around tremendously in the early years. They are not stable. They start to become more stable after age five. You also need to think about this in terms of analyses on a genetic study.

Happé: I agree entirely. From the small study that Uta Frith and I did (Scheuffgen et al 2000), there is some evidence that individuals who are scoring very poorly on standard IQ tests actually show good potential for processing information of various sorts at speed. We need to look below the surface of IQ.

Rutter: Do you want to comment on, what seems to me, a similar issue: namely, specific talents? I am struck with the frequency of these in autism. In the follow up study that Sue Goode and I did, about one in six had marked talents, although not at the extreme savant level. These seem not to occur at all commonly in individuals with a broader phenotype. The issue of why some autistic people have these extraordinary skills seems to me as puzzling as why there is mental retardation. The expectation of a different explanation for each one seems unhelpfully complicated.

Happé: The key thing is that we don't know at what rate some kind of special or isolated ability occurs in the relatives. This would be an interesting question. In the small study that we did (Briskman et al 2001) relatives were identifying themselves as exceptionally good proofreaders, for example. These are not things we will find out about unless we ask specific questions. The other kind of explanation, that Alan Snyder (e.g. Snyder & Thomas 1997) would favour, is that you have to have a somewhat impaired system in order to get this sort of runaway skill. Personally, I wouldn't favour that, but we do need the data.

Bolton: I'd like to make a general comment that leads on from the discussion we had earlier about our concepts about neurodevelopmental disorders. We need to ensure that we keep in our minds the fact that genetic effects are often pleiotropic and developmental. The effect is that we could wrongly assume that a neurodevelopmental abnormality is important in pathophysiology when in fact it is simply a correlated feature. For example, although brain changes in autism seem to arise early on during prenatal brain development, this doesn't necessarily mean that these processes are the ones that underlie the development of the syndrome. They may just be a marker of the pleiotropic and developmental effect of the genes and it is really later postnatal events that determine the development of autism. Therefore we shouldn't really confine ourselves to thinking that the putative causal processes may be purely prenatal. We have evidence from various other lines of enquiry that suggests that there might be postnatal events that can lead to the development of an autistic syndrome. It may be that postnatal environmental events, in interaction with the genetic effects are relevant. We have to remember that genes turn on and off during development: it may be that the genetic risk factors perturb early prenatal brain development and also later aspects of postnatal development, and it is these later changes that are relevant to the emergence of autism. This might be a consideration, for example, in thinking about the development of macrocephaly in children with autism. The genetic risk factors could also interact with postnatal environmental factors to cause autism.

Rutter: What postnatal influence have you in mind?

Bolton: Well, there is evidence from your own studies of Romanian adoptees who experienced extreme early deprivation in early infancy suggesting that they are quite prone to developing a quasi-autistic-like syndrome. There is also evidence from Peter Hobson's research that children with congenital blindness are prone to developing an autistic-like syndrome. In addition, the research that I have been doing on the development of autism spectrum disorders in children with tuberous sclerosis indicates that the emergence of epilepsy in early infancy is a marker of risk for autism-spectrum disorder. This raises the possibility that the abnormal electrophysiology that is going on at that time somehow perturbs brain development and the establishment of the cognitive representations that might be important for social development.

Folstein: This might be a good time to tell you about my experience in Dar es Salaam, Tanzania last summer (2001). I had a wrong-headed idea that I could find a more homogeneous genetic group by recruiting from a particular tribe. I knew that genetic variation in Africa is huge, but I thought that within Tanzania, where we had some contacts, we might be able to find a mostly Bantu population. This didn't work out: there were lots of different tribes in Dar es Salaam, which is the largest city in Tanzania. We went to a unit for autistic children that a friend of mine had started when she was a child psychiatrist at Mohimbili University at Dar es Salaam. We examined all the children that attended that school or who were on the waiting list. These were all singletons. We translated the ADI-R into Swahili, and as I was interviewing the mother of the first case, I asked her when she first noticed something wrong. She gave me a very specific date in December of some year. I decided I had better take a proper paediatric history. The long and short of it is that of the 16 children who met criteria for autism, half had onset following their first episode of cerebral malaria. Most of these children were non-verbal, but I was amazed that they really did meet criteria. In particular, they had compulsive behaviour that is so typical of autism. If it was quite severe—and some of these children were unconscious for several days—they even lost all their motor skills, but this was only temporary; the social abnormalities were permanent. The parents provided clear evidence that the children had normal language and social interactions before their malarial encephalitis. Several developed epilepsy following their encephalitis. It is a tiny sample, though.

Charman: What was the range of ages when they got the cerebral malaria?

Folstein: Between 18 and 40 months.

Bishop: Are you saying this is an alternative aetiology, or did these people also have a higher rate of broader phenotype in relatives.

Folstein: I tried to look at this, along with a few other factors. Head size was more likely to be big in the ones that had not had malaria. The sex ratio was a little different, with an equal male–female similarity in the ones who had had malaria. The cases that had a milder form of cerebral malaria—they hadn't been unconscious for extended periods—suffered a little pre-encephalitis delay in their language. It was very difficult for me to ask about the broader phenotype except for language.

Lipkin: Did any of these children have a Parkinsonian syndrome or tremor?

Folstein: No. I did examine them for that.

C. Frith: It is interesting that this increase in head size keeps coming up. It is apparently not present at birth. So it sounds as if something abnormal is happening with brain growth during the first decade of life. We know that at this time there is a huge reorganization of the brain in terms of synaptogenesis and pruning. This is a period when there could well be environmental effects on brain organization.

Pericak-Vance: Did you know anything about the siblings? Did they have malaria?

Folstein: Everyone gets malaria there, although there was one child who came from a middle-class family who had never had malaria. But none of the other children had any change in their autism with their first malarial episode. Most had clear symptoms before their first malaria and overall they had much milder cases, without encephalitis. I hypothesize that the subsequent outcome is related how young the child is when first infected and the parasite load of that initial infection.

Fombonne: How long were they unconscious for?

Folstein: It varied. One child was unconscious for a week.

Bauman: I know it's a long shot, but were any imaging studies done?

Folstein: No. I could get CAT scans on these children if I sent the money. Even just finding the records of every child that survived cerebral malaria at that age range is challenging. The records are very poor.

Bauman: Has anyone looked at postmortem brains of these children?

Folstein: No.

Bauman: Does anyone know what the neuropathology of malaria is in the brain?

Folstein: I've looked this up, and there isn't a great deal.

Lipkin: The neuropathology is chiefly intravascular coagulation and strokes.

Folstein: And this seems to be more common in the temporal areas. It may not be anything to do with malaria as such.

Lipkin: Malaria is also associated with profound hyperthermia and cytokine release. These indirect effects of infection could also be important to the pathobiology of an autistic syndrome.

Sigman: I have been working in Africa, in rural Kenya, since 1985. For a while I couldn't find any children with autism. When I asked my Kenyan collaborators if they knew of cases, they would introduce me to mildly shy children. More recently, we have found a school in our community with a fair number of children with autism. Given the high prevalence of malaria, you would expect to see a lot of affected children if malaria was a cause of autism. In fact, I've seen very few children with autism. If you would like to come into my community you could look, because we have been there for such a long time and have followed many children.

Folstein: I was there for a short time, and the head teacher at the school invited whomever they thought was autistic. The other thing, at least in this school, is that they didn't call people autistic if they had reasonable language. There were a couple of autistic children who did turn up with good language and the nurse tried to turn them away. In this particular place they had a very narrow view of what autism was.

Sigman: In the 1980s, there were very few children with developmental disabilities in the schools. Now, a classroom has been set up in one of the schools for children with developmental delays. None of those children are autistic. This surprised me because I would have expected to see some children with autism in this group.

Charman: Have you got a general interpretation of why you think you are finding so few children with autism among the thousands that you have seen?

Sigman: I don't know. We have certainly seen lots of children with malaria. This is really interesting and needs to be followed up.

Fombonne: I don't know if you are aware that there are areas in West Africa where it is not possible to find individuals with Down syndrome. When families have children with Down syndrome they routinely practice infanticide by abandoning them in the bush. This might be an explanation for why autism is rare in some areas, or at least this possibility should be entertained and/or ruled out.

Sigman: That is a possibility. It is also possible that a different set of criteria are used for identification and diagnosis, as Susan Folstein has mentioned. Malaria is endemic in Africa and one would expect to see huge numbers of cases.

Folstein: You would predict that there would be a higher prevalence of autism in Africa than elsewhere, for that reason.

Fombonne: Did these people have specific clinical characteristics or different ADI scores?

Folstein: One didn't meet the criteria on count of age of onset, but that's it. They were low functioning.

References

Briskman J, Happé F, Frith U 2001 Exploring the cognitive phenotype of autism: weak 'central coherence' in parents and siblings of children with autism. II. Real-life skills and preferences. J Child Psychol Psychiatry 42:309–316

Chakrabarti S, Fombonne E 2001 Pervasive developmental disorders in preschool children. J Am Med Assoc 285:3093–3099

Gilmour J, Hill B, Place M, Skuse DH 2003 Social communication deficits in conduct disorder: a clinical and community survey. J Child Psychol Psychiatry, in press

Le Couteur A, Bailey AJ, Goode S et al 1996 A broader phenotype of autism: the clinical spectrum in twins. J Child Psychol Psychiatry 37:785–801

Rapin I 1996 Preschool children with inadequate communication: developmental language disorder, autism, low IQ. Clinics in Developmental Medicine 139. MacKeith Press, London

Scheuffgen K, Happé F, Anderson M, Frith U 2000 High 'intelligence', low 'IQ'? Speed of processing and measured IQ in children with autism. Dev Psychopathol 12:83–90

Snyder AW, Thomas M 1997 Autistic artists give clues to cognition. Perception 26:93–96

Strategies for autism candidate gene analysis

G. Barnby and A. P. Monaco[1]

Wellcome Trust Centre for Human Genetics, Roosevelt Drive, Headington, Oxford OX3 7BN, UK

Abstract. The identification of autism susceptibility genes has moved a step closer over the last four years with the completion of eight whole genome screens for linkage. Several overlapping areas of linkage have been reported, most notably on chromosomes 7q22-31 and 2q32. These regions of replicated linkage provide a focus to search for candidate genes whose normal functions in neurodevelopment are altered to increase the risk for autism. Strategies that aim to narrow further the rather broad size of these linkage regions, such as high density single nucleotide polymorphism (SNP)-based association studies, currently suffer from practical and statistical limitations. Alternatively, positional candidate genes can be screened for deleterious variants in autistic individuals selected from large samples such as those collected by the International Molecular Genetic Study of Autism Consortium (IMGSAC). Targeted genotyping of candidate gene variants in this large multiplex family sample will then be performed to confirm association with autism.

2003 Autism: neural basis and treatment possibilities. Wiley, Chichester (Novartis Foundation Symposium 251) p 48–69

It was hoped that autism would be one of the most tractable of the multigenic psychiatric disorders to unravel based on its high heritability. Siblings of probands with autism were estimated to be around 100 times more likely to be autistic compared with the general population (Szatmari et al 1998). However, recent epidemiological studies have revised the population prevalence of autism upwards from 4 in 10 000 to 12–17 in 10 000 (Chakrabarti & Fombonne 2001). One possible consequence of this increase in population prevalence is a revision downwards of the sibling relative risk, markedly reducing the familial clustering of autism. Explanations of this apparent increase in population prevalence are varied and include many environmental triggers such as diet and pollutants, but these must be balanced with considerations of improved diagnosis and awareness

[1]This paper was presented at the symposium by A. P. Monaco to whom correspondence should be addressed.

of autism. Comparison of previous and current prevalence rate investigations are also confounded by changes in screening procedures and methodological differences between studies. Despite these recent findings, autism is still considered highly heritable and has a strong familial clustering compared to other psychiatric disorders such as schizophrenia ($\lambda s = 10$)(Williams et al 1999), attention deficit hyperactivity disorder ($\lambda s = 5$) (Fisher et al 2002) and bipolar disorder ($\lambda s = 7$) (Kelsoe et al 2001).

The core triad of behaviours characteristic for autism are impairments in verbal and non-verbal communication, reciprocal social interaction and restricted activities and repetitive movements. The refinement of autism diagnostic tools, such as the Autism Diagnostic Interview—Revised (ADI-R) and the Autism Diagnostic Observation Schedule—Generic (ADOS-G) continues to contribute to delineating the autistic phenotype and reducing heterogeneity within study samples. Improved understanding and characterization of the broader phenotype within multiplex families will facilitate analysis of genetic data with the aim of disentangling locus specific contributions from different facets of the autistic phenotype.

Autism is found in three times more males than females suggesting a possible role for the sex chromosomes in its genetic aetiology. This has only been mildly supported by two genome screen reports (Liu et al 2001, Shao et al 2002). Although several hypotheses have been proposed to account for such a skew in sex ratio (Skuse 2000), no genes have been implicated as yet. Additionally, multivariate analysis of phenotypic expression in relatives by sex does not support X-linkage under a multifactorial threshold model (Pickles et al 2000).

Both chromosomal abnormalities and disorders of known genetic aetiology are associated with autism. The most frequent chromosomal region affected is 15q11-q13 which overlaps with the Prader–Willi/Angelman critical region (PWACR). For reviews of chromosomal abnormalities associated with autism see Gillberg (1998) and Wassink et al (2001a). Cases of fragile-X syndrome, neurofibromatosis and tuberous sclerosis are also elevated in autistic individuals (Feinstein & Reiss 1998, Smalley 1998, Mbarek et al 1999). Their increased co-morbidity with autism may indicate some contribution to the autistic phenotype, but this only accounts for a small proportion of total cases of autism, and as such they are unlikely to be major susceptibility genes. Assessment of affected individuals for linkage studies generally includes testing to exclude individuals with known medical and genetic disorders, and abnormal karyotypes.

Autism genome screens

Eight whole genome screens for autism have been completed to date with loci implicated on all chromosomes except 11, 12, 14, 20 and 21 (Fig. 1).

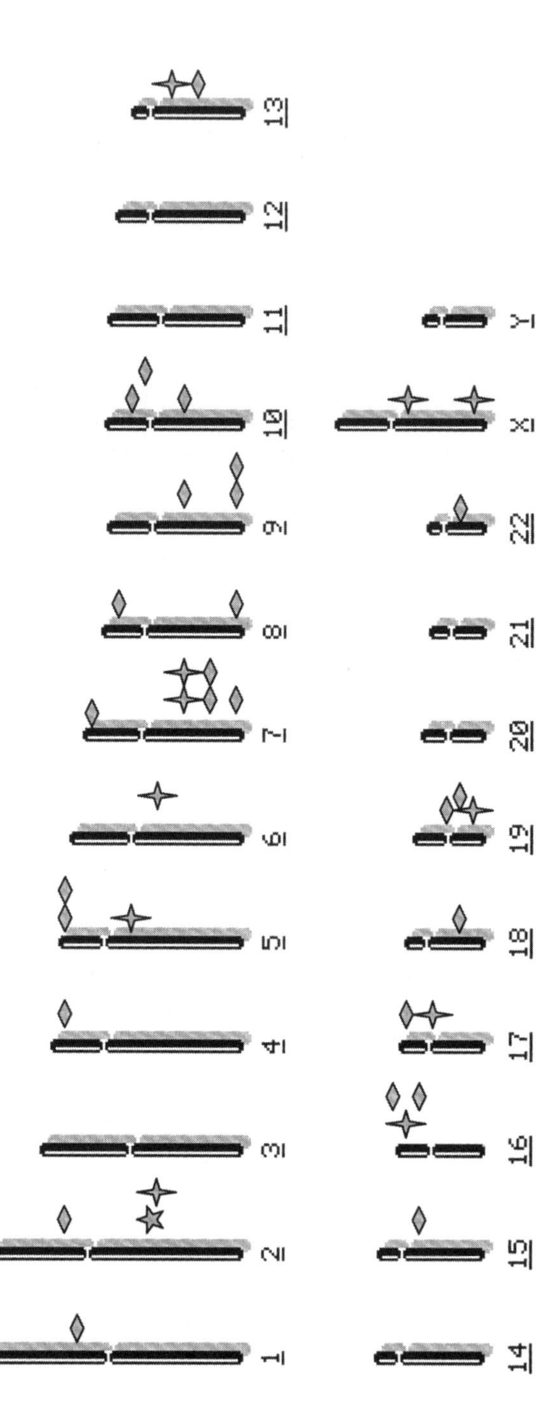

FIG. 1. Autism genome screen results.

Undoubtedly a number of these loci will prove to be false positives and given the relatively small number of affected sibling pairs being analysed (between 50 and 150) a number of true loci of small effect could remain unidentified. Establishing a realistic cut off level for significance remains problematic for complex diseases. Traditionally, LOD scores >3 are used to declare significant linkage for monogenic disorders. This has been revised for genome wide screens in complex diseases to a significant linkage LOD score threshold >3.6 i.e. statistical evidence for linkage that would be expected to occur one time at random in 20 genome screens (Lander & Kruglyak 1995). However, limitations in available sample sizes and issues such as heterogeneity, where affected individuals have different underlying genotypes at different disease loci, has led to guidelines for suggestive linkage of a maximum LOD score (MLS) >2.2 i.e. statistical evidence for linkage that would be expected to occur one time at random in a genome screen (Lander & Kruglyak 1995).

Of the linkages reported for autism, significant linkage has been found at only one locus on chromosome 2q32 with an MLS of 3.74 at D2S2118. Eight loci have been reported with MLS above the suggestive linkage threshold: D7S477 with an MLS of 3.2, D16S3102-2.93, HTTINT2-2.34 (IMGSAC 2001a), D6S283-2.23, D7S1813-2.2 (Philippe et al 1999), D13S800-3.0 (CLSA 2001), D2S364-2.25 (Buxbaum et al 2001) D5S2494-2.55 DXS1047-2.56 (Liu et al 2001) D19S714-2.53 and DXS6789-2.54(Shao et al 2002) (Fig. 1 and Table 1). Concentrating on overlapping areas of linkage detected by multiple independent studies could also

TABLE 1 Maximum LOD scores (MLS) above 2.2 reported in autism whole genome screens

Chromosome	*Marker*	*MLS*	*Reference*
2	D2S2118	3.74	(IMGSAC 2001b)
7	D7S477	3.2	(IMGSAC 2001b)
16	D16S3102	2.93	(IMGSAC 2001b)
17	HTTINT2	2.34	(IMGSAC 2001b)
6	D6S283	2.23	(Philippe et al 1999)
7	D7S1813	2.2	(Philippe et al 1999)
13	D13S800	3.0	(CLSA 2001)
2	D2S364	2.25	(Buxbaum et al 2001)
5	D5S2494	2.55	(Liu et al 2001)
X	DXS1047	2.56	(Liu et al 2001)
19	D19S714	2.53	(Shao et al 2002)
X	DXS6789	2.54	(Shao et al 2002)

help distinguish between chance findings and the true location of susceptibility genes.

Meta analysis

A meta analysis was carried out on four published autism genome screens by Badner & Gershon (2002). They used multiple scan probability statistics (MSPs), which combine reported *P* values and test whether they occur more often than expected by chance in the same region, and found that the frequency and magnitude of reported linkage to 7q and 13q was significant (Badner & Gershon 2002). However, this analysis did not include all the genome screens reporting linkage to 2q32. The authors note that this type of regional meta analysis of published data suffers from publication bias of positive overlapping results and from sample heterogeneity between studies. Proposed meta analysis of raw genotypic data also suffers from the problem of between study sampling differences which is further compounded by potential genetic heterogeneity and epistasis and the use of different genetic markers.

Meta analysis of genotype data of complex diseases has been carried out and has proved fruitful in inflammatory bowel disease (IBD), demonstrating the utility of such large collaborations on analysis. An international collaboration pooled data from 12 microsatellites in 613 affected sibling pairs (ASPs). A multipoint MLS of 5.79 on chromosome 16 was generated for Crohn's disease (Cavanaugh 2001). The underlying susceptibility gene has now been identified within this region (Hugot et al 2001, Ogura et al 2001).

Fine mapping by linkage analysis

Fine mapping regions of linkage with high-density maps of microsatellite markers was initially seen as an effective method for narrowing the search for candidate genes. Following disappointing findings with this strategy and revisions and improvements in statistical genetics, a re-assessment of fine-mapping strategies may be necessary. Figure 2 shows the changes in multipoint MLS across chromosome 7 as increasing numbers of microsatellite markers were genotyped and ASPs were added to the IMGSAC sample (IMGSAC 2001a). Analysis of seven microsatellite markers in 91 ASPs using ASPEX generated a multipoint MLS of 2.53 between D7S530 and D7S684 at 144.7cM cM. Following the addition of 62 ASPs and 72 microsatellite markers, analysis using ASPEX gave a multipoint MLS of 3.37 at D7S477 positioned at 119.6cM. The change in both the position and width of the linkage curve is striking although it has been suggested that this is not unexpected, due to chance variations in the detection of linkage even with large numbers of multiplex families (Roberts et al 1999, Cordell 2001).

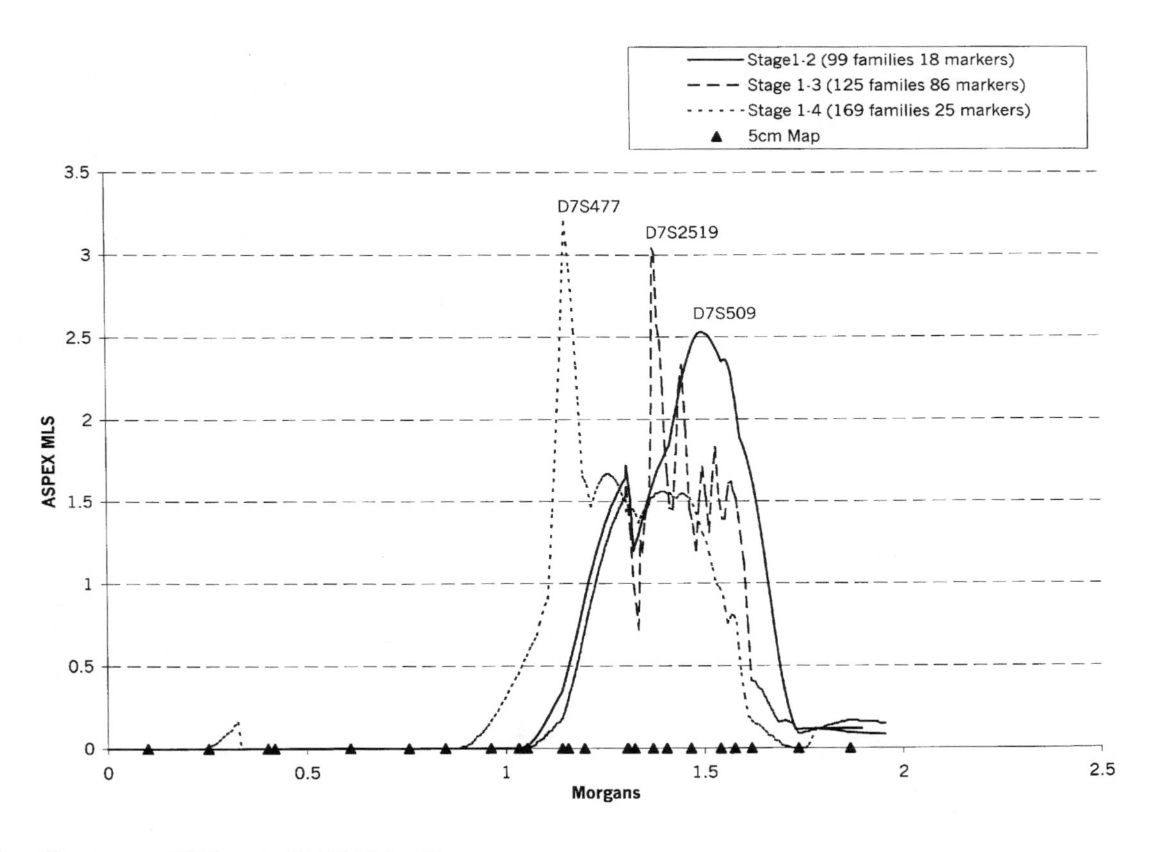

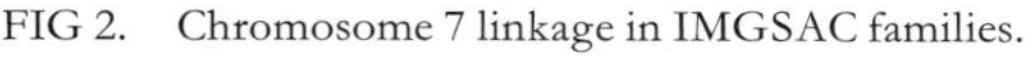

FIG 2. Chromosome 7 linkage in IMGSAC families.

Analysis strategies based on stringent phenotypic definitions, to reduce sample heterogeneity, and conditioning one locus on another have been suggested for improving linkage location estimates (Cordell 2001). A number of criteria have been used by different groups to select more homogeneous groups of families. These include criteria based on the presence or absence of speech, delay in speech or phrases, IQ level, ADOS and ADI scores and nationality. Analysis of IMGSAC data based on severity of autism criteria has been carried out at 7 loci. Although different subsets of affected individuals seem to be responsible for linkage at different loci there was no significant narrowing of the linkage region (IMGSAC 2001b). However, this type of approach remains problematic because power is lost as of the number of families being analysed is reduced.

Association studies

Analysis of IMGSAC genome screen data for association using the family based transmission disequilibrium test (TDT) (Spielman et al 1993) was carried out on chromosome 7q. Three markers were found to be associated with autism in 170 families; D7S477 (119.6cM), D7S2437 (146.59cM) and D7S2533 (148.89cM) (IMGSAC 2001a). D7S2533 was also found to be associated in an independent German sample of 86 trios. However no linkage disequilibrium (LD) was found between the associated alleles of the three markers. Additionally three markers between D7S2437 and D7S2533 also failed to show any association although a number of other markers in the general region show a trend toward being associated both in the German and IMGSAC samples. Microsatellite loci separated by many megabases have been found to detect LD between markers where single-nucleotide polymorphism (SNP) surveys have not, possibly due to greater power from multi-allelic markers (Pritchard & Przeworski 2001). Fine-mapping can potentially narrow regions of linkage and detect association, but the benefit of adding extra markers is also limited by increasing map error and saturation of informativeness. A 5 cM map of markers is sufficient for extracting linkage data whereas a much higher density of markers may be necessary to detect association (Feakes et al 1999).

Association studies aim to detect differences in the frequency of a marker allele (either from a microsatellite or SNP) in cases compared to controls. The family-based TDT is an extension of this which uses data from heterozygous parents to determine the transmission of an allele identical by descent to affected offspring, thus controlling for population stratification effects. The hypothesis under consideration is that the allele being transmitted is the disease-causing variant itself or is in LD with the causative mutation. Debate concerning the extent of LD in the genome is ongoing and will hopefully be resolved definitively in the near future as whole chromosome LD maps are generated. The outcome of this debate and current research is essential for guiding strategies for identifying positional candidate genes. The underlying assumption for successful use of association strategies to identify susceptibility genes, is that a disease will be caused by common gene variants and these variants will be in LD with other common polymorphisms and detectable through them.

Linkage disequilibrium in the genome

Early predictions that LD stretches up to 100 kb have been questioned by simulations which found that useful LD was unlikely to extend beyond 3–5 kb in human populations (Kruglyak 1999). This has several implications. Firstly, any systematic screen for association throughout the whole genome will need to type ~500 000 SNPs. Secondly, any targeted SNP typing around positional candidate genes must be densely packed to avoid missing true associations. Finally, such large amounts of data will require appropriate statistical handling to correct for multiple testing (Kruglyak 1999). The assumptions of human demographic history underlying these simulations have been challenged. Incorporating population bottlenecks or observed levels of nucleotide diversity and assuming steady population size in simulations gave a less conservative estimate of one SNP per 20 kb as sufficient to detect LD on average. The heterozygosity of SNPs chosen also affects the distance at which LD can be detected with older and more heterozygous SNPs being in weaker LD (Collins et al 1999, Ardlie et al 2002).

The diversity of findings from simulation studies are comparable to the diversity of empirical findings. For a recent review see Ardlie et al (2002). The size of regions surveyed ranged from less than 10 kb to entire chromosomes, study sizes ranged between 10 and 962 individuals and between 4 and 24 000 markers were typed. Comparisons are difficult to make between the studies because different measures of LD were used and different populations were studied (Ardlie et al 2002).

An example of the variation in LD found in the genome is illustrated by SNP typing carried out around apolipoprotein E (*APOE*), a well established susceptibility gene for late-onset Alzheimer's disease (AD) (Martin et al 2000). Analysis of 60 SNPs, in 220 AD cases and 220 controls, within a 1.5 Mb region around *APOE* found sixteen SNPs to be associated with AD (Martin et al 2000). Seven of these SNPs clustered within 40 kb of the gene with the strongest association 16 kb away. Interestingly, several SNPs closer to the *APOE* susceptibility allele showed no evidence for association. The authors concluded that associations can be detected for a complex disease gene, but in order to do so the choice of SNPs is crucial.

Variation in LD detected in empirical studies is substantial and unpredictable. Physical distance appears to account for below 50% of variation in LD. The remainder can be ascribed to genetic drift, demographic factors and variable rates of mutation, recombination hot and cold spots and gene conversion. Association studies relying on LD face serious challenges in terms of both initial detection of susceptibility loci and identification of candidate genes (Abecasis et al 2001).

Candidate gene association studies

A number of functional candidate genes based on biochemical and neurological findings in autistic individuals have been tested for association. The usefulness of this type of approach has been limited by the dearth of exceptional and abundance of possible candidate genes. However, emerging consensus in neurobiological findings and improved sensitivity of diagnosis may guide candidate gene choice in the future.

Variants in the serotonin transporter (*5HTT*) gene have been tested for association with autism following the observation that approximately one-third of autistic individuals have elevated levels of whole blood or platelet serotonin. Conflicting results from typing the functional deletion/insertion in the promoter region (5HHTLPR) of *5HTT* have been published. A recent study and review found no association and concludes that the *5HTT* gene is unlikely to play an important role in the genetic aetiology of autism (Betancur et al 2002). However, a more detailed study of 27 polymorphic markers near *5HTT* genotyped in 81 autism trios, does find evidence for association with four markers, although not with the 5HHTLPR polymorphism (Kim et al 2002). These results are not in themselves surprising within the context of association studies in other complex diseases which have frustratingly failed to replicate initial positive findings (Dahlman et al 2002).

Candidates within the PWACR on chromosome 15q11-13 have also been examined including the γ-aminobutyric acid (GABA) receptor β3 gene (*GABRB3*). GABA is the chief inhibitory neurotransmitter in the brain and association with autism was found in a sample of 138 probands, but this has not been replicated by a number of studies. For a review of this and other candidate gene studies see Folstein & Rosen-Sheidley (2001).

Other negative association findings include testing known polymorphisms in 10 positional functional candidate genes tested in 39 multiplex families by Philippe et al (2002): proenkephalin (*PENK*), prodynomorphin (*PDYN*), proprotein convertase subtilisin/kexin type2 (*PCSK2*), dopamine receptor D2 (*DRD2*) and D5 (*DRD5*), tyrosine hydroxylase (*TH*) and monoamine oxidases A (*MAOA*) and B (*MAOB*), brain derived neutrophic factor (*BDNF*) and neural cell adhesion molecule (*NCAM*).

Studies screening candidate genes

The initial appeal of large-scale association studies has become diminished by the above theoretical and practical issues. An alternative approach based on screening the entire coding sequence of candidate genes in affected individuals is becoming increasingly attractive. One such study by Wassink et al (2001b) screened 135 affected individuals and 160 controls across the entire *WNT2* (wingless-type MMTV integration site family member 2) gene coding sequence. Two non-conservative missense mutations were found in conserved residues segregating in two independent families and association was found with a SNP in the 3′UTR. *WNT2* is an interesting candidate gene for a number of reasons. It is within the 7q31-33 candidate region, is important in the development and patterning of the vertebrate central nervous system and a knockout mouse of dishevelled (*Dvl*), involved in WNT signalling, shows reduced social interaction (Wassink et al 2001b). Variants found in the homeobox (HOX) genes *HOXA1* and *HOXB1*, involved in cortical to hindbrain development, have also been tested for association with some evidence for association to a SNP in *HOXA1* (Ingram et al 2000).

A number of recent publications have failed to support these initial positive findings in larger samples, for example *WNT2* (McCoy et al 2002) and *HOXA1* and *HOXB1* (Li et al 2002).

Co-localization to chromosome 7q31 of a gene mutated in severe speech and language disorder, forkhead box P2 (*FOXP2*) (Lai et al 2001), and the autism susceptibility locus *AUTS1* has led to the suggestion that this represents a good candidate gene, especially given the high frequency of speech impairments in autism. Screening of the entire coding sequence of *FOXP2*, has been carried out in IMGSAC and Specific Language Impairment Consortium (SLIC) families (Newbury et al 2002). No evidence for association was found in 169 IMGSAC families using intragenic microsatellite markers and two SNPs tested, and no coding changes were detected in 48 affected autistic individuals whose coding sequence was screened.

A number of candidate genes beneath the peaks of linkage of the IMGSAC genome screen, are now being considered for screening. The three highest multipoint MLS' are at: D2S2188 on chromosome 2q with a multipoint MLS of 3.74, D7S477 on chromosome 7q22 reaching a multipoint MLS of 3.20 and at D16S3102 on chromosome 16p with a multipoint MLS of 2.93 (IMGSAC 2001b). Two Distal-less homeobox genes (*DLX1* and *DLX2*) are located on 2q32, interestingly their homologues *DLX5* and *DLX6* are located on 7q22 and have been found to be cross regulated in mouse. Other good candidate genes in the region include T Brain 1 (*TBR1*), Neuropilin (*NRP2*), and cAMP response element-binding protein 1 and 2 (*CREB1* and *CREB2*) as well as α-1-chimaerin (*α1CHIM*) and the alternatively spliced α2CHIM, a GTPase-activating protein (GAP).

The imprinting status of the 7q32 region has been examined along with four adjacent candidate genes; mesoderm specific transcript homologue (*PEG1/MEST*), coatomer protein complex subunit gamma 2 (*COPG2*) and the carboxypeptidases A1 and A5 (*CPA1* and *CPA5*). The results do not suggest a major role for imprinting or these genes in the genetic aetiology of autism at the chromosome 7 susceptibility locus (Bonora et al 2002). Reelin on 7q22 has also been suggested as a candidate gene for autism following TDT analysis of a polymorphic trinucleotide repeat in the 5′ UTR (Zhang et al 2000, Persico et al 2001).

Candidates to consider on chromosome 16 include an *N*-methyl-D-aspartate receptor (*NMDA*) channel subunit gene. The NMDA channel is formed by two glutamate-receptor channel subunits which when knocked out have been found to cause spatial learning deficits in mice. The CREB binding protein (CREBBP) which activates transcription of cAMP-responsive genes is also beneath the peak of linkage. Micro deletions of the gene are found in Rubinstein–Taybi syndrome (RTS) whose features include mental retardation and characteristic facial features. Autistic features have also been reported in a case of RTS (Hellings et al 2002) Tuberous Sclerosis 2 (*TSC2*) is also located distally to the peak of linkage.

Discussion

The conflict between expediency and thoroughness can be seen in the variety of methods being used to examine candidate genes for autism susceptibility. As discussed above, the choice of SNPs typed can determine whether any association is detected, making it difficult to exclude candidate genes on the basis of a single marker. Even given a SNP density of one per 20–50 kb we will not be able to assume that we have captured the underlying haplotypic structure between variants. 'Islands of linkage disequilibrium' have been found to exist in various regions of the genome, including the class II region of the MHC, 5q31, chromosome 21 and 22q12.1, enabling haplotypic tagging. This exploits known LD between SNPs allowing one SNP to be used as a proxy for many others, reducing the volume of genotyping needed (Daly et al 2001, Eisenbarth et al 2001, Goldstein 2001, Jeffreys et al 2001, Johnson et al 2001, Patil et al 2001, Rioux et al 2001). However the variation in LD found by these studies indicates that an empirical assessment of LD in a region should be made before embarking on high marker density disease mapping studies. Extrapolating the extent of LD from another region or on the basis of simulation studies could lead to inappropriate density of markers being typed (Ardlie et al 2002).

Three hypotheses can be considered for autism genetic susceptibility: (1) common population variants occur in an individual and if these reach a threshold number an autistic phenotype is expressed, (2) common population variants are present and rare variants are also necessary, and (3) rare variants only are sufficient to cause an autistic phenotype. If the rare variant hypothesis is true, from a population genetics perspective we would not expect to detect association, because there will only be a small number or an absence of shared haplotypes in the affected population. This suggests that a more exhaustive screening of candidate genes is necessary than can be accomplished using association study techniques, as was undertaken for *FOXP2*.

Simulations show that strong selection, as seen in autism, will keep disease alleles at a low frequency and LD distances short (Pritchard 2001). Simulations on the basis of autism genome screen data (Risch et al 1999) also found that the distribution of λs is likely to be skewed with the top 15 loci having a λs range of 3.89–1.04 and the remainder contributing little to population risk. Linkage data will detect differences in sharing IBD at a locus regardless of the frequency of the underlying mutation. This data can be utilized to select families who show increased sharing in an attempt to reduce genetic heterogeneity and focus candidate gene screening efforts on a subset of individuals.

An important and related issue to the identification of rare disease-causing variants is discerning between significant and expected differences in their frequency in coding sequence. Analysis of sequence and amino acid conservation and known protein structure may indicate whether coding variants are likely to affect the structure and function of a candidate gene. Concurrent screening of control groups is crucial before variants can be labelled 'disease causing mutations'. There are few detailed assessments of the prevalence and distribution of rare variants within genes. One study by Glatt et al (2001) screened 450 individuals for variants in the serotonin transporter SLC6A4 and the monoamine transporter SLC18A2. In total 41 variant sites were found with SLC6A4 having significantly more coding variants than SLC18A2. The low frequency of coding SNPs (cSNPs) found suggests that a cSNP database would be incomplete unless a large reference population was used. Furthermore the gene specific frequency of coding variants supports extensive screening of control groups to determine the baseline variation in a gene. Even though the frequency of any individual variant may be low, the aggregated frequency of these high risk mutations may be significantly different from control samples (Glatt et al 2001). In genes where there is a low frequency of coding variants screening for more frequent non-coding variants affecting expression patterns in regulatory regions may be necessary.

Positive reports of association with autism and detection of coding variants in candidate genes must be seen within the context of the limitations of LD mapping approaches and the paucity in knowledge of underlying rare variant frequencies.

Current research strategy

We are currently pursuing the following strategy for candidate gene analysis. Gene selection will be based on position in the region of linkage and known function, and followed by screening of the entire coding sequence. Screening will be narrowed to a subset of autistic individuals from families showing increased sharing at that locus and carried out by either direct sequencing or denaturing high-performance liquid chromatography (DHPLC). This should enable the identification of all variants found in the coding sequence and surrounding genomic sequence. Subsequent typing of selected variants with a range of heterozygosities in the entire IMGSAC sample and 200 controls will provide data for association analysis. Any coding variants will be analysed for evolutionary conservation and potential functional impact using bioinformatics tools. It will be essential for any findings to be replicated in an independent sample to confirm their importance and ultimately functional data will be required.

Each disease has its own unique evolutionary history, and it is difficult to predict whether susceptibility genes for autism have only recently become significant players in human development. The aggregation of mutations in these genes challenges the plasticity of human neurobiological development to such an extent that they result in an autistic phenotype. Unlike chronic late onset diseases, where susceptibility alleles may have been previously selectively neutral and only 'uncovered' with changes in life expectancy and lifestyle, the elimination of mutations leading to autism by decreased procreativity in severe cases will have prevented these variants becoming high in frequency (Weiss & Terwilliger 2000). A combination of strategies may be necessary to detect both low frequency rare variants and higher frequency common variants and to successfully identify autism susceptibility genes.

Acknowledgements

This work would not have been possible without the co-operation of both the individuals with autism and their families and the many referring professionals. We also thank the secretarial support staff at the Wellcome Trust Centre and Janine Lamb, Simon Fisher and Angela Marlow for a critical reading of this manuscript. This work has been funded by the UK Medical Research Council, The Wellcome Trust, BIOMED 2 (CT-97-2759), EC Fifth Framework (QLG2-CT-1999-0094), Telethon—Italy (E.1007), the Janus Korczak Foundation, Deutsche Forschungsgemeinschaft, Fondation France Télécom, Conseil Régional Midi-Pyrénées, Danish Medical Research Council, Sofiefonden, the Beatrice Surovell Haskells Fund for Child Mental Health Research of Copenhagen, the Danish Natural Science Research Council (9802210), the National Institute of Child Health and Development (5-P01-HD-35482) and the National Institutes of Health (MO1 RR06022 GCRC NIH, NIH K05 MH01196, K02 MH01389), particularly the Collaborative Programs for Excellence in Autism research. A.P.M. is a Wellcome Trust Principal Research Fellow.

References

Abecasis GR, Noguchi E, Heinzmann A et al 2001 Extent and distribution of linkage disequilibrium in three genomic regions. Am J Hum Genet 68:191–197

Ardlie KG, Kruglyak L, Seielstad M 2002 Patterns of linkage disequilibrium in the human genome. Nat Rev Genet 3:299–309

Badner JA, Gershon ES 2002 Regional meta-analysis of published data supports linkage of autism with markers on chromosome 7. Mol Psychiatry 7:56–66

Betancur C, Corbex M, Spielewoy C et al 2002 Serotonin transporter gene polymorphisms and hyperserotonemia in autistic disorder. Mol Psychiatry 7:67–71

Bonora E, Bacchelli E, Levy ER et al 2002 Mutation screening and imprinting analysis of four candidate genes for autism in the 7q32 region. Mol Psychiatry 7:289–301

Buxbaum JD, Silverman JM, Smith CJ et al 2001 Evidence for a susceptibility gene for autism on chromosome 2 and for genetic heterogeneity. Am J Hum Genet 68:1514–1520

Cavanaugh J 2001 International collaboration provides convincing linkage replication in complex disease through analysis of a large pooled data set: Crohn disease and chromosome 16. Am J Hum Genet 68:1165–1171

Chakrabarti S, Fombonne E 2001 Pervasive developmental disorders in preschool children. J Am Med Assoc 285:3093–3099

CLSA Collaborative Linkage Study of Autism 2001 An autosomal genomic screen for autism. Am J Med Genet 105:609–615

Collins A, Lonjou C, Morton NE 1999 Genetic epidemiology of single-nucleotide polymorphisms [see comments]. Proc Natl Acad Sci USA 96:15173–15177

Cordell HJ 2001 Sample size requirements to control for stochastic variation in magnitude and location of allele-sharing linkage statistics in affected sibling pairs. Ann Hum Genet 65:491–502

Dahlman I, Eaves IA, Kosoy R et al 2002 Parameters for reliable results in genetic association studies in common disease. Nat Genet 30:149–150

Daly MJ, Rioux JD, Schaffner SF, Hudson TJ, Lander ES 2001 High-resolution haplotype structure in the human genome. Nat Genet 29:229–232

Eisenbarth I, Striebel AM, Moschgath E, Vogel W, Assum G 2001 Long-range sequence composition mirrors linkage disequilibrium pattern in a 1.13 Mb region of human chromosome 22. Hum Mol Genet 10:2833–2839

Feakes R, Sawcer S, Chataway J et al 1999 Exploring the dense mapping of a region of potential linkage in complex disease: an example in multiple sclerosis. Genet Epidemiol 17:51–63

Feinstein C, Reiss AL 1998 Autism: the point of view from fragile X studies. J Autism Dev Disord 28:393–405

Fisher SE, Francks C, McCracken JT et al 2002 A genomewide scan for loci involved in attention-deficit/hyperactivity disorder. Am J Hum Genet 70:1183–1196

Folstein SE, Rosen-Sheidley B 2001 Genetics of autism: complex aetiology for a heterogeneous disorder. Nat Rev Genet 2:943–955

Gillberg C 1998 Chromosomal disorders and autism. J Autism Dev Disord 28:415–425

Glatt CE, DeYoung JA, Delgado S et al 2001 Screening a large reference sample to identify very low frequency sequence variants: comparisons between two genes. Nat Genet 27: 435–438

Goldstein DB 2001 Islands of linkage disequilibrium. Nat Genet 29:109–111

Hellings JA, Hossain S, Martin JK, Baratang RR 2002 Psychopathology, GABA, and the Rubinstein–Taybi syndrome: a review and case study. Am J Med Genet 114:190–195

Hugot JP, Chamaillard M, Zouali H et al 2001 Association of NOD2 leucine-rich repeat variants with susceptibility to Crohn's disease. Nature 411:599–603

IMGSAC International Molecular Genetic Study of Autism Consortium 2001a Further characterization of the autism susceptibility locus AUTS1 on chromosome 7q. Hum Mol Genet 10:973–982

IMGSAC International Molecular Genetic Study of Autism Consortium 2001b A genomewide screen for autism: strong evidence for linkage to chromosomes 2q, 7q, and 16p. Am J Hum Genet 69:570–581

Ingram JL, Stodgell CJ, Hyman SL et al 2000 Discovery of allelic variants of HOXA1 and HOXB1: genetic susceptibility to autism spectrum disorders. Teratology 62:393–405

Jeffreys AJ, Kauppi L, Neumann R 2001 Intensely punctate meiotic recombination in the class II region of the major histocompatibility complex. Nat Genet 29:217–222

Johnson GC, Esposito L, Barratt BJ et al 2001 Haplotype tagging for the identification of common disease genes. Nat Genet 29:233–237

Kelsoe JR, Spence MA, Loetscher E et al 2001 A genome survey indicates a possible susceptibility locus for bipolar disorder on chromosome 22. Proc Natl Acad Sci USA 98:585–590

Kim SJ, Cox N, Courchesne R et al 2002 Transmission disequilibrium mapping at the serotonin transporter gene (SLC6A4) region in autistic disorder. Mol Psychiatry 7:278–88

Kruglyak L 1999 Prospects for whole-genome linkage disequilibrium mapping of common disease genes. Nat Genet 22:139–144

Lai CS, Fisher SE, Hurst JA, Vargha-Khadem F, Monaco AP 2001 A forkhead-domain gene is mutated in a severe speech and language disorder. Nature 413:519–523

Lander E, Kruglyak L 1995 Genetic dissection of complex traits: guidelines for interpreting and reporting linkage results [see comments]. Nat Genet 11:241–247

Li J, Tabor HK, Nguyen L et al 2002 Lack of association between HoxA1 and HoxB1 gene variants and autism in 110 multiplex families. Am J Med Genet 114:24–30

Liu J, Nyholt DR, Magnussen P et al 2001 A genomewide screen for autism susceptibility loci. Am J Hum Genet 69:327–340

Martin ER, Lai EH, Gilbert JR et al 2000 SNPing away at complex diseases: analysis of single-nucleotide polymorphisms around APOE in Alzheimer disease. Am J Hum Genet 67:383–394

Mbarek O, Marouillat S, Martineau J, Barthelemy C, Muh JP, Andres C 1999 Association study of the NF1 gene and autistic disorder. Am J Med Genet 88:729–732

McCoy PA, Shao Y, Wolpert CM et al 2002 No association between the WNT2 gene and autistic disorder. Am J Med Genet 114:106–109

Newbury DF, Bonora E, Lamb JA et al 2002 FOXP2 Is not a major susceptibility gene for autism or specific language impairment. Am J Hum Genet 70:1318–1327

Ogura Y, Bonen DK, Inohara N et al 2001 A frameshift mutation in NOD2 associated with susceptibility to Crohn's disease. Nature 411:603–606

Patil N, Berno AJ, Hinds DA et al 2001 Blocks of limited haplotype diversity revealed by high-resolution scanning of human chromosome 21. Science 294:1719–1723

Persico A, D'Agruma L, Maiorano N et al 2001 Reelin gene alleles and haplotypes as a factor predisposing to autistic disorder. Molecular Psychiatry 6:150–159

Philippe A, Martinez M, Guilloud-Bataille M et al 1999 Genome-wide scan for autism susceptibility genes. Paris autism research international sibpair study [published erratum appears in Hum Mol Genet 8:1353]. Hum Mol Genet 8:805–812

Philippe A, Guilloud-Bataille M, Martinez M et al 2002 Analysis of ten candidate genes in autism by association and linkage. Am J Med Genet 114:125–128

Pickles A, Starr E, Kazak S et al 2000 Variable expression of the autism broader phenotype: findings from extended pedigrees. J Child Psychol Psychiatry 41:491–502

Pritchard JK 2001 Are rare variants responsible for susceptibility to complex diseases? Am J Hum Genet 69:124–137

Pritchard JK, Przeworski M 2001 Linkage disequilibrium in humans: models and data. Am J Hum Genet 69:1–14
Rioux JD, Daly MJ, Silverberg MS et al 2001 Genetic variation in the 5q31 cytokine gene cluster confers susceptibility to Crohn disease. Nat Genet 29:223–228
Risch N, Spiker D, Lotspeich L et al 1999 A genomic screen of autism: evidence for a multilocus etiology. Am J Hum Genet 65:493–507
Roberts SB, MacLean CJ, Neale MC, Eaves LJ, Kendler KS 1999 Replication of linkage studies of complex traits: an examination of variation in location estimates. Am J Hum Genet 65: 876–884
Shao Y, Wolpert CM, Raiford KL et al 2002 Genomic screen and follow-up analysis for autistic disorder. Am J Med Genet 114:99–105
Skuse DH 2000 Imprinting, the X-chromosome, and the male brain: explaining sex differences in the liability to autism. Pediatr Res 47:9–16
Smalley SL 1998 Autism and tuberous sclerosis. J Autism Dev Disord 28:407–414
Spielman RS, McGinnis RE, Ewens WJ 1993 Transmission test for linkage disequilibrium: the insulin gene region and insulin-dependent diabetes mellitus (IDDM). Am J Hum Genet 52:506–516
Szatmari P, Jones MB, Zwaigenbaum L, MacLean JE 1998 Genetics of autism: overview and new directions. J Autism Dev Disord 28:351–368
Wassink TH, Piven J, Patil SR 2001a Chromosomal abnormalities in a clinic sample of individuals with autistic disorder. Psychiatr Genet 11:57–63
Wassink TH, Piven J, Vieland VJ et al 2001b Evidence supporting WNT2 as an autism susceptibility gene. Am J Med Genet 105:406–413
Weiss KM, Terwilliger JD 2000 How many diseases does it take to map a gene with SNPs? Nat Genet 26:151–157
Williams NM, Rees MI, Holmans P et al 1999 A two-stage genome scan for schizophrenia susceptibility genes in 196 affected sibling pairs. Hum Mol Genet 8:1729–1739
Zhang H, Zhang C, Robitaille S et al 2000 The Reln gene as a candidate locus for autism spectrumdisorders. Poster presentation at the 50th Annual Meeting of the American Society for Human Genetics. October 2000, Pennsylvania. View the abstract at *http://www.faseb.org/cgi-bin/ashg00/f2007.htm*

DISCUSSION

U. Frith: You mentioned earlier that you are involved in genetic studies of dyslexia. Is the situation comparable? I'd assume that the diagnosis is slightly simpler and we know that there might be strong genetic transmission from parent to child. Does this make a difference?

Monaco: In our dyslexia study we use measures which are normally distributed in the population and our probands are usually in the lower percentiles on multiple measures. Then in our families we test the difference between the siblings quantitatively and use this as the approach for linkage analysis. The linkage we have found in dyslexia is on chromosomes 2, 6 and 18 and these loci have been replicated independently. We do not see any overlap with the linkages for autism.

U. Frith: Are you closer to identifying the actual genes in the case of dyslexia than in the case of autism?

Monaco: No. We still have the same problem. There are regions of linkage that are spread over tens of centimorgans. What I was stating earlier was that we are getting away from the individual measures and using multivariate analysis to increase our power of gaining information from all the measures at a locus. But then we still have the same problem of screening candidate genes and searching for linkage disequilibrium. The only difference I would say is that in dyslexia and language impairment, since they affect 5% of all schoolchildren, we do feel that the common variant hypothesis may play more of a role. Those genes are not lost in the population since dyslexic individuals have normal reproductive fitness. In autism, the same assumptions may not hold and new mutations or variants may play a more significant role.

Folstein: When you added the other cases in your follow-up study, you got a broader peak than the first time. But by now you must have screened every conceivable marker in that interval.

Monaco: We screened 70 markers at about 500 kb spacing. Besides looking for association it just made the linkage curves more difficult to interpret. We did this mostly to look for association.

Folstein: So in narrowing that interval the usual rules are not applying.

Rutter: Why might this be the case?

Monaco: Some people would say that the wideness of the curve is evidence that there is actually a true linkage there. When you get these spikes that go way up and come way down very quickly, this may have less meaning than a broad curve.

Pericak-Vance: Dan Weeks (University of Pittsburgh, PA) did some of the simulations with children. Narrow peaks were more likely to be false positives than the broader peaks.

Monaco: I do not know why adding more families has not narrowed the region. The answer would probably be that we have to add thousands of families to get localization.

Pericak-Vance: I don't even think that will work; we see the same pattern with every other complex disease we work with. This is even with several hundred thousand families. Sometimes if we can get some clarification of phenotype stratification this helps, but it is a major problem. I am beginning to agree with you that there has not been a common variant in autism, unlike Alzhiemer's. There might be a series of mutations. In terms of chromosome 15, we find some really defined areas of linkage with some phenotypes, but we see no evidence of association in that group of families.

Rutter: What you are both saying suggests that we shouldn't follow the current fashion of favouring association strategies.

Monaco: If you did just an association strategy you may find something but you may also miss any new mutations or lower frequency variants in the heterogeneous case.

Pericak-Vance: Association is in vogue now because of the technology that allows rapid SNP genotyping. People forget that this might not answer all the questions.

Folstein: But you wouldn't want to not do it.

Pericak-Vance: That's true, but you can't assume that this is going to give you everything.

Bolton: The model that you presented earlier, based on the analyses that Andrew Pickles has done, was for multiple interacting genes. What evidence is there that the linkage in these families is shared across loci? Wouldn't another way of trying to test whether or not certain functional polymorphisms in these regions are involved in pathogenesis be to condition your analysis dependent on the presence of identity by descent at one of the other loci?

Monaco: We did do some analysis of results from 153 families. None of it was really strongly significant, but there was a trend for a relationship of families sharing on 2, 7 and 16. Interestingly, 15 seemed to be the odd one out. This might follow from our case type analysis. It requires larger numbers of families for a more powerful test.

Folstein: I'm having a little trouble with Mendelian mechanisms. The idea is that because autism is a lethal, if it were a single-gene disorder it would have to be maintained by new mutations. But if it involves several genes working together, those are transmitted by the other siblings and other people in the family. I guess I have never really thought about it that way.

Monaco: You have raised a good point. You cannot explain autism as a monogenic disorder. However, I feel uncomfortable following just an association approach, because one has to assume that rare variants or new mutations may contribute. From the monogenic models one can argue that they should contribute. But I agree: when multiple genes are involved, there is that complexity.

Folstein: Also, it seems that if you take the Pickles hypothesis that each family does not necessarily have the same set of genes, even though there may be only a few for each individual, the total number must be large, and therefore likely to be very common. If, say, three genes are needed per case, and the prevalence of the idiopathic type of autism is 5 per 10 000, the prevalence of each individual contributing allele must be quite high.

Monaco: The other way to view it is that in singleton cases, perhaps one of the reasons that the percentage of severely affected siblings is low is because a lot of these cases are due to new mutations. It may not be passed through the other siblings because the mutation may have occurred in the parent's germline.

Folstein: We should probably compare maternal and paternal age in the singletons versus the multiplex families. This would be easy to do.

Bailey: Maternal age is unremarkable in the IMGSAC multiplex sample (unpublished data).

Folstein: People have reported it in earlier studies, which were most likely singletons.

Bailey: But there's quite a lot of variation in the findings with regard to maternal age, although very clearly a subset of studies find an elevation.

Folstein: You have to also look at paternal age, if you are looking for point mutations.

Bishop: Am I correct in thinking that the severity in multiplex families varies with birth order?

Bailey: There's no particular evidence in this sample for a birth order effect.

Rutter: It would be helpful to know what the fertility is of individuals with the broader phenotype.

Folstein: We both think it is low.

Rutter: Are there good data on this?

Folstein: I don't think so. None of the co-twins had children.

Rutter: No, they certainly didn't. They were at an age where they might have been beginning to have children. Once one gets broader in the relatives, I'm not sure whether the same would apply.

Bailey: Tony Monaco, I wondered whether you wanted to say something about using relatives as an aid to identifying susceptibility alleles.

Monaco: Tony Bailey and his group are looking at the milder phenotypes in the parents and other siblings. One possibility in families with children who share alleles at a particular locus is if the parent is known to have some affection status, we could use that information to see which chromosome haplotypes in that region are being transferred to the children to look for recombination events. If you cannot get a narrow region by statistically averaging things over hundreds of families, you could start to at least see where the recombination events are occurring in individual families. It is dangerous territory, but it might be enlightening.

Skuse: You mentioned that a number of other studies using similar strategies to yours had found linkage to chromosome 7, for example. How convincing are those data as replications? I believe no one has found a LOD score as high as you have on 7q.

Monaco: No, but I think in replication one does not require a genome-wide significant linkage for it to be considered replication. *A priori* you already have linkage results from that region and you are just looking for independent evidence that this did not occur by chance.

Pericak-Vance: We have LOD scores over 2 on 7, even though at the onset I would have said that our study was underpowered. This was based on a genome-wide scan.

Monaco: They gave guidelines to what a significant result is from a genome-wide scan. But are there guidelines for what you would call evidence for replication of an original result? I did this fairly arbitrarily.

Pericak-Vance: I think most people do have guidelines. A LOD score of 1+ as the cut off for initial evidence that an area is interesting and warranting follow-up. In follow-up we look for LOD scores that increase with additional data. However, true replication is difficult. In general you will need a much larger data-set to successfully replicate your initial finding in a genetically heterogeneous disease.

Folstein: Since there are so many non-geneticists here it might be useful to explain why this is. There is a good reason for it. If there are several genes that you are looking for, and you find one, then the next time you look for those same genes, the probability is that you will find a different one. You need a much bigger sample the next time you look if you expect to find the same gene again. We are all very underpowered for replication.

Bailey: To put it another way: the observed degree of replication is surprising given the sample sizes we are dealing with.

Folstein: Especially compared with bipolar disorder and schizophrenia. I am very pleased with the way it is going.

Bailey: If anything, the replication on 2 is even more convincing.

Monaco: Especially in terms of location. It is almost exactly in the same region.

Pericak-Vance: On chromosome 2 all studies are within 15 cM of each other. It is very difficult, because it is almost impossible to replicate given the number of families that we have. At the present rate for any individual site we would be collecting families for the next 30 years before we get anywhere.

Folstein: We have a signal there too, although it is not very big. We just have to believe and then look for the genes that are in our sample and the intervals that give each of us the best LOD scores.

Monaco: Another way this has been done is meta-analysis. This involved taking linkage results for a region from multiple groups, pulling out the strongest result, and then adding the LOD scores from the other studies. They said that 7 and 13 were significant but they did not have the results from chromosome 2 in this study. Outside of merging all the actual data, this is not an unreasonable way of looking at the problem.

Folstein: 13 was particularly interesting because everyone said that they didn't have a signal there when they tried to replicate the CLSA data, but 13 emerged from analysing the data together.

Skuse: How do you know that what you have linkage to is an autistic feature, and not just some other feature of mental retardation?

Folstein: By subsetting.

Monaco: In many of our regions that case type 1/type 2 families were contributing most to the linkage.

Bailey: I would argue that seeing mental retardation as separate isn't useful from this particular perspective. Most groups have found that it is almost impossible to cut the data in such a way that linkages are consistently increased across different chromosomes. What happens is that the linkage goes up in one region by selecting a particular subgroup of patients and it drops somewhere else. The safeguard is what Tony Monaco and I have always done, which is to always analyse all families. We never simply pick an arbitrary group that happens to give the biggest signal. If all families are included this gives some guarantee that one is not generating a false-positive result. Of course, in the IMGSAC sample IQ is much higher than in the epidemiological studies of singletons. This is a very high-functioning group once one selects for multiplicity. Do you find this Susan Folstein?

Folstein: Yes. A lot of people doing this say that they don't know what the relative risk is any more because of prevalence. Remember, we don't include in these genetic studies a lot of the cases that account for the more recently reported high prevalence. All these new cases, that have genetic disorders or specific medical conditions that are increasing in prevalence are excluded at the beginning. This decreases the prevalence. Also, I don't think anyone is ascertaining on the spectrum. We might include the spectrum in family members, but we try to ascertain on autism. This also decreases the prevalence. I am not sure that the risk ratio is very much different from what we said it was 10 years ago.

Bishop: I could see why you would want to exclude all these people with other conditions. But wouldn't there be a case for saying that if these are conditions that increase the risk of autism but don't definitely always give you autism, they might be interesting conditions to look at? They might need another mutation somewhere to push them over the edge.

Folstein: You would have to have an awful lot of them to be able to test that hypothesis. It is not unreasonable, though.

Rutter: This comes back to the query as to why there is an association. One could argue that with tuberous sclerosis or fragile X, it provides a lead for where the susceptibility genes for autism might be located. Alternatively, it could mean only that both conditions involve disrupted brain systems, and that it is that disruption that predisposes to autism through some route unconnected with genetics.

Bishop: If you had a candidate gene, such as the *RELN* gene, it might be interesting to see whether these mutations are also present in people with tuberous sclerosis and autism. Perhaps everything is adding together.

Monaco: At this point, it is not a lot of work to test those single variants in other populations.

Rutter: Tony Monaco, do you want to revisit the point Patrick Bolton raised earlier about genes switching on and off?

Monaco: I would not restrict this to the idea of switching genes on: I would ask when their function is most critical. Sometimes it is difficult to turn a gene off completely or turn it on specifically, outside of very early development. In particular, if genes take on different functions at different time points, they might have a specific function very early and then later have a different function in brain development after birth. Given the consistent increase in brain volume seen in autism we are also looking at genes that might be involved in apoptosis.

Folstein: It would be helpful to focus on genes that are active in post-fetal life but before age 4. We need more neuroscience knowledge.

Monaco: The mouse might be useful in this respect.

Lipkin: As a non-geneticist I am wondering how you go about prioritizing what you are going to do. You showed us a 250 kb segment: you were looking at the promoter regions, the 3′ UTRs and coding sequences. There are a lot of genes there. What is your strategy given what you now have available? Are you planning to pursue sequence analysis?

Monaco: No, you would not sequence that entire region in many individuals. What we did with that gene is about as exhaustive as most groups would get at this point. There are blocks of sequence in the genome called haplotypes which are related to each other through population history. By testing a couple of variants in one of those blocks, you can test the whole block. In some ways, if you get those variants, genotype them in your large families, and find out how these blocks are organized across the chromosome region, you can genotype a lot fewer markers to test for association.

Lipkin: Can you create chromosomal arrays where you break down individual regions of chromosomes and then do hybridization studies with RNAs extracted from different regions? Is that feasible?

Monaco: It can be a fishing expedition. It is very difficult to understand how to take all that gene expression information forward.

Lipkin: It does not sound like that was a targeted study.

Monaco: The major problem is getting RNA samples from brain for microarray analysis.

Folstein: If these genes were functioning in fetal life, what you are getting from adult brain RNA may be completely irrelevant.

Lipkin: I was thinking more in terms of peripheral samples, which you could get at an earlier time point. Rather than doing a huge fishing expedition I was envisaging concentrating on chromosomes 2, 7, 16 and 17.

Monaco: It does not make sense to focus because the microarrays you can buy have all 33 000 genes on them. You use bioinformatic analysis to check out the genes in your region.

How might genetic mechanisms operate in autism?

Susan E. Folstein, Michael Dowd, Raymond Mankoski and Ovsanna Tadevosyan

Department of Psychiatry, Tufts-New England Medical Center, 750 Washington Street, #1007, Boston, MA 02111, USA

Abstract. Twin and family studies provide strong evidence that autism has a largely genetic aetiology. The pattern of familial aggregation suggests that in individual families, a small number of genes act together to cause the phenotype. However, it is unlikely that the same genes act in all families. Thus, the total number of genes involved could be large. One key to finding genes for disorders with considerable locus heterogeneity is to detect genetically more homogeneous subsamples. There exist several traits in families who have a child with autism—biochemical, physical, or behavioural—that are likely to reflect underlying genetic heterogeneity and can thus be used to divide families into more homogeneous subsets. These traits (1) show variation in autism samples; (2) are found in non-autistic family members more often than controls; (3) aggregate in particular autism families; and (4) result in increased signals when used in linkage analysis to define 'affected'.

2003 Autism: neural basis and treatment possibilities. Wiley, Chichester (Novartis Foundation Symposium 251) p 70–83

The present situation

Several linkage signals on numerous chromosomes have been reported, but few are in exactly the same place, as shown in Fig. 1, and none are large. The field of autism genetics has reached what has proved to be a difficult point for investigators of disorders that have 'complex' genetic mechanisms. It is not easy to replicate linkage findings in multiplex disorders, even when loci are truly present (Suarez et al 1994), and the ability to localize true signals is poor (Altmuller et al 2001).

How can we make further progress?

There are several possible next steps, all of which may need to be pursued iteratively in order to localize the linkage signals enough to decrease the number of candidate genes to a reasonable number, although what is a 'reasonable number' will increase with advances in technology.

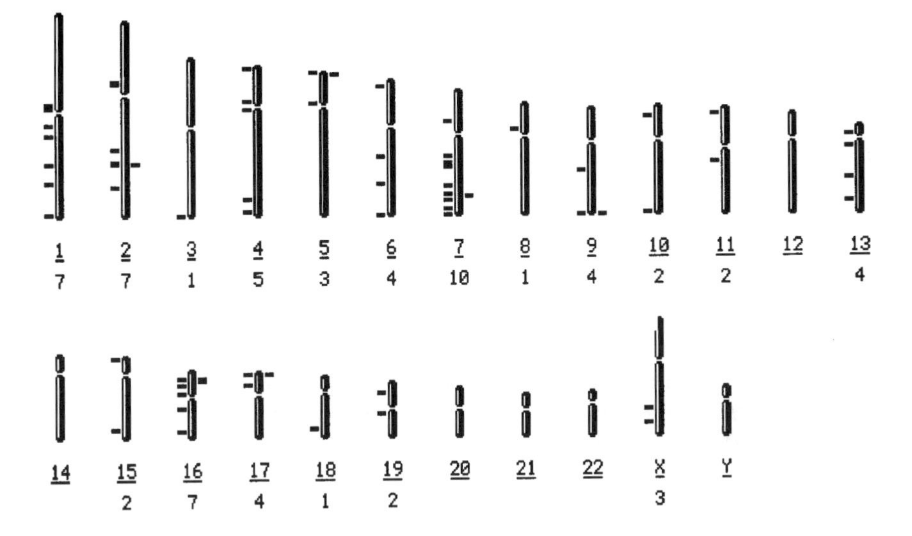

FIG. 1. 'Suggestive' signals from the published genome screens. (Each perpendicular line indicates a reported signal; numerals below the chromosome numbers indicate total number of signals reported on that chromosome.)

As a first step, *we need larger samples*. Several consortia are now working toward this end. It may be possible to tease larger and more precise localization of signals from a 'mega-analysis' of several hundred families although this approach is limited when the genotypes have been carried out in different laboratories and using different sets of markers. It is possible that even using very large samples, locus heterogeneity will limit the power of such analyses.

Thus, a second step is to carefully reconsider all included families in order to *minimize phenocopies of the autism phenotype*. As the definition of autism has become broader, many of the cases no longer resemble the highly familial disorder Kanner described in 1943. He did not diagnose any child who had dysmorphic features, and he also required a rich array of autistic features, particularly 'a preoccupation with the maintenance of sameness' (Kanner 1943). The children thus diagnosed had a highly familial disorder (Piven et al 1990). Miles & Hillman (2000) made a similar point. When they compared the families of autistic probands with and without dysmorphic features, a positive family history of autism and the broader phenotype was three–four times more common in the families ascertained through probands without dysmorphic features. Dysmorphic features are particularly common in trios, but they also occur in sib pairs. Omitting such families from linkage and association studies should reduce the phenocopy rate somewhat, but much heterogeneity will remain.

Third, therefore, *we need take advantage of the phenotypic heterogeneity* in autism, rather than be hampered by it. Evidence is accumulating that (a) autism is not the best phenotype to use to find its genes, and that (b) it is possible to dissect autism into genetically relevant components. For some years, considerable research has focused on defining genetically relevant phenotypes that can increase the power to find susceptibility loci for autism. These can be noticed as variation within the autism phenotype, and some are also evident in non-autistic family members.

Some of these phenotypes are manifested as traits in the family members of autistic children. Eisenberg (1957) reported this in the 1950s in his description of fathers, and it was rediscovered in the first twin study of autism (Folstein & Rutter 1977). Monozygotic twins with autism are often concordant for autism, but when they are discordant, the co-twin often has features that are conceptually reminiscent of autism. In a subsequent series of studies, these features, collectively known as the broader autism phenotype (BAP), were also found more often in the parents and siblings of autistic children than in controls. They include abnormalities of language development, social reticence, and a preference for (and difficulty with change in) daily routines (Bolton et al 1994, Landa et al 1992, Piven et al 1994, 1995, 1997). An excess of mental retardation is present only in families ascertained through a profoundly retarded autistic proband (Folstein et al 1999, Starr et al 2001).

We made an initial test of the genetic value of dissecting the phenotype by subsetting the CLSA families based on proband language development and including as 'cases' those parents with language abnormalities. The linkage signals improved on 7q, 13q and 2q (2q data are unpublished). This has been replicated on 7q and 2q (Liu et al 2001, Buxbaum et al 2001, Shao et al 2002). The same marker that gave the peak score on 13q (D13S800) provided a LOD of more than 3 in families with developmental language disorder (Bartlett et al 2001). Other traits are also candidates for subsetting families, including symptom clusters documented in the Autism Diagnostic Interview-Revised (ADI-R), head size, platelet serotonin and a history of autoimmune disease.

Choosing genes to study in detail

Plausible candidate genes for autism are frighteningly numerous. However, integration of evidence from a variety of fields provides increased clarity for such considerations. Evidence comes from the following types of studies. (1) Biochemical analysis of peripheral blood from patients and their families have repeatedly implicated the serotonin system and, more recently other neurotransmitter systems and neuropeptides (e.g. oxytocin). (2) Positron emission tomography (PET)-based studies in children with autism have provided more direct evidence that serotonin synthesis is altered and $GABA_A$

receptor binding is decreased in various structures in probands (Chugani et al 1999, 1997) and in parents with traits resembling autism (Goldberg et al 2001). (3) Studies of neuroanatomy and neuropathology, both from magnetic resonance imaging and autopsy studies point to developmental abnormalities in the limbic system and cerebellum, as well as large head size. The brain appears to overgrow early in life, and then grow too slowly. Neuronal size, packing, and migration issues have all been identified. Correlation of these findings with observations of similar neuropathology in mouse models is helpful in identifying genes and systems that may play a similar role in the human. (4) Radiolabelled ligand binding analysis of autopsy material suggests abnormalities in GABA receptors in hippocampus. (5) Studies of animal social behaviour have provided clues. Differences in social behaviour between montane and prairie voles is caused by a variation in the promoter of the oxytocin receptor (OXTR). Social abnormalities in the *Dishevelled* mouse (segregating a naturally occurring mutation in the *Dvl1* gene) points to genes in that system. (6) Cytogenetic abnormaliies, which occur throughout the genome, but most commonly involve maternal duplications on 15q11-q13. Translocations and deletions are identifying areas on 7q, 2q and Xq among others that may harbour genes whose disruption leads to autism phenotypes.

When all of these these data are combined with evidence for linkage and information from linkage based on phenotypically defined subsamples, a far more manageable number of candidates emerges that are reasonable to study. We will discuss a few examples. Autopsy studies of brains of persons with autism have found consistent profound decreases in Purkinje cells in cerebellar hemispheres (Kemper & Bauman 1998, Ritvo et al 1986). Naturally occurring mouse mutants, including reeler and lurcher, have developmental abnormalities of these neurons. Lurcher, which has a mutation in the glutamate receptor subunit $\delta 2$ (*Grid2*) gene shows cell autonomous apoptosis of cerebellar Purkinje cells during postnatal development. The PARIS group found a strong linkage signal at a marker in another glutamate receptor gene (*GLUR6*) on 6q. Reelin (*RELN*), in contrast, is a secreted glycoprotein that acts as a signalling molecule with significant roles in neurodevelopment. A positive association has been reported between the *RELN* gene on human 7q and autism (Persico et al 2001). Other examples are relevant to our own studies.

WNT2

It had been observed that the mouse knockout of the *Dvl1* gene displays abnormal social behaviour — no grooming of cage mates and failure to sleep in a communal pile. The interval with the strongest linkage signal on 7q in autism families contains a gene, *WNT2*, which depends for its function on the DVL family of proteins.

Wassink et al (2001) reported two families with non-conservative mutations as well as significant association in the larger sample to an allelic variation in the gene.

5HTT

For the WNT2 study, we used only one variable on the ADI, onset of phrase speech, to divide the families. In this example, we took advantage of the rich variety of symptoms and behaviours documented on the ADI. Thus I will first describe the development of factors based on items of the Autism Diagnostic Interview (ADI) and Autism Diagnostic Interview—Revised (ADI-R).

We speculated that we could identify more homogeneous subsets for genetic analysis by developing factors of the many items in the ADI. We developed these factors without the imposition of pre-existing diagnostic concepts. We used ADIs (the original version of the interview) and ADI-Rs from 293 subjects (only one proband was used per family) who met criteria for autism. The best solution included six factors (Table 1). Items were retained in each factor when $r^2=0.40$ with other items in the factor and had low correlation with the next 'closest' factor. We then validated these factors using an independent sample of 70 autistic children who had been given, in addition to ADI-R, extensive tests of language, cognition, functioning and psychiatric symptoms. The factors, except for Social Intentionality showed modest but highly significant within-sib pair correlation, suggesting that they are genetically relevant.

We then used these factors in a series of linkage analyses. First we selected several chromosomal intervals that provided HLODs of $\geqslant 1$ when we used Autism as the phenotype and that were interesting in at least one other respect. One of these was a small signal at 17q (HLOD$=1.11$, $\alpha=0.22$) in an interval containing the serotonin transporter gene. Linkage has also been reported for this region in the IMGSAC families (IMGSAC 2001). The serotonin transporter (SLC6A4 or 5HTT) has long been considered a candidate because of consistent findings of abnormalities in the serotonin system. The selective serotonin reuptake inhibitors (SSRIs) improve mood and reduce compulsions and anxieties associated with autism (e. g. Awad 1996). Association studies of the *5HTT* gene have been mixed, although this is of limited significance as only two polymorphisms have been examined in the vast majority of studies. One of these markers is an insertion/deletion polymorphism in the promoter (5HTTLPR), speculated to affect gene expression (Cook et al 1997). One detailed study (Kim et al 2002) analysed a number of SNP and microsatellite markers, and corresponding haplotypes, across the transcriptional unit; they found significant transmission disequilibrium effects at three markers.

For this illustration (Fig. 2), we used a simple approach. For each factor, we determined the mean score for each pair of probands. The families were then divided into two subsamples, split at the median factor score. One simple

TABLE 1 ADI factors

Factor	*Validating test*	*Correlation*	*Partial correlation*	*Sib–sib correlation*
1. Spoken Language	Expressive Vocabulary Test	$r = -0.28$, $P = 0.020$	Controlled for IQ, $r = -0.30$, $P = 0.013$	$r = 0.21$, $P = 0.008$
2. Social Intentionality	Vineland Adaptive Behavior Scales socialization standard score	$r = -0.37$, $P = 0.002$		$r = 0.11$, $P = 0.142$
3. Compulsions	KSADS Impairment and time spent on compulsions	$r = 0.57$, $P = 0.000$		$r = 0.25$, $P = 0.001$
4. Developmental Milestones	Vineland total score Full scale IQ	$r = -0.27$, $P = 0.026$ $r = -0.26$, $P = 0.029$		$r = 0.29$, $P = 0.0001$
5. Splinter skills	Difference between NVIQ and VIQ	$r = -0.01$, $P = 0.951$		$r = 0.39$, $P < 0.0001$
6. Sensory Aversions	Phobia score derived from KSADS ADOS unusual sensory reactions	$r = 0.26$ $P = 0.054$ $r = -0.27$, $P = 0.001$	Controlled for V IQ, $r = -0.39$, $P = 0.002$	$r = 0.26$, $P = 0.002$

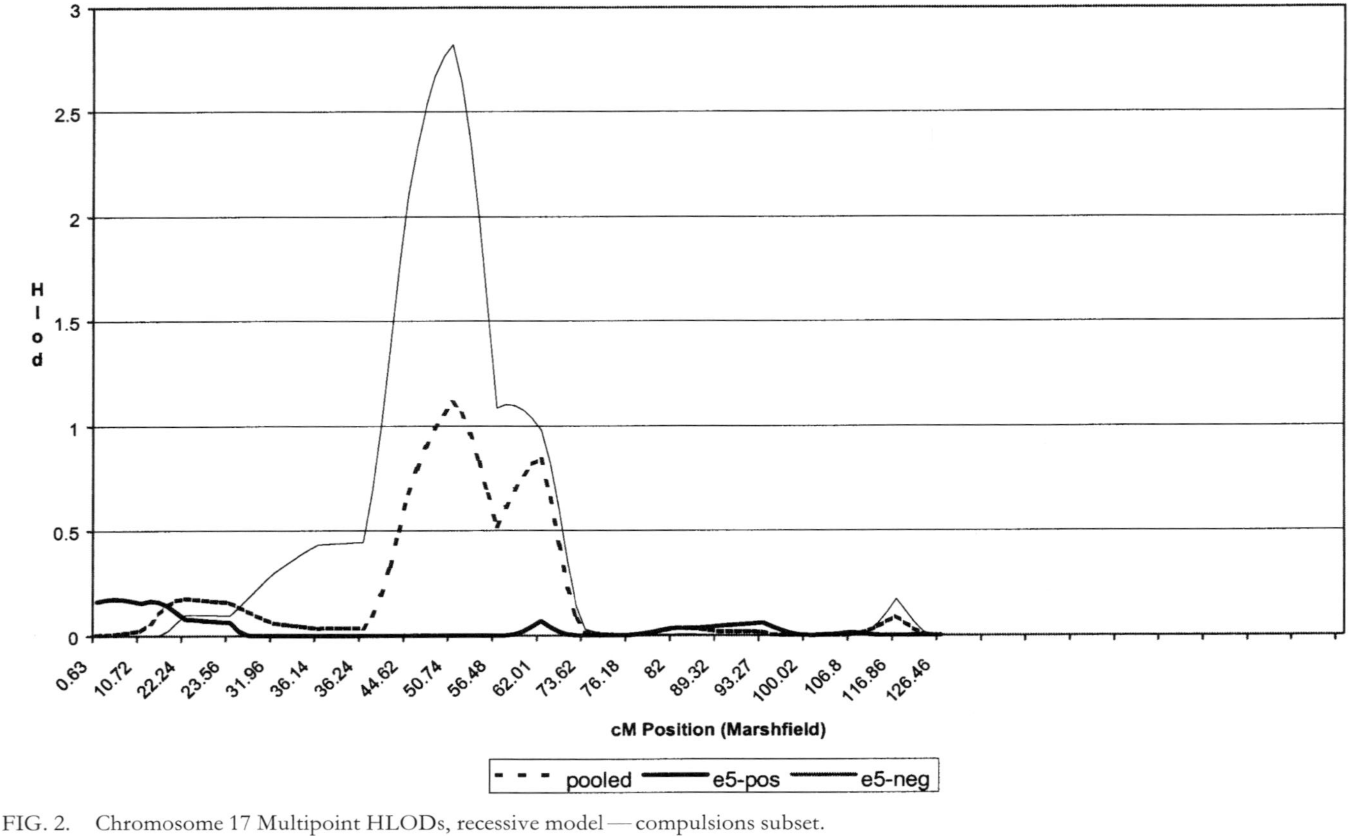

FIG. 2. Chromosome 17 Multipoint HLODs, recessive model—compulsions subset.

dominant and one simple recessive model were used. The HLOD in the Boston/AGRE data set ($n=97$ for this interval) was 1.11 ($\alpha=0.2$) at D17S1294 under a recessive model. Analysis of the subset of families with high factor scores on 'Compulsions' ($n=52$) revealed a recessive HLOD of 2.82 ($\alpha=0.44$) and NPL of 2.82 ($P<0.003$) at (51 cM). This result was particularly heartening since one of the main effects of the SSRIs on autistic symptoms is to decrease the driven, compulsive quality of their repetitive behaviours.

Head circumference: another example of an autism-related phenotype

In his original description of autism, Kanner noted that the children tended to have large heads. This has been rediscovered in recent years, and documented. The head size in probands is normally distributed, but skewed to the right and more children than expected meet criteria for macrocephaly (Deutsch et al 2003, Lainhart et al 1997, Piven et al 1996, 1995). Deutsch et al (2003) has shown that probands' heads also tend to be wider than expected, and, using our Boston multiplex families, that their sibs and parents' heads also tend to be large. In 46 families where head circumference (HC) was measured in both parents both probands, macrocephaly (HC=97th percentile) was more common in probands when one of their parents had macrocephaly.

Possible epigenetic effects

The field has focused mainly on direct genetic effects in the aetiology of autism; that is, effects on development that result from the genes that are present in the child. However, epigenetic effects are also being explored. For example, the level of maternal dopamine-β hydroxylase (DBH) has been reported to be lower in the mothers of multiplex autism families than in controls (Robinson et al 2001). DBH is an enzyme that converts dopamine to noradrenaline. The dopaminergic system has been implicated in a number of psychiatric disorders including schizophrenia, social anxiety disorder, addiction and attention deficit/hyperactivity disorder. Maternal DBH levels may play a role in the development of autism by altering the prenatal intra-uterine concentrations of neurotransmitters and morphogens — specifically, dopamine and noradrenaline. DBH levels are in part regulated by two promoter alleles: a 19 base-pair insertion/deletion (+/−) polymorphism and a −1021 C/T transition (Zabetian et al 2001). The 19 bp +/+ allele and the −1021 C/C allele are associated with increased levels of DBH.

We examined DBH levels in 100 mothers of children with autism. We have found that the children of mothers with the 19 bp +/+ allele have statistically significant worse outcomes with respect to obsessive-compulsive behaviours than children of mothers with the 19 bp −/− allele. Children of mothers with the −1021 C/C genotype may have worse outcomes with respect to spoken language than did the children of mothers with the −1021 T/T genotype.

Data from several groups showing more severe disorder in second born autistic children also suggest possible fetal–maternal effects. We replicated this using our factor scores: second born probands had worse scores on the spoken language factor and were less likely to have splinter skills ($P < 0.001$ for each). However, it is possible that these data are biased. Perhaps if the firstborn child has severe autism and is diagnosed early in life, the parents would be less likely to have a second child. If the autism of the firstborn was milder, it may not be diagnosed until subsequent childbearing decisions had already been made.

What is the likely nature of the genes and alleles involved in causing autism?

The model most frequently suggested is the one described at the outset of this paper: a few interacting genes in an individual family, but with different combinations of genes acting in different families. When several to many genes interact to cause a moderately common phenotype, the relevant alleles of these genes will be very common in the population, and thus not likely to cause disease unless they are paired with other susceptibility alleles. This idea is consistent with the traits that make up the broader autism phenotype. They are traits in the normal range, albeit at the extreme of that range, and they usually do not interfere enough with functioning to result in clinical referral. Indeed, several studies of the siblings of children with autism have indicated that they may have unusually high intelligence, and there are many examples of parents who make important contributions to their fields. It is thus of interest that variations in the promoter regions have been implicated in several of the candidates for which some positive evidence of involvement has been reported. The promoter includes sequences that control regulation of gene expression patterns, including timing and amount of message produced.

However, while this theory of mechanism is satisfying because of its conceptual unity, other complicating genetic factors are likely. For example, in Wassink's study of *WNT2* on Chromosome 7q31 (Wassink et al 2000), he found two families with mutations in conserved coding regions that may themselves be sufficient to cause autism. In both families, one parent and both probands had the mutation. But in addition to the two families, he also found positive association with a different allele, suggesting a more general role for the gene as a susceptibility factor, as opposed to a causative factor. The findings in the DBH gene point to epigenetic effects, as do the birth order data.

References

Altmuller J, Palmer LJ, Fischer G, Scherb H, Wjst M 2001 Genomewide scans of complex human diseases: true linkage is hard to find. Am J Hum Genet 69:936–950.

Awad GA 1996 The use of selective serotonin reuptake inhibitors in young children with pervasive developmental disorders: some clinical observations. Can J Psychiatry 41:361–366

Bartlett CW, Flax J, Tallal P, Brzustowicz LM 2001 Linkage analysis of chromosome 3 in families selected for specific language impairment. Am J Hum Genet 69:506

Bolton P, Macdonald H, Pickles A et al 1994 A case–control family history study of autism. J Child Psychol Psychiatry 35:877–900

Buxbaum JD, Silverman JM, Smith CJ et al 2001 Evidence for a susceptibility gene for autism on chromosome 2 and for genetic heterogeneity. Am J Hum Genet 68:1514–1520

Chugani DC, Muzik O, Rothermel R et al 1997 Altered serotonin synthesis in the dentatothalamocortical pathway in autistic boys. Ann Neurol 42:666–669

Chugani DC, Muzik O, Behen M et al 1999 Developmental changes in brain serotonin synthesis capacity in autistic and nonautistic children. Ann Neurol 45:287–295

Cook EH Jr, Courchesne R, Lord C et al 1997 Evidence of linkage between the serotonin transporter and autistic disorder. Mol Psychiatry 2:247–250

Deutsch CK, Folstein SE, Gordon-Vaughn K et al 2003 Macrocephaly and cephalic disproportion in autistic probands and their first-degree relatives. Am J Med Genet, in press

Eisenberg L 1957 The fathers of autistic children. Am J Orthopsychiatry 127:715–724

Folstein S, Rutter M 1977 Infantile autism: a genetic study of 21 twin pairs. J Child Psychol Psychiatry 18:297–321

Folstein SE, Santangelo SL, Gilman SE et al 1999 Predictors of cognitive test patterns in autism families. J Child Psychol Psychiatry 40:1117–1128

Goldberg J, Szatmari P, Zwaigenbaum L, Nahmias C 2001 Brain serotonin 2A 5HT2A receptor density in parents of autistic probands. International Meeting for Autism Research IMFAR, November 2001, San-Diego, CA, *http://www.imfar.org/L_ABSTRACTS.doc*

IMGSAC 2001 A genomewide screen for autism: strong evidence for linkage to chromosomes 2q, 7q, and 16p. Am J Hum Genet 69:570–581

Kanner L 1943 Autistic disturbances of affective contact. Nervous Child 2:217–250

Kemper TL, Bauman M 1998 Neuropathology of infantile autism. J Neuropathol Exp Neurol 57:645–652

Kim SJ, Cox N, Courchesne R, Lord C et al 2002 Transmission disequilibrium mapping at the serotonin transporter gene (SLC6A4) region in autistic disorder. Mol Psychiatry 7:278–288

Lainhart JE, Piven J, Wzorek M et al 1997 Macrocephaly in children and adults with autism. J Am Acad Child Adolesc Psychiatry 36:282–290

Landa R, Piven J, Wzorek MM, Gayle JO, Chase GA, Folstein SE 1992 Social language use in parents of autistic individuals. Psychol Med 22:245–254

Liu J, Nyholt DR, Magnussen P et al 2001 A genomewide screen for autism susceptibility loci. Am J Hum Genet 69:327–340

Miles JH, Hillman RE 2000 Value of a clinical morphology examination in autism. Am J Med Genet 91:245–253

Persico AM, D'Agruma L, Maiorano N et al 2001 Reelin gene alleles and haplotypes as a factor predisposing to autistic disorder. Mol Psychiatry 6:150–159

Piven J, Gayle J, Chase GA et al 1990 A family history study of neuropsychiatric disorders in the adult siblings of autistic individuals. J Am Acad Child Adolesc Psychiatry 29:177–183

Piven J, Wzorek M, Landa R et al 1994 Personality characteristics of the parents of autistic individuals. Psychol Med 24:783–795

Piven J, Arndt S, Bailey J, Havercamp S, Andreasen NC, Palmer P 1995 An MRI study of brain size in autism. Am J Psychiatry 152:1145–1149

Piven J, Arndt S, Bailey J, Andreasen N 1996 Regional brain enlargement in autism: a magnetic resonance imaging study. J Am Acad Child Adolesc Psychiatry 35:530–536
Piven J, Palmer P, Jacobi D, Childress D, Arndt S 1997 Broader autism phenotype: evidence from a family history study of multiple-incidence autism families. Am J Psychiatry 154:185–190
Ritvo ER, Freeman BJ, Scheibel AB et al 1986 Lower Purkinje cell counts in the cerebella of four autistic subjects: initial findings of the UCLA-NSAC Autopsy Research Report. Am J Psychiatry 143:862–866
Robinson PD, Schutz CK, Macciardi F, White BN, Holden JJ 2001 Genetically determined low maternal serum dopamine beta-hydroxylase levels and the etiology of autism spectrum disorders. Am J Med Genet 100:30–36
Shao Y, Raiford KL, Wolpert CM et al 2002 Phenotypic homogeneity provides increased support for linkage on chromosome 2 in autistic disorder. Am J Hum Genet 70:1058–1061
Starr E, Berument SK, Pickles A et al 2001 A family genetic study of autism associated with profound mental retardation. J Autism Dev Disord 31:89–96
Suarez B, Hampe C, Van Eerdewegh P 1994 Problems of replicating linkage claims in psychiatry. In: Gershon E, Cloninger C (eds) Genetic approaches to mental disorders. American Psychiatric Publishing Inc, Washington, DC, p 23–46
Wassink TH, Piven J, Vieland VJ et al 2001 Evidence supporting WNT2 as an autism susceptibility gene. Am J Hum Genet 105:406–413
Zabetian CP, Anderson GM, Buxbaum SG et al 2001 A quantitative-trait analysis of human plasma-dopamine beta-hydroxylase activity: evidence for a major functional polymorphism at the DBH locus. Am J Hum Genet 68:515–522

DISCUSSION

Bailey: One of the striking results was that when you added the parents in with language delay, this really doesn't increase the linkage by very much.

Folstein: We have fewer families and a higher score.

Bailey: If we had a perfect measure, so that we were able to identify an affected parent perfectly, does anyone have a feel for how much including accurately identified parents would increase the power to detect linkage?

Folstein: Obviously family history is extremely insensitive.

Bailey: But if we had a hypothetically perfect measure, what gain in linkage would we get from identifying affected parents?

Pericak-Vance: It depends on what the underlying genetic model is.

Sigman: Dan Geshwind has findings with multiplex families using the AGRE database.

Folstein: He was able to replicate our work on chromosome 7, but he couldn't do it with phrased speech. They only got it with first words, but they didn't have any parents' phenotypes for language.

Sigman: I believe in the current study that he is finding a relationship with age of phrased speech.

Bishop: In the families in which you have found amino acid changes, and you say that both probands and one parent had this change, was this parent in any way abnormal?

Folstein: The *WNT2* paper has been published (Wassink et al 2001). What we have is in there. In one family the father definitely had the broad phenotype, and I think he had a language problem. In the other family they hadn't examined the mother.

Monaco: When you did this phenotyping that creates smaller subsamples and you present linkage analysis at a locus, one worries about the drop in sample size. For one locus you showed a sample of 15 families.

Folstein: I was profligate with my power. I decided to do it.

Monaco: One way around this would be to simulate your groupings. Peggy Pericak-Vance has done this on chromosome 2. If you subset your families into 15 families in one group and so many in a second group and simulate at random, you can see how often you will get this result by chance.

Folstein: We have done this and the answer is hardly ever. It is hard to know what the best way to do this is.

Monaco: It is difficult to draw conclusions from 15 families.

Folstein: I agree. I was just throwing this out.

Dawson: I am assuming that when you are using your factor scores, you are using them to code the children as affected versus non-affected.

Folstein: We have so far, but you can use it on a quantitative trait.

Pericak-Vance: We tried it both ways. We did it as a qualitative trait and we also used it as a QTL. We found the same results using the QTL analysis.

Dawson: We have been focusing intensely on developing this quantitative measure that can be used with both parents and children. We have administered it to over 200 families, and we have done a factor analysis. We have some nice factors that are showing variability and distributions over the probands and the family members. I am curious to see how important the geneticists think it will be to use this approach versus a sub-typing approach.

Folstein: The classical rule about QTLs is that a normally distributed trait is required. This is probably what prompted you to do it across the whole family. But I think people are really getting results just using it within disorder as well.

Dawson: To rephrase my question, how important is it going to be to try to think about these in a dimensional way, and also to try to develop measures that look at the supernormal range? For a QTL analysis you want both ends of the distribution. Is it important to develop these measures, or will other approaches get us there as effectively? It is a huge investment of resources to develop such measures.

Pericak-Vance: I think we don't know. Someone has to do it. This is sort of what Tony Monaco is doing with dyslexia.

Monaco: Yes, but there we are starting with a true quantitative measure. Autism is a clinical diagnosis and you are trying to make a QTL out of the clinical description.

Dawson: In the history of learning disability, people went from saying that an individual has a reading disability or not, to having nice quantitative measures. I'm asking whether we should push to develop these kinds of measures in autism as well.

Pericak-Vance: That will be critical if we want to finely dissect autism.

Bailey: I'd like to ask Geraldine Dawson's question in a slightly different way. There are two challenges. One is to identify the rough location of all potential susceptibility loci; the other is then to narrow the critical region in order to identify the susceptibility gene. Using these time-consuming techniques of phenotyping other relatives, have we any evidence that they may be useful in terms of narrowing regions of linkage, as opposed to identifying the location of further loci of weak effect?

Folstein: That is an empirical question. Since everyone has the ADI, let's do something really easy first. If we could increase the power to detect signals without phenotyping the parents, that would be a big saving.

Dawson: I just wondered whether there is enough historical evidence from other disorders, or perhaps learning disability is a good heuristic example. The scientists doing the phenotyping need guidance from the scientists doing the genotyping with respect to where to put resources and what are successful strategies.

Bailey: Tony Monaco, can I ask you a more direct question? Are the peaks of linkage in dyslexia, where you are essentially using a QTL approach, significantly narrower than those seen in autism?

Monaco: They are just as broad.

Bishop: But reading is a skill that you have to be taught. We know that how well you do it depends not just on what your brain is like, but also the education system.

Dawson: However, in autism there are intervention effects also.

Folstein: Except for the language and reading area, I can't decide what may be the best way to measure the impairments in family members. There are so many different things that can be measured in terms of psychological tests, as well as social skills using measures such as Constantino's social reciprocity scale. I don't know how this relates to what it is we think of as the same trait in autism

Monaco: One source of power that the quantitative approach might have in looking for susceptibility genes is that by assessing the broader phenotype you are using the variation in scores and whether they get certain alleles or not as adding more power than just having the single proband or the affected sibling pair.

Folstein: It should add power if you get it right, which is more transparent to me in reading and language than it is in these social traits.

Bishop: I wouldn't be so confident about language: it involves a complex set of functions and there are a lot of different things that can be measured.

Bailey: Let me ask a more provocative question: given that there is reasonable evidence that there are genes on chromosomes 7 and 2, should phenotypers stop what they are doing with other relatives, and is there anything that they can do that would help to identify those genes?

Monaco: In the background one can keep a number of genes being screened for mutations in the proband set that you have chosen, and you hope to get lucky one day and pick the right gene. But in terms of localization, using the information about the parents and unaffected siblings and starting to look at the segregation pattern in a particular family, then you can actually look at the recombination points.

Bishop: If we are moving away from just clinical cases we need to think about good old-fashioned behaviour genetics methods, such as twin studies, where you wouldn't just be concentrating on autistic twins. This would probably be less expensive than rushing in and doing whole genome scans. You could then identify which traits were heritable and which were reasonably good candidates for molecular genetics.

Pericak-Vance: I think almost every field in complex genetics is looking at phenotype stratification.

Sigman: For those of us who spend a great deal of effort on sensitive assessments of language skills, it is surprising that parental report of the children's ages when they spoke their first word should be such a powerful marker.

Pericak-Vance: Could it be a surrogate for something else?

Folstein: If we look at whether those with late onset of speech continue not to have speech, the answer is that some do and some don't. So the analysis could be done an entirely different way: by dividing the families according to whether the probands ever got speech or not, and also to use the language factor from the ADI.

Rutter: In passing, you commented that QTL approaches presuppose a normal distribution. The critics of behaviour genetics as a field are keen to criticise it on the same grounds. However, distributions are hugely sensitive to measurement influences; changes in measurement readily produce whatever distribution you want. Is the QTL assumption of normality just a statistical convenience or is it saying something meaningful about reality?

Folstein: There are two issues with the distribution. People have said that a normal distribution is needed in the population for a QTL, but I wonder whether this is just because of the kinds of traits that they have been interested in studying. Most of the ADI factors are more or less normally distributed. We just took the median in some cases. Peggy Pericak-Vance has got exactly the same result using an ordered subset with the ADI-related factors that have nothing like a normal distribution in the population.

Reference

Wassink TH, Piven J, Vieland VJ et al 2001 Evidence supporting *WNT2* as an autism susceptibility gene. Am J Med Genet 105:406–413

X-linked genes and the neural basis of social cognition

David Skuse

Behavioural and Brain Sciences Unit, Institute of Child Health, 30 Guilford Street, London WC1N 1EH, UK

Abstract. The neural basis of social cognition is subject to intensive research in both humans and non-human primates. Autism is the archetypal disorder of social cognitive skills, and increasing interest is being paid to the role played by efferent and afferent connectivity between the amygdala and neocortical brain regions, in predisposing to this condition. Such circuits are now known to be critical for the processing of social information. Recent research suggests a sub-cortical neural pathway, routed through the amygdala, may turn out to be a key player in the mystery of why humans are so prone to disorders of social adjustment. This pathway responds to certain simple classes of potential threat, including direct eye contact and, in humans, arousal evoked by this exquisitely social stimulus is modulated and controlled by a variety of specific (largely frontal) neocortical regions. Dysfunction of these modulating circuits can occur in the context of developmental disorders that are associated with haploinsufficiency of one or more classes of X-linked genes, lacking Y-homologues, which may be sexually dimorphic in expression.

2003 Autism: neural basis and treatment possibilities. Wiley, Chichester (Novartis Foundation Symposium 251) p 84–108

Our social interactions with other people are distinctively coloured by our emotions, in ways that are both obvious and subtle. Neuroscientists are making rapid progress in understanding the interface between our processing of emotions, feelings and social cognition — that set of rules and responses that makes for a well-adjusted individual. Research on emotion processing is illuminating our understanding of how social competence develops and how it is maintained.

In a recent review Dolan (2002) defined emotions as representing 'complex psychological and physiological states that, to a greater or lesser degree, index occurrences of value'. By that, he meant that emotions are psychological events that influence our behaviour by making some activity more desirable, more likely to be rewarding, and other activities less desirable in that they are unlikely to be associated with reward or are associated with an adverse and unpleasant outcome.

He points out that the range of emotions to which an organism is susceptible will importantly reflect the complexity of its adaptive niche. Higher order primates, in particular humans, live in a complex social world. For that reason, emotional regulation will have a profound effect upon our social behaviour, and any perturbation in our ability to regulate our emotions is likely to have adverse consequences for our social adaptation.

'Unlike most psychological states emotions are embodied and manifest in uniquely recognizable, and stereotyped, behavioural patterns of facial expression, comportment, and autonomic arousal' (Dolan 2002). Most of us can 'read' other people's emotional states without effort, and our ability to respond appropriately to such states has huge significance for our ability successfully to rear our young and to find a mate with whom to reproduce ourselves. When the ability to read other's emotional states is significantly impaired, we appear at the very least socially gauche or we may have difficulty in responding appropriately in any social situation, a characteristic feature of autistic conditions.

Recognition of facial expressions

We gain critically important information about how to respond appropriately in social encounters by monitoring the expression on another's face, which provides information about that person's emotional state. In certain circumstances, those emotional expressions can evoke that emotion in oneself—happiness is one obvious example. Haxby et al (2002) proposed that there are dedicated systems for processing emotion expressions in other's faces, in which the amygdala and insula play a crucial role. When we process the emotional content of a face we take into account a wide range of visual cues. These include whether we know the individual or not (face recognition memory), the facial configuration (for example, whether the mouth is open or shut, whether the eyes are wide or narrowed), and in particular eye gaze (is this person looking at us?). Accurate perception of emotional expression involves the coordinated participation of regions for the visual analysis of expression and regions for the representing and producing emotions, in which the amygdala apparently plays a significant role. Studies from both humans with congenital or acquired damage to the amygdalae (Calder et al 2001) and from primates in which lesions have been induced (Amaral 2002) show that this subcortical structure influences our ability to gain and to maintain socially appropriate behaviour. Whether its functional integrity is critical for normal social cognitive development in humans, is still an open question. It may not be so for macaque monkeys (Amaral et al 2003).

The amygdala and perception of fear

We observe, in functional magnetic resonance imaging (fMRI), amygdala activation in response to facial expressions, and find there is greater activation in the amygdala (in terms of a BOLD response) when we perceive *fear* compared with other emotional faces (Morris et al 1998a). When the amygdala is bilaterally ablated the perception of fear is selectively impaired (Adolphs et al 1999). We do not fully understand why this is so, but recent evidence suggests that the amygdala responds specifically to eye contact, and that it is maximally activated by exaggerated wide-open eyes, such as are associated with a fearful expression (Morris et al 2002). This response may occur because direct gaze can indicate potential threat (equivalent to our innate reaction to a striking snake). Consequently, direct eye contact elicits an instinctive 'fear response' in humans, and in primates too. Behavioural studies in monkeys have shown that eye contact is a critical component of threatening and fear-related displays (Nahm 1997). A simple stare is often the most effective stimulus in evoking a fight or flight response in non-human primates (Emery 2000).

The perception of direct gaze from a face that is neither threatening nor fearful also elicits a response in the amygdala in humans (Kawashima et al 1999). Several studies have now shown that, in monkeys too, there are cells in the amygdala that respond selectively to eye gaze (Sato & Nakamura 2001). Neural interactions between the amygdala and neocortical regions that are engaged by visual stimuli of faces is enhanced if those faces have direct gaze orientations (George et al 2001).

The question of whether the amygdala's response to *fearful faces* is specifically related to processing of information from *eyes* had not been addressed before Morris and colleagues looked at this issue (Morris et al 2002). They hypothesize that there is no specificity in the response of the amygdala to fear, but that fearful eyes are characteristically wide open, and thus exaggerated representations of eye contact in general. (Note, in contrast, that happy faces are characterized by narrowed eyes.)

Eye contact evokes amygdala responses that are processed simultaneously by both conscious (explicit) and non-conscious (implicit) neural mechanisms. Implicit processing of fearful expression and other fear-related stimuli engages *subcortical* visual pathways that are routed directly to the amygdala, without passing through the visual cortex first (see Morris et al 1998a, 1998b, 2001). Consequently, they can evoke a very rapid physiological response — before the neocortex has had time to consider the information and decide on an appropriate course of action (Morris et al 2001). Because of the rather low spatial resolution of the subcortical circuit (via the pulvinar nucleus of the thalamus and the superior colliculus), it can only discriminate threats that are associated with simple visual stimuli (besides eye contact, this could include snakes or spiders; Dolan 2002).

There is increasing anatomical evidence in support of the 'subcortical visual pathway' hypothesis. For example, the superior colliculus receives its visual input primarily from magnocellular retinal ganglion cells—which have large and rapidly conducting axons. The principal projection of the magnocellular pathway is the pulvinar nucleus in the posterior thalamus (Elgar & Campbell 2001). In turn, there are direct projections from the pulvinar to the amygdala in monkeys and in rats (Morris et al 2002). High-resolution representations of objects are processed via an alternative neocortical pathway, leading ultimately to conscious perceptions, and this pathway is much slower, probably engaging the geniculostriate parvocellular system (Elgar et al 2002). Analogous parallel cortical and subcortical fear-recognition pathways, involving responses to threatening auditory stimuli, have been described in rats (LeDoux 2000).

Morris et al (2002) propose that the low-spatial resolution subcortical pathway provides a potential route by which neural responses to the threat posed by 'fearful eyes' (and by implication, eye contact in general) can reach the amygdala independently of the geniculostriate neocortical system. They found, using an ingenious fMRI-based investigation, that fearful eyes alone are sufficient to evoke increased neural responses in this non-conscious circuit (Morris et al 1999), which is probably the remnant of a simple neural system that had survival value throughout the animal kingdom (Emery 2000).

Which cortical circuits modulate, and are modulated by, amygdala activity evoked by eye contact? Because the amygdala is an essential and central component of a threat-detection system, with extensive neocortical and subcortical connections that are crucial for the automatic non-conscious responses to a threatening stimulus (e.g. fight and flight), appropriate social responses require complex cortical processing of such stimuli. Modulation of those instinctive responses depends on the social context in which it occurs, therefore our response to a stimulus that *could* be a threat is modulated by complex neocortical connections. The outcome of interactions between the aroused amygdala and social cognition processing centres in the neocortex permits appropriate responses in a social encounter. We hypothesize that, in humans, a crucial component of the modulating circuitry is the recruitment of language centres, and the conscious processing of a 'feeling' response, which is important especially in social interactions with strangers. However, there are limits to the modulating ability of these higher cortical circuits. Our instinctive response of discomfort when engaged in conversation with someone suffering from thyrotoxicosis, with exophthalmos, is attributable to this limitation. Wide open eyes, in which the sclera is visible over the iris, appear 'fearful' and thus activate the 'threat detection circuit' particularly efficiently. Nevertheless, humans are in general more able to tolerate direct eye contact for longer than other primates.

Several studies of patients with bilateral lesions of the amygdala have shown they suffer impairment in their ability to recognize negative emotions, *particularly fear* (Adolphs et al 1999, Broks et al 1998, Calder et al 1996). This deficit is associated with other impairments specifically related to failure to accurately monitor eye contact, including impaired detection of eye gaze (Young et al 1995). These individuals are also impaired, apparently as a consequence of the lesions, at making social judgements (Stone et al 2003, Adolphs et al 1995). This finding provides supportive evidence for the hypothesis that the amygdala plays a more general role in processing information relevant to social cognition (Adolphs 2002). Recent evidence suggests our ability to make social judgements, such as whether someone is trustworthy or not, are critically dependent on the integrity of reciprocal connectivity between the amygdala and the ventromedial prefrontal cortex (Winston et al 2002). In addition, the amygdala is activated along with neural circuits that are important for the development of theory of mind (Siegal & Varley 2002). If such circuits are damaged, competence in this social-cognitive function becomes impaired (Stone et al 2003).

Amygdala dysfunction and autism

In people with high-functioning autism, in which social cognitive skills are seriously impaired, there is a relatively specific failure to recognize—in the sense of being able consciously to label accurately—specifically *fearful* facial expressions (Howard et al 2000), a deficit found too in paranoid schizophrenia (Phillips et al 1999). Such individuals also suffer from impaired theory of mind skills (Frith 2001, Sarfati et al 1999), and are unable to make rational judgements about the trustworthiness of strangers (Adolphs et al 2001, Tenyi et al 2002). Because functional and structural anomalies of the amygdala are found in some, if not many, cases of autism (Sweeten et al 2002, Shenton et al 2001), we hypothesize that very same neural circuits that developed for the perception of fearful expressions in conspecifics are being used for another purpose altogether—a purpose that some have argued is characteristically, if not uniquely, human (Hare et al 2001, 2002). Our current research is addressing the question: how has the coincidence between the development of fear recognition skills and the development of theory of mind come about, and what genetic mechanisms influence their function?

Eye contact and social cognition

The ability to follow and respond to the direction of gaze of a conspecific is a crucial skill in humans, shared with some, but not all, primate species (Perrett et al 1985, Emery 2000, Jellema et al 2000). Our ability to meet and to follow another's gaze is present during early infancy and is associated with a growing appreciation of salient events in a socially structured world (Allison et al 2000). There are two components to the developed skill, which we have termed 'allocentric' and 'egocentric' (Elgar et al 2002).

In allocentric gaze-monitoring the directional aspect of perceiving and processing gaze becomes recruited for the purpose of following the intentional gaze of another, and is thus critical for the development of joint attention (Leekam & Moore 2000). Allocentric gaze means paying attention to the salience and significance of events extrinsic to the viewer; to things happening 'out there'. It enables the viewer to engage effectively with external and potentially distant events, orienting the observer to the appropriate location, and it uses extrinsic spatial co-ordinates. Its development normally occurs during the first few years of postnatal life, and we become increasingly accurate at determining where someone else is looking with experience (Langton et al 2000).

Allocentric gaze skill can be contrasted with direct engagement of gaze with the onlooker, which we have labelled 'egocentric' (Elgar et al 2002). In contrast to the relatively slow development of allocentric gaze sensitivity, infants show egocentric gaze sensitivity from birth (Batki et al 2000). The young infant responds actively to being looked at, exhibiting a range of teasing and smiling behaviours of increasing complexity, suggesting an early developing ability to engage with conspecifics by facial acts involving direct gaze with the interactant (Aitken & Trevarthen 1997). This interest in the gaze of others upon herself (Farroni et al 2002) is accompanied by increasing precision in the child's ability to detect when she is being looked at (Lee et al 1998). The main function of egocentric gaze, relates primarily to the viewpoint of the perceiver and onlooker, rather than to the spatial relations of the extrinsic world. Its goal is to engage the viewer and to control interaction.

Disruption of gaze monitoring

Neural circuitry that involves the amygdala, the orbito-frontal cortices and the superior temporal sulcus constitutes a probable basis for the development of gaze monitoring, which is critically involved in the perceptual processing of a range of social behaviours (Brothers 1990). This network is preferentially activated when viewing faces and especially eye regions (Allison et al 2000, Calder et al 2002). Individuals with bilateral disruption to amygdala-related circuits, typically have impairments of gaze monitoring (Young et al 1995). Failure to make direct eye contact when in dyadic communication, as adults, has profound consequences for our interpretation of our social partner's mental health, or their trustworthiness. The importance different cultures place on the appropriate role of eye contact in social interactions (Kleinke 1986) does not detract from that conclusion — rather, it emphasizes its validity.

Face processing deficits in autism

Face recognition memory

A failure to attend to other's faces is the best single discriminating feature of children who are developing autistic disorders, at one year of age (Dawson et al 1998, 2002). Dawson et al (2002) found that children between 3 and 5 years with an autistic spectrum disorder did not in general show differential event-related potential (ERP) activity to their mother's face, compared with an unfamiliar face. This was in striking contrast to carefully chosen comparisons. In general, there is a selective impairment of face recognition memory in autistic individuals (Klin et al 1999, Boucher et al 1998) compared with others of comparable non-autistic learning disabilities. In contrast, they retain a good word-recognition memory (Minshew & Goldstein 2001) and there is some evidence that they are relatively less impaired at the matching of unfamiliar faces than in facial recognition tasks (Minshew & Goldstein 2001, Davies et al 1994).

Emotion expression detection

Impaired recognition of emotional expressions has been reported for many years in autism (Hobson et al 1988), although it has until recently been unclear whether this is a generalized deficit in the ability of autism individuals to name facial expressions in general, or to differentiate one from another, or whether it is a more specific deficit than that. There is emerging evidence, from a series of replicated observations, that autistic individuals have reduced perception of *fear* specifically in relation to other facial emotions (Howard et al 2000, Pelphrey et al 2002).

Egocentric gaze monitoring

People with autism are notoriously poor at making appropriate eye contact in social conversations, such contact usually being either excessively direct or deviant (Calder et al 2002). It was nearly 40 years ago that Hutt & Ounsted (1966) first proposed that individuals with autism averted their gaze because it served to reduce arousal. It is surprising that, in the many thousands of research studies done since then, that the functional integrity of neural circuits mediating autonomic arousal in relation to direct gaze have so rarely been objectively assessed in autistic individuals (Hirstein et al 2001). There appears to be an associated difficulty in the detection of gaze direction, and a striking difference from normal individuals in the monitoring of eyes in social interactions (Klin et al 2002). People with autism commonly possess deficits in the perception of direction and meanings of gaze. They may experience difficulties with both explicit egocentric tasks ('Is s/he looking at me?'; Howard et al 2000), as well as integrative tasks, which make use of detection of direction of gaze in order to infer the other's intentions or attitude ('What does she want/feel/mean?'). By contrast, the ability to follow line of gaze ('Where is s/he looking?') appears intact (Leekam et al 1997). Incidentally, a similar dissociation has been observed in adult patients with schizophrenia (Rosse et al 1998, Franck et al 2002).

Theory of Mind abilities

Remarkably, and importantly for the development of our hypothesis concerning a link between fear perception and theory of mind skills, arousal of amygdala-linked neural circuits by fearful faces is possible with non-conscious stimuli (Morris et al 1998b). Deficits in the interpretation of facial expressions, face recognition memory, and the impaired interpretation of gaze direction are associated, in autistic individuals, with poor performance on a range of theory of mind tasks—from 'reading the language of the eyes' (Calder et al 2002) to attributing intention to the apparently purposeful movement of abstract shapes in a cartoon (Castelli et al 2002).

We propose that the reason such associations are present is because the same neural circuitry is being used both for monitoring socially relevant cues, especially eye contact, and for the development of theory of mind skills. A system that originally evolved to alert us to the presence, and the intention, of a predator that was 'eyeing us', is now modulated by higher cortical centres (especially in the medial prefrontal cortex and cingulate). The autonomic arousal and the associated 'feelings' evoked by eye contact are now being processed for such diverse purposes as enhancing attachment in infancy, pair-bonding in adulthood, and our ability to 'read the mind' of others.

At the heart of this arousal system is the subcortical neural pathway that alerts us to direct eye contact by others and which automatically signals 'beware, threat!' The same complex neural systems we have developed to inhibit the automatic response of fight or flight, are activated by a simple theory of mind task such as the Castelli et al (2002) cartoons (J. S. Morris, R. J. Dolan, D. H. Skuse, unpublished data). We hypothesize there is a delicate balance between the arousal engendered by this phylogenetically ancient system, with its associated memory, gaze monitoring, and threat detection roles, and the secondary processing of such arousal by neocortical systems that link ultimately to our theory of mind abilities. Because the system is delicately balanced, a variety of congenital and acquired disorders will result in literal disintegration of the harmonious interplay of the constituent elements. Playing centre stage in this neural network is the amygdala. The effects of dysfunction within that structure will therefore have especially widespread and pervasive consequences upon the development or maintenance of appropriate social cognitions and behaviours.

Structural and functional anomalies of the amygdala in autism

We might, in light of the evidence for the involvement of the amygdala in the processing of socially relevant face-related cognitions, anticipate that there are functional if not structural anomalies of the amygdala and its related neocortical pathways in autism. There are now many studies of this matter, and the results of functional, if not structural, anomalies are remarkably consistent. In general, there appears to be a lack of activation (in terms of fMRI) of the amygdala in response to social stimuli involving face perception, especially the perception of eyes (Baron-Cohen et al 1999). Diminished amygdala activation (as well as other unusual patterns of neural activity) has been found when autistic individuals visualize faces alternating with meaningless shapes (Pierce et al 2001). This latter study is of particular interest for it concluded that, compared with normal individuals, autistic individuals 'see' faces utilizing different neural systems, with each patient doing so via 'unique neural circuitry'.

Disorders of amygdala structure and function in X-monosomy

We have been studying the neural basis of a social-cognitive deficit in Turner syndrome (TS), a sporadic disorder of human females in which all or part of one X chromosome is deleted. The usual karyotype is 45,X, and this contrasts with the normal female complement of 46,XX chromosomes. Creswell & Skuse (2000) reported that autistic features are frequently associated with TS, and that the risk of autism is increased up to 500-fold. Such difficulties may reflect haploinsufficiency of X-linked gene products, which are needed in two copies for normal 46,XX female development (Zinn & Ross 1998). We have recently reported reliable deficits in the recognition of faces and in the identification of a 'fearful' facial expression, in women of normal verbal intelligence and TS (Elgar et al 2003). Because these deficits were reminiscent of those reported in people with autism (e.g. Howard et al 2000) we hypothesized that, in view of the increased risk of autistic behaviours in 45,X females, they would possess other anomalies in socio-perceptual processing. The processing of gaze was one such feature that interested us, because children with autism and those at risk for developing autism show less eye contact and a reduced ability to follow the gaze of another, especially when the attention of the other is directed to an event of social interest (e.g. Ruffman et al 2001). We confirmed that women with TS had difficulty ascertaining gaze direction from face photographs showing small lateral angular gaze deviations. They also had difficulty discriminating the detection of egocentric gaze, and the detection of allocentric gaze (Elgar et al 2002). These findings were indicative of an anomaly in the processing of facial information, in particular that involving the eyes, and implicated functional anomalies in the amygdala (Elgar et al 2003). We have therefore conducted a range of structural and functional imaging studies of the amygdala in TS, the results of which confirm this hypothesis. The structural studies show that the size of the amygdala is larger in this condition than in matched comparison 46,XX females (Good et al 2003). Our functional imaging analyses are still undergoing analysis. In other (unpublished) findings of behavioural studies in TS, which focused on the amygdala's role in fear conditioning, we found most 45,X women had impaired habituation and excessive SCR responses in a well-studied conditioning paradigm (Morris et al 1998a). The fact that a deficit in the perception of fear in another's face can be associated with excessive reactivity (rather than hypo-reactivity) of the amygdala in fear conditioning is a remarkable dissociation that demands further analysis. The strong implication is that in this condition there is anomalous modulation by the amygdala of cortical circuits concerned with face processing and other aspects of social cognition, and of the amygdala itself by frontal cortical regions. We have also recently shown that the ability to classify fear in facial expressions is correlated with face recognition skill in women, but not in men (Campbell et al 2002). This intriguing dissociation between the sexes may reflect sexual dimorphism in the mnemonic functions of the amygdala (Cahill et al 2001), and could in turn have relevance to the observation that males are more vulnerable to disorders of social cognition than females (Skuse 2000).

Intriguingly, not all 45,X females shared these deficits in fear perception, gaze monitoring, and fear-conditioning. About one third were severely affected, and the remainder had a range of impairments distributed around a median that was low-normal. Examination of the data plotted graphically suggested a bimodal distribution. It is not at this stage clear just what mechanism or mechanisms relating to X-monosomy are responsible for our findings. However, the implication is that, directly or indirectly, haploinsufficiency of X-linked genes that normally escape X-inactivation (and are not imprinted) causes maldevelopment and dysfunction of the amygdala and related circuits that are essential for processes relating to social cognition. In view of the parallels with deficits that have been reported in association with autism, we also assessed our 45,X subjects with a cartoon-based task that measures theory of mind skills (Castelli et al 2002). Our hypothesis that many 45,X women would score in the autistic range on aspects of this task, was supported. As we found in their other cognitive deficits, failure appropriately to attribute intention to the movement of animated abstract shapes, in cartoons, occurred in about one-third of the young adult 45,X subjects investigated (unpublished data).

Conclusions

In a variety of neurodevelopmental and acquired conditions, in which social cognitive deficits are predominant (the most striking example of which is autism), there is an association between a triad of impairments in cognitive processes that appear superficially to be unrelated. First, and foremost, there is a specific deficit in the ability to distinguish the facial expression of fear from other facial expressions, or to name that facial expression accurately. Second, there is a failure to track eye gaze accurately, to tell whether someone is making direct eye contact or not, and an associated difficulty with joint attention. Third, both these deficits are associated with higher order social cognitive deficits, including poor theory of mind abilities and a tendency to be disinhibited or overly trusting in social interactions with strangers. The neural basis for this triad of impairments seems to be functional abnormalities of the amygdala and related neural circuitry. A consensus is emerging among developmental cognitive neuroscientists that the functional integrity of the amygdala, and its connections with a distributed neural system that encompasses both neocortical and subcortical structures, is critical for the development and maintenance of social cognitive skills. Remarkably, fear perception is a critical indicator of the integrity of this process. Our competence at consciously detecting fear in another's face in turn reflects our theory of mind abilities. Discovering the neural basis of this association may shed light not only upon the aetiology of autism, but it could be critical for a broader understanding of human vulnerability to disorders of social cognition.

We have evidence that the development of these neural systems is influenced, at least in part, by a specific class of X-linked genes. Such genes escape X-inactivation and are thus expressed in two copies in normal females. In 46,XY males, dosage equivalence to 46,XX females is attained only if there is an equivalent copy of the gene on the Y-chromosome (true for a small proportion). Alternatively, there may be up-regulation of the male's single copy of the gene to a female-equivalent dosage, but how this regulation is achieved is as yet unknown. We hypothesize that male vulnerability to disorders of social cognition such as autism is related to haploinsufficiency (relative to females) of one or more classes of X-linked genes, the nature and position of which on the X-chromosome we are attempting to define.

Acknowledgements

The ideas expressed in this article have arisen largely out of discussions between the author, John Morris and Ray Dolan, with helpful contributions from Kate Lawrence (née Elgar) and Ruth Campbell. This is an abbreviated version of a similar article based on the Jack Tizard Memorial Lecture 2002, to be published in *Child and Adolescent Mental Health*, 2003. Original research by John Morris, Ray Dolan and David Skuse supporting many of the hypotheses outlined in this review is currently being prepared for submission for publication.

References

Adolphs R 2002 Neural systems for recognizing emotion. Curr Opin Neurobiol 12:169–177
Adolphs R, Tranel D, Damasio H, Damasio AR 1995 Fear and the human amygdala. J Neurosci 15:5879–5891
Adolphs R, Tranel D, Hamann S et al 1999 Recognition of facial emotion in nine individuals with bilateral amygdala damage. Neuropsychologia 37:1111–1117
Adolphs R, Sears L, Piven J 2001 Abnormal processing of social information from faces in autism. J Cogn Neurosci 13:232–240
Aitken KJ, Trevarthen C 1997 Self/other organization in human psychological development. Dev Psychopathol 9:653–677
Allison T, Puce A, McCarthy G 2000 Social perception from visual cues: role of the STS region. Trends Cogn Sci 4:267–278
Amaral DG 2002 The primate amygdala and the neurobiology of social behavior: implications for understanding social anxiety. Biol Psychiatry 51:11–17
Amaral DG, Capitanio JP, Jourdain M, Mason WA, Mendoza SP, Prather M 2003 The amygdala: is it an essential component of the neural network for social cognition? Neuropsychologia 41:235–240
Baron-Cohen S, Ring HA, Wheelwright S, Bullmore ET, Brammer MJ, Williams SC 1999 Social intelligence in the normal and autistic brain: an fMRI study. Eur J Neurosci 11: 1891–1898
Batki A, Baron-Cohen S, Wheelwright S, Connellan J, Ahuwalia J 2000 Is there an innate gaze module? Evidence from human neonates. Infant Behav Devel 23:223–229
Boucher J, Lewis V, Collis G 1998 Familiar face and voice matching and recognition in children with autism. J Child Psychol Psychiat 39:171–181
Broks P, Young AW, Maratos EJ et al 1998 Face processing impairments after encephalitis: amygdala damage and recognition of fear. Neuropsychologia 36:59–70
Brothers L 1990 The social brain: a project for integrating primate behaviour and neurophysiology in a new domain. Concepts Neurosci 1:27–51
Cahill L, Haier RJ, White NS et al 2001 Sex-related difference in amygdala activity during emotionally influenced memory storage. Neurobiol Learn Mem 75:1–9
Calder AJ, Young AW, Rowland D, Perrett DI, Hodges JR, Etcoff NL 1996 Facial emotion recognition after bilateral amygdala damage: differentially severe impairment of fear. Cognit Neuropsychol 13:699–745
Calder AJ, Lawrence AD, Young AW 2001 Neuropsychology of fear and loathing. Nat Rev Neurosci 2:352–363
Calder AJ, Lawrence AD, Keane J et al 2002 Reading the mind from eye gaze. Neuropsychologia 40:1129–1138
Campbell R, Elgar K, Kuntsi J et al 2002 The classification of 'fear' from faces is associated with face recognition skill in women. Neuropsychologia 40:575–584
Castelli F, Frith C, Happe F, Frith U 2002 Autism, Asperger syndrome and brain mechanisms for the attribution of mental states to animated shapes. Brain 125:1839–1849
Creswell C, Skuse D 2000 Autism in association with Turner syndrome: implications for male vulnerability. Neurocase 5:511–518
Davies S, Bishop D, Manstead AS, Tantam D 1994 Face perception in children with autism and Asperger's syndrome. J Child Psychol Psychiatry Allied Discipl 35:1033–1057
Dawson G, Meltzoff AN, Osterling J, Rinaldi J 1998 Neuropsychological correlates of early symptoms of autism. Child Dev 69:1276–1285
Dawson G, Carver L, Meltzoff AN, Panagiotides H, McPartland J, Webb SJ 2002 Neural correlates of face and object recognition in young children with autism spectrum disorder, developmental delay, and typical development. Child Dev 73:700–717

Dolan RJ 2002 Emotion, cognition, and behavior. Science 298:1191–1194
Elgar K, Campbell R 2001 Annotation: the cognitive neuroscience of face recognition: implications for developmental disorders. J Child Psychol Psychiatry 42:705–717
Elgar K, Campbell R, Skuse D 2002 Are you looking at me? Accuracy in processing line-of-sight in Turner syndrome. Proc R Soc Lond B Biol Sci 269:2415–222
Elgar K, Kuntsi J, Campbell R, Coleman M, Skuse DH 2003 Face and emotion recognition deficits in Turner syndrome: implications for male vulnerability to pervasive developmental disorders. Neuropsychology, in press
Emery NJ 2000 The eyes have it: the neuroethology, function and evolution of social gaze. Neurosci Behav Rev 24:581–604
Farroni T, Csibra G, Simion F, Johnson MH 2002 Eye contact detection in humans from birth. Proc Natl Acad Sci USA 99:9602–9605
Franck N, Montoute T, Labruyere N et al 2002 Gaze direction determination in schizophrenia. Schizophr Res 56:225–234
Frith U 2001 Mind blindness and the brain in autism. Neuron 32:969–979
George N, Driver J, Dolan RJ 2001 Seen gaze-direction modulates fusiform activity and its coupling with other brain areas during face processing. Neuroimage 13:1102–1112
Good CD, Lawrence K, Thomas NS et al 2003 Dosage sensitive X-linked locus influences the development of amygdala and orbito-frontal cortex, and fear recognition in humans. Brain, in press
Hare B, Call J, Tomasello M 2001 Do chimpanzees know what conspecifics know? Anim Behav 61:139–151
Hare B, Brown M, Williamson C, Tomasello M 2002 The domestication of social cognition in dogs. Science 298:1634–1636
Haxby JV, Hoffman EA, Gobbini MI 2002 Human neural systems for face recognition and social communication. Biological Psychiatry 51:59–67
Hirstein W, Iversen P, Ramachandran VS 2001 Autonomic responses of autistic children to people and objects. Proc R Soc Lond B Biol Sci 268:1883–1888
Hobson RP, Ouston J, Lee A 1988 What's in a face? The case of autism. Brit J Psychol 79: 441–453
Howard MA, Cowell PE, Boucher J et al 2000 Convergent neuroanatomical and behavioural evidence of an amygdala hypothesis of autism. Neuroreport 11:2931–2935
Hutt C, Ounsted C 1966 The biological significance of gaze aversion with particular reference to the syndrome of infantile autism. Behav Sci 11:346–356
Jellema T, Baker CI, Wicker B, Perrett DI 2000 Neural representation for the perception of the intentionality of actions. Brain Cogn 44:280–302
Kawashima R, Sugiura M, Takashi K et al 1999 The human amygdala plays an important role in gaze monitoring: a PET study. Brain 122:779–783
Kleinke CL 1986 Gaze and eye contact: a research review. Psychol Bull 100:78–100
Klin A, Jones W, Schultz R, Volkmar F, Cohen D 2002 Visual fixation patterns during viewing of naturalistic social situations as predictors of social competence in individuals with autism. Arch Gen Psychiatry 59:809–816
Langton SR, Watt RJ, Bruce I 2000 Do the eyes have it? Cues to the direction of social attention. Trends Cogn Sci 4:50–59
LeDoux JE 2000 Emotion circuits in the brain. Annu Rev Neurosci 23:155–184
Lee K, Eskritt M, Symons LA, Muir D 1998 Children's use of triadic eye gaze information for 'mind reading'. Dev Psychol 34:525–539
Leekam S, Moore C 2000 The development of attention and joint attention in children with autism. In: Burack JA, Charman C, Yirmiya N, Zelazo PR (eds) The development of autism: perspectives from theory and research. Earlbaum, New York, p 105–130

Leekam S, Baron Cohen S, Perrett D, Milders M, Brown S 1997 Eye-direction detection: a dissociation between geometric and joint attention skills in autism. Brit J Dev Psychol 15:77–95

Minshew NJ, Goldstein G 2001 The pattern of intact and impaired memory functions in autism. J Child Psychol Psychiatry 42:1095–1101

Morris JS, Friston KJ, Buchel C et al 1998a A neuromodulatory role for the human amygdala in processing emotional facial expressions. Brain 121:47–57

Morris JS, Ohman A, Dolan RJ 1998b Conscious and unconscious emotional learning in the human amygdala [see comments]. Nature 393:467–470

Morris JS, Ohman A, Dolan RJ 1999 A subcortical pathway to the right amygdala mediating 'unseen' fear. Proc Natl Acad Sci USA 96:1680–1685

Morris JS, Buchel C, Dolan RJ 2001 Parallel neural responses in amygdala subregions and sensory cortex during implicit fear conditioning. Neuroimage 13:1044–1052

Morris JS, de Bonis M, Dolan RJ 2002 Human amygdala responses to fearful eyes. Neuroimage 17:214–222

Nahm FK 1997 Heinrich Kluver and the temporal lobe syndrome. J History Neurosci 6:193–208

Pelphrey KA, Sasson NJ, Reznick JS, Paul G, Goldman BD, Piven J 2002 Visual scanning of faces in autism. J Autism Dev Disord 32:249–261

Perrett DI, Smith PA, Potter DD et al 1985 Visual cells in the temporal cortex sensitive to face view and gaze direction. Proc R Soc Lond B Biol Sci 223:293–317

Phillips ML, Williams L, Senior C et al 1999 A differential neural response to threatening and non-threatening negative facial expressions in paranoid and non-paranoid schizophrenics. Psychiatry Res 92:11–31

Pierce K, Muller RA, Ambrose J, Allen G, Courchesne E 2001 Face processing occurs outside the fusiform 'face area' in autism: evidence from functional MRI. Brain 124: 2059–2073

Rochat P, Morgan RCM 1997 Young infants' sensitivity to movement information specifying social causality. Cogn Dev 12:441–465

Rosse RB, Schwartz BL, Johri S, Deutsch SI 1998 Visual scanning of faces correlates with schizophrenia symptomatology. Prog Neuropsychopharmacol Biol Psychiatry 22:971–979

Ruffman T, Garnham W, Rideout P 2001 Social understanding in autism: eye gaze as a measure of core insights. J Child Psychol Psychiatry 42:1083–1094

Santos LR, Hauser MD 1999 How monkeys see the eyes: cotton-top tamarin's reactions to changes in visual attention and action. Anim Cogn 2:131–139

Sarfati Y, Hardy-Bayle MC, Brunet E, Widlocher D 1999 Investigating theory of mind in schizophrenia: influence of verbalization in disorganized and non-disorganized patients. Schizophr Res 37:183–190

Sato N, Nakamura K 2001 Detection of directed gaze in rhesus monkeys (*Macaca mulatta*). J Comp Psychol 115:115–121

Shenton ME, Dickey CC, Frumin M, McCarley RW 2001 A review of MRI findings in schizophrenia. Schizophr Res 49:1–52

Siegal M, Varley R 2002 Neural systems involved in 'theory of mind'. Nat Rev Neurosci 3: 463–471

Skuse DH 2000 Imprinting, the X-chromosome, and the male brain: explaining sex differences in the liability to autism. Pediatr Res 47:9–16

Stone VE, Baron-Cohen S, Calder A, Keane J, Young A 2003 Acquired theory of mind impairments in individuals with bilateral amygdala lesions. Neuropsychologia 41: 209–220

Sweeten TL, Posey DJ, Shekhar A, McDougle CJ 2002 The amygdala and related structures in the pathophysiology of autism. Pharmacol Biochem Behav 71:449–455

Tenyi T, Herold R, Szili IM, Trixler M 2002 Schizophrenics show a failure in the decoding of violations of conversational implicatures. Psychopathology 35:25–27

Winston JS, Strange BA, O'Doherty J, Dolan RJ 2002 Automatic and intentional brain responses during evaluation of trustworthiness of faces. Nat Neurosci 5:277–283
Young AW, Aggleton JP, Hellawell DJ, Johnson M, Broks P, Hanley R 1995 Face processing impairments after amygdalotomy. Brain 118:15–24
Zinn AR, Ross JL 1998 Turner syndrome and haploinsufficiency. Curr Opin Genet Dev 8:322–327

DISCUSSION

Schultz: You are speculating that there is interconnectivity between these cortical regions and the amygdala. Have you done any region-of-interest analyses to look for correlations in the amount of amygdala activation across your sample of Turner syndrome (TS) women?

Skuse: We haven't yet. This work was only recently completed and we haven't had a chance to look formally at interactions between these areas, although that work is in progress. It is being done by John Morris, in my Unit.

Schultz: Did you say that the activity of the amygdala was observed in a contrast of conditioned versus unconditioned neutral face?

Skuse: That's right. Amygdala activity was observed in two different tasks, in our study of TS women using fMRI. One was a CS+/CS− contrast in a fear-conditioning paradigm, in which the UCS was a very loud noise and the CS were two neutral faces. The other difference in activity, we observed was in the contrast of amygdala responses to a series of morphed faces which contrasted 'fearful' eyes and 'neutral' eyes using a paradigm developed by John Morris and Ray Dolan (Morris et al 2002). There was an exaggerated amygdala response to the CS+ and the fearful eyes respectively, in the TS sample compared with a group of normal females.

Schultz: Do you know whether there would be over activity in the amygdala independent of that relative difference if you compared, for example, the CS+ to a neutral object?

Skuse: We have evidence suggesting that there is probably over activity, even to the CS−.

Amaral: I need some clarification. You said that there was a bimodal distribution in your TS patients' ability to perceive fear in a face. Could you review the bimodality in relation to amygdala volume?

Skuse: About one-third of the TS females have fear recognition skills within the normal range, and one-third have seriously deficient skills—similar to those of patients with a bilateral amygdalectomy. There appears to be bimodality in the distribution, but we have not yet been able to correlate that bimodality with amygdala volume.

C. Frith: I'm assuming that you haven't done a regions-of-interest analysis of the amygdala volume in parallel with voxel-based morphometry (VBM).

Skuse: We wanted to validate the VBM results by a formal region-of-interest assessment of the volume of the amygdala. As you know, this is a very difficult structure to define anatomically. Our neuroimaging colleagues said the procedure would be a waste of time, the information we would gain would be no more valid than the estimated volume from VBM analyses.

C. Frith: What I predict is that you might find no difference with the regions-of-interest analysis, but you could find a difference in the VBM measure. This would be very informative as to what is actually taking place.

Skuse: That is a very interesting prediction! VBM is an automated technique that uses mathematical functions to predict a local average amount of grey matter within each voxel, but there has been controversy about the correct interpretation of these data. VBM does seem to show a reasonable correlation with region of interest (ROI) techniques in the measurement of amygdala volume (Good et al 2002).

Rutter: Can you speculate on the implications of the different possibilities?

C. Frith: As I understand it, the amygdala is not really a single 'thing'; it is a collection of many separate nuclei, and is a junction for many pathways. In a sense, the volume is not the interesting parameter, but rather the proportions of grey and white matter.

Skuse: That's right, and this was one of the arguments against measuring it with an ROI technique because it is very difficult to tell where its boundaries are.

C. Frith: I would suggest that this is showing that the problem is something to do with connectivity rather than just the size.

Skuse: I agree that concentrating on the volumetric aspect of our findings could be a red herring, so far as understanding the deficit in fear recognition is concerned. Potential anomalies in the amygdala's connectivity with other brain regions represent a far more interesting issue. I don't want to dwell on the volumetric differences of the amygdala in 45,X Turner syndrome, except to say that there is an intriguing inverse relationship between the number of the X chromosomes and the size of this structure. 46 XX females and 47 XXY males have similar sized amygdalae, which are smaller than those of normal 46,XY males, and these in turn are smaller than the amygdalae of 45,X females (Patwardhan et al 2002).

Amaral: I agree with Chris Frith that it is controversial what volume-based measurements are actually showing. The other controversy is whether in autism the amygdala is getting larger or smaller. Ultimately, it may depend on the age at which you sample your population. There are nice data showing that there is an age-related change in the size of the amygdala postnatally, from the ages of about 5–15. We have replicated this work in the age range of 8–18. There is a 50% increase.

Skuse: All of our subjects were adults. Everyone was over 16 and the vast majority were over 18.

Amaral: The hypothesis is very interesting, but it turns out that the pathology that has been demonstrated in the amygdala thus far is in the nuclei that aren't communicating with the cortex. We haven't seen any differences in lateral or basal nuclei. The amygdala consists of thirteen different nuclei, so you have to be careful to look at exactly which nuclei are showing changes. My guess is that at this point we simply lack the data to know this.

Skuse: I agree; the scanner that we used didn't have the resolution for us to look at individual nuclei. Hopefully, as we get better resolution we will be able to distinguish between the medial and lateral nuclei, for example. We suspect that there is a disconnection within the amygdala between the activity of those different nuclei.

Bailey: Dr Simon Wallace has just completed a very careful study of face processing in autism and Asperger's syndrome for his PhD (unpublished thesis, University of London). Twenty-eight able individuals and very well matched controls performed a range of face processing tasks, including tasks where image presentation was restricted to between 40 and 100 ms giving tight control over the strategies that participants could use. He found significant impairments in face recognition, emotion recognition and judging eye gaze direction, but unremarkable non-face object processing. The clinical group were impaired at recognizing most facial expressions, but particularly fear and disgust. Secondly, Bob Schultz (Schultz et al 2000) and Eric Courchesne (Pierce et al 2001) have nice data showing that individuals are not activating fusiform gyrus when they see faces. We have similar data using magnetoencephalography (unpublished data). It turns out that when people see pictures of faces, there is a very fast pathway to anterior temporal regions, and probably the fusiform gyrus, that is activated at 30–60 ms and which precedes activation of primary visual cortex (Bräutigam et al 2001). There is a different pattern of activity in that fast pathway in the affected individuals, and no evidence of activation of fusiform gyrus. In terms of the neuropathology, we have some preliminary evidence for abnormal connectivity, but in a number of separate brain regions. We can see abnormal fibre tracts in the caudate nucleus in one case, in the cerebellum in another and in the pons (Bailey et al 1998) in the only child, and also an abnormal myelinated plexus in the cerebral cortex in one case. I suggest that this is probably a connectivity problem, but it seems plausible that this is not tightly confined to one anatomical structure. The challenge will be to sort out which of these localized abnormalities are correlated with particular behavioural impairments.

Dawson: Susan Folstein's student, Raphe Bernier, who currently working with me, is using a fear-conditioning paradigm with adults with autism. We are collecting data on this now. It will be interesting to see whether the amygdala is hyperactive or hypoactive. I wanted to mention Steve Dager's findings (Sparks et al 2002) from our group at the University of Washington on three year olds with autism. Like most people, he found an enlarged brain. He looked at several regions of interest and found that the amygdala was proportionally larger, even relative to brain size. It was the only structure that was larger relative to overall brain volume. It was more accentuated in children with autistic disorder compared to children with pervasive developmental disorder. Size of the amygdala was positively correlated with severity of impairment on a number of behavioural measures. It is a surprising finding.

Rogers: I have a comment about the methods of looking at abnormalities in social function and TS. Can we assume that all social deficits are autism? When you showed the AQ data, the mean of people with normal development was about 14, the TS group was 16 and the autism group was 45 points. This doesn't convince me that people with TS are autistic. We don't have a very good way of categorizing social deficit. It is important to ask whether all social dysfunction is autism, or is there something very specific about the particular social deficits that make up autism.

Skuse: You are right. These women were selected because they were well adjusted; they weren't selected because they had autistic features. Nevertheless, they did have some degree of social and cognitive impairment of an autistic type. On the self-rated autism scale I presented (The Autism Quotient Self-report questionnaire; Baron-Cohen et al 2001) women with TS had significantly more autistic-like traits than a control group of typically developing females. Autistic-like behaviours were most marked for the social skills subscale of this questionnaire. Many of the women, for example, indicated that they did not find it easy to work out what someone was feeling from looking at their face. Some autistic-like impairment in social interaction is common in TS, although sometimes it is subtle. We have found more solid evidence for impairments of an autistic type in their cognitive processing of social cues than in measures at a behavioural level, which are necessarily rather crude. Not only are there the face-processing deficits we have discussed, but there are striking impairments on a recently published theory of mind task (Castelli et al 2002). I strongly suspect we would find a similar set of cognitive or emotion-processing deficits in a larger proportion of the general population than are diagnosed with a disorder on the autistic spectrum. These 'endophenotypes' may turn out to be more informative phenotypes for linkage or association studies than the disorder as conventionally described.

Rogers: We don't know how many different groups would have fear-conditioning problems or problems on the social intentionality test. There may be a variety of different people with different disabilities who have no social features of autism.

Skuse: The point of looking at this group of 45,X females, in order to discover neural systems that could be dysfunctional in autism, is that they are haploinsufficient for X-linked genes, and this is why they have the problem. We know that their physical and psychological phenotype must be due to the fact that they only have one X chromosome, and this narrows down our search for genes that predispose to autistic aspects of cognitive functioning to the X-chromosome. We can use deletion mapping of people who have only lost bits of the X chromosome to narrow down the area of the search even further (Good et al 2003).

Rogers: You generalized a little to address sensory reactivity, and suggested that there was hyperresponsivity in autism in sensory measures in general. Our autism and fragile X data are relevant here. Work published by Don Rojas (Rojas et al 2001) and Lucy Miller (Miller et al 2001), contrasting autism and fragile X with normal development, demonstrate lack of appropriate habituation and hyper-responsivity to a variety of different sensory inputs in people with fragile X syndrome. In contrast, people with autism show hypo-responsivity and more rapid habituation in those paradigms. The idea of hyper-sensitivity or hyper-sensory responsiveness in autism is not yet convincingly established.

Skuse: No published studies have examined the functional integrity of the amygdala in autism, in a classic fear-conditioning experiment. The nearest I could find was by Hirstein et al (2001), which showed both hyper- and hypo-responsivity of the amygdala to social cues in autistic subjects. The group they were most interested in were the hyper-responsive ones. In fragile X, invariably the male subjects have quite severe mental retardation. Autism is certainly proportionately more common in males, and in people with very low abilities. In contrast, we have been studying a sample of 45,X females with at least average IQ—a group in whom it is all the more surprising that we find so many autistic features. But we must be cautious. If you take a group of people with moderate to severe mental retardation, you might find autonomic hyper-reactivity is non-specifically associated with learning difficulties in general.

Bishop: I was struck by the fact that the only time I've heard people talking about conditioning in relation to disorder of empathy is that low autonomic reactivity goes with problems in being empathic. What I would have predicted with high reactivity is social anxiety. It is counterintuitive that over-reactivity would lead you to be socially insensitive.

Dawson: Fear conditioning is not the same as oversensitivity to stimuli. They are two different things.

Bishop: What's the difference, then?

Dawson: Theoretically, you could have a person who is hypo-responsive to someone showing a fear face and at the same time, establish fear conditioning quickly. Fear conditioning involves a well worked out neural circuit involving the amygdala, and someone should look at it in autism. It may be that for some reason, people with autism establish fear conditioning quickly and have trouble habituating.

Bishop: There is a wonderful book called *Boo!* (Simons 1996), about people who are very jumpy or over-reactive to stimuli such as loud noises. There are people for whom this just does not condition. But they are not autistic.

Dawson: Fear conditioning is not something we would expect to be specific to autism.

Skuse: Can I emphasize that we are talking about the interaction between a different brain areas that are interconnected, and it is something to do with impairments in the interconnectivity of these different areas that leads to the autistic phenotype. Taking each element of dysfunction in isolation can lead to misleading conclusions. For example, you can show that people who are socially anxious avert their gaze from other's faces (Chen et al 2002). They are not autistic; they are socially anxious. We suspect there is something about the eyes that they find aversive, but they are not autistic. You can't take any single element of this and say that this is the key to understanding the aetiology of autism. It is much more complex.

Charman: In the past you have talked about the difference between TS subjects who inherit the X from the mother or father. Is this still part of the picture? Could you also comment on the phenotypic overlap between TS and autism?

Skuse: We did describe excessive autistic features among 45,X females who have a single maternally derived X as opposed to a paternal X chromosome (Skuse et al 1997). We have not found any significant differences between these subgroups in terms of brain structure, which is compatible with structural imaging by other groups (Brown et al 2002). Seven of the people we imaged had a paternal X and 14 had a maternal X.

Charman: From your previous work it should have been the maternals who were the most impaired.

Skuse: That is true, we certainly expected clearer evidence for differences in brain structure than we found. However, at this stage we have only conducted our functional brain-imaging experiments with 45,X females in whom the single X is maternal.

Rutter: It seems that we need to back-off from the imprinting mechanism that you put forward originally.

Skuse: No, I think not. We followed the sample on whom the imprinted influence on social behaviour was originally described (Skuse et al 1997) from childhood to adulthood, and we had them rate themselves as adults on various dimensions of autistic behaviour (Baron-Cohen et al 2001). The results hold up: there is still significantly more self-reported autistic behaviour among the $45,X^{M}$s than the $45,X^{P}$s, which is what the original hypothesis stated. We had expected to find that we would also distinguish these two groups on the cognitive measures that we are using. We have not been able to find that distinction in terms of any single measure. On the other hand, we do have an interaction effect that is intriguing. If you look at the correlation between fear recognition and face recognition memory in a normal population of males, there is none at all (Campbell et al 2002). In normal females the correlation is about 0.5. We split our 45,X sample into those with a paternal X and those with a maternal X. The latter group are like normal males in that they have a single maternal X-chromosome. Those with a single paternal X are like normal females in this respect. Within the $45,X^M$ group there was no correlation between fear recognition and face recognition memory (comparable to normal males), whereas in the $45,X^P$ group the correlation is identical to normal females. This finding suggests that an imprinted gene, which is expressed from the paternal X-chromosome, influences emotion processing circuits relevant to recognition memory.

Lord: How does that fit with your bimodal distribution?

Skuse: There is no evidence that the bimodality in respect of fear-recognition from faces is influenced by an imprinted X-linked gene.

Bailey: I have no trouble with accepting that some of the deficits you have found in the TS group, particularly the interesting data on amygdala volume and fear conditioning, might be quite similar to some of the pathology and mechanisms occurring in idiopathic autism. The question is whether the genetic influences are the same. I'd like to bring you back to what you wrote a few years ago, arguing that the TS data represented evidence that there was a locus on the X chromosome that was likely to be implicated in idiopathic autism. One of the difficulties with this hypothesis is that if you look at milder phenotypes in other family members, the most frequently affected groups were male siblings and fathers. This is incompatible with transmission of an X locus from the affected father to an autistic male. Where is your thinking at the moment in terms of genetic mechanisms? Is the TS model most useful from a neurobiological perspective rather than looking for genes?

Skuse: I agree with your cautionary comments about the limited value of our work in tracking down genes that predispose to autism in the general population. I have never suggested that there is an imprinted locus that actually causes autism. Nevertheless, I proposed that their lack of a paternally derived X-chromosome may have rendered males more vulnerable to autism due to the influence of specific autosomal loci (Skuse 2000). Today, I have presented a revised and rather more complex theory. Our latest data are consistent with the idea that male vulnerability is due to X-linked genes, which influence the development and function of neural circuits involved in social-cognitive processing. The new twist to the theory is that there appear to be non-imprinted X-linked genes involved too, and these influence emotion processing from social cues such as faces. We hypothesize that such genes are functionally polymorphic, escape X-inactivation, and have no Y-homologue. We now know there are many such genes on the X-chromosome, mainly on the short arm (Disteche 1999), although for the most part their function is uncertain. To ensure dosage equilibrium between males and females males are going to up-regulate a great many of those X-linked genes. Remarkably little work has been done on this, but sexual dimorphism could result if they do not up-regulate all of them. A functionally polymorphic X-linked allele could act in a simple X-linked way to enhance male vulnerability to autism in the same way that lack of expression of a imprinted gene might do so.

Bailey: How would you marry that with Asperger's statement that this was a genetic disorder passed from father to son? Are you suggesting that this modifier locus is not operating in the affected fathers; that they have some unrelated cause of their mild autism-like traits?

Monaco: David Skuse, I think things are getting confused. You said that there was an imprinted locus that escapes X inactivation. So what you are saying is, I think, that all males are equally susceptible because there is no variation at this locus. Because we are males we have this bad luck of having this gene that is not turned on normally, and which makes us susceptible.

Skuse: I am suggesting there could be two X-linked mechanisms involved in increasing male vulnerability to autism. Incidentally, both are likely to interact with endogenous sex steroid influences upon gene function. One is an imprinted X-linked locus (which is not expressed at all in males therefore is not subject to variation that could be mapped in them). The second mechanism involves a functionally polymorphic X-linked locus that must be passed from mothers to sons. This locus appears to escape X-inactivation. We do not know the relative proportions of these polymorphic variants in the general population, but one variant appears to be associated with very poor fear recognition, according to our 45,X data. Females would of course be largely heterozygous for these variants, thus any phenotype that is associated with carrier status is likely to be subtle in them.

Monaco: If there was allelic variation we should have picked it up in the linkage studies.

Skuse: That may be true, but it is likely the variant in which we are interested is probably in epistasis with other loci that are not X-linked. This is going to make mapping difficult unless you have a clearer idea of what phenotype to map, or much larger samples than are currently available. The best example of a genetic mechanism analogous to what I am proposing, has been called 'triallelic inheritance' (Katsanis et al 2001). Whether autism is due to three, or to more than three genes, the principle is similar—if one of these alleles is X-linked the expression of the phenotype could be sexually dimorphic.

Bishop: It could depend on the genetic background it is expressed against.

Skuse: That is exactly the point. For the full expression of the autistic phenotype a particular genetic background is necessary, with several other contributory genetic variants, which may be autosomal. On the other hand, there may be a polymorphism of the non-imprinted X-linked locus I have described, which increases vulnerability to the impact of those other variants on social cognitive processing. We have deletion mapped a locus that appears to influence the development of the amygdala and its functionality to Xp11.4 (Good et al 2003). Females with one intact and one partial X chromosome, who have deleted that region tend to have a large amygdala that is dysfunctional (they lack normal fear-recognition from faces). If their second X-chromosome is deleted distal to that region, they have normal sized amygdalae and normal functionality.

Bishop: I'm not saying that it is not the X: I'm just saying that if it is increasing your susceptibility it is not the only factor determining function, but works in combination with other genes.

Bailey: Can I check that I have understood this? Are you saying that the locus on the X is a general risk factor that is not autism specific?

Skuse: Yes, at the moment. There are probably one or more X-linked alleles that are susceptibility factors for phenotypes that, if could measure them, we could detect in people with neither autism nor TS.

Bailey: How do those particular alleles get from father to son?

Folstein: They don't have to if they are general susceptibility genes.

Bailey: I thought David said that there is a general risk, but also a more specific gene. Are you arguing that there is any autism-specific association with the X, or is it simply that males are more vulnerable to social difficulties?

Skuse: For the most part, 45,X females don't have autism: they have some features of dysfunction, which are autistic-like. I don't know what this has to do with a general susceptibility to autism, but most of us suspect that autism is a genetically heterogeneous condition. Nevertheless, males are certainly more vulnerable than females. I propose there is a *general* male susceptibility (due to an imprinted X-linked gene) and in addition a *specific* vulnerability that is X-linked and due to functional variation at a non-inactivated locus. The latter will increase susceptibility in some males, and in a substantially smaller proportion of females. My interest here is not primarily to find the genetic cause of autism. I am trying to understand which neural circuits might be dysfunctional in people who have autistic features of behaviour. If we could get a genetic hold on why these neural circuits are dysfunctional, even though that may not be the reason why most autistic people are autistic, we could begin to develop animal models to investigate that dysfunction further. We have begun to do this already with colleagues at the Babraham Institute in Cambridge (Isles & Wilkinson 2000).

Bishop: What makes it very interesting is that the TS girls are not at risk for language impairment. On structural language skills they are fine. You are really saying that what this is increasing the risk for is pragmatic problems and social communication deficits. The structural language problems, which also show a massive sex difference (e.g. Robinson 1991) are not increased in TS. There is a real pulling apart of two things that in autism would normally go together.

Skuse: It is not quite true to say that females with normal intelligence and TS have normal language. We have used your own Children's Communication Checklist to investigate this matter (Bishop 1998). Children and adolescents with TS were found to exhibit relatively poor pragmatic language skills in a number of domains. For instance, in comparison with typically developing age and IQ-matched females, their conversation was significantly more stereotyped, with inappropriate initiations, poor conversational rapport and a lack of coherence (Akers et al 2003).

Rutter: Is it just coincidence that there is a male preponderance for all these neurodevelopmental disorders or might there be a common mechanism? The overall pattern of features in TS does not seem to suggest commonality with other neurodevelopmental disorders.

Folstein: This is what made us all so excited about the original paper: it gave a plausible explanation for the male preponderance of a lot of traits. How could we follow-up that idea? If we see increased amygdala size in autism and TS, what does the amygdala size look like in ADHD?

Dawson: We also have to consider development. A study by Elizabeth Aylward and Nancy Minshew (Aylward et al 1999) suggests that by adulthood the amygdala is smaller.

Rutter: Tony Monaco, your work spans language disorders, dyslexia and autism. Do you see a common feature that might account for the male preponderance?

Monaco: We haven't seen anything on the X-chromosome in any of those disorders. In autism, if it holds that there is very little variation at this locus and we cannot detect linkage in male–male pairs, we might be able to see something in female–female pairs. Unfortunately, the numbers are very low.

References

Akers R, Lawrence K, Campbell R, Skuse DH 2003 The development of pragmatic language skills in Turner syndrome. In preparation

Aylward EH, Minshew NJ, Goldstein G et al 1999 MRI volumes of amygdala and hippocampus in non-mentally retarded autistic adolescents and adults. Neurology 53:2145–2150

Bailey A, Luthert P, Harding B et al 1998 Clinicopathological study of autism. Brain 121:889–905

Baron-Cohen S, Wheelwright S, Skinner R, Martin J, Clubley E 2001 The autism-spectrum quotient (AQ): evidence from Asperger syndrome/high-functioning autism, males and females, scientists and mathematicians. J Autism Dev Disord 31:5–17

Bishop DV 1998 Development of the Children's Communication Checklist (CCC): a method for assessing qualitative aspects of communicative impairment in children. J Child Psychol Psychiatry 39:879–891

Bräutigam S, Bailey A, Swithenby S 2001 Task dependent early latency (30–60 ms) visual processing of human faces and other objects. Neuroreport 12:1531–1536

Brown WE, Kesler SR, Eliez S et al 2002 Brain development in Turner syndrome: a magnetic resonance imaging study. Psychiatry Res 116:187–196

Campbell R, Elgar K, Kuntsi J et al 2002 The classification of 'fear' from faces is associated with face recognition skill in women. Neuropsychologia 40:575–584

Castelli F, Frith C, Happe F, Frith U 2002 Autism, Asperger syndrome and brain mechanisms for the attribution of mental states to animated shapes. Brain 125:1839–1849

Chen YP, Ehlers A, Clark DM, Mansell W 2002 Patients with generalized social phobia direct their attention away from faces. Behav Res Ther 40:677–687

Disteche CM 1999 Escapees on the X chromosome. Proc Natl Acad Sci USA 96:14180–14182

Good CD, Scahill RI, Fox NC et al 2002 Automatic differentiation of anatomical patterns in the human brain: validation with studies of degenerative dementias. Neuroimage 17:29–46

Good CD, Lawrence K, Thomas NS et al 2003 Dosage sensitive X-linked locus influences the development of amygdala and orbito-frontal cortex, and fear recognition in humans. Brain, in press

Hirstein W, Iversen P, Ramachandran VS 2001 Autonomic responses of autistic children to people and objects. Proc R Soc Lond B Biol Sci 268:1883–1888

Isles AR, Wilkinson LS 2000 Imprinted genes, cognition and behaviour. Trends Cogn Sci 4:309–318

Katsanis N, Ansley SJ, Badano JL et al 2001 Triallelic inheritance in Bardet–Biedl syndrome, a Mendelian recessive disorder. Science 293:2256–2259

Miller LJ, Reisman J, McIntosh D, Simon J 2001 An ecological model of sensory modulation: performance of children with fragile X syndrome, autism, attention deficit disorder with hyperactivity and sensory modulation dysfunction. In: Smith-Roley S, Imperatore-Blanche E, Schaaf RC (eds) Understanding the nature of sensory integration with diverse populations. Therapy Skill Builders, San Antonio, TX

Morris JS, de Bonis M, Dolan RJ 2002 Human amygdala responses to fearful eyes. Neuroimage 17:214–222

Patwardhan AJ, Brown WE, Bender BG, Linden MG, Eliez S, Reiss AL 2002 Reduced size of the amygdala in individuals with 47,XXY and 47,XXX karyotypes. Am J Med Genet. 114:93–98

Pierce K, Muller RA, Ambrose J, Allen G, Courchesne E 2001 Face processing occurs outside the fusiform 'face area' in autism: evidence from functional MRI. Brain 124:2059–2073

Robinson RJ 1991 Causes and associations of severe and persistent specific speech and language disorders in children. Dev Med Child Neurol 33:943–962

Rojas DC, Benkers T, Rogers SJ, Teale PD, Reite ML, Hagerman RJ 2001 Auditory evoked magnetic fields in adults with fragile X syndrome. Neuroreport 12:2573–2576

Schultz RT, Gauthier I, Klin A et al 2000 Abnormal ventral temporal cortical activity during face discrimination among individuals with autism and Asperger syndrome. Arch Gen Psychiatry 57:344–346

Simons RC 1996 Boo! Culture, experience, and the startle reflex. Oxford University Press, New York

Skuse DH 2000 Imprinting, the X-chromosome, and the male brain: explaining sex differences in the liability to autism. Pediatr Res 47:9–16

Skuse DH, James RS, Bishop DV et al 1997 Evidence from Turner's syndrome of an imprinted X-linked locus affecting cognitive function. Nature 387:705–708

Sparks BF, Friedman SD, Shaw DW et al 2002 Brain structural abnormalities in young children with autism spectrum disorder. Neurology 59:184–192

General discussion I

Amaral: I have a general question for the geneticists. Is the field moving ahead as rapidly as possible to get to the genetic determinants of autism? If not, what is holding it up? In neuropathology it is clear what is impeding progress: the lack of brains and more people to look at the brains. But I would like a sense of what momentum there is in the genetics of autism.

Monaco: One of the things holding it up is the sheer number of genes we have to test at these loci, because they are so large. A lot of us test genes but we do not publish the negative results. The *RELN* data are publishable, but a number of the other genes we have tested will never see the light of day. One way to prevent each of us from testing the same genes would be to put our negative results on our websites. This would be valuable and it is easy to do. We do this within our consortium, but this could be done more broadly. Otherwise, the bottleneck is that if you want to screen genes exhaustively you cannot do more than one a month.

Bolton: In the study of other disorders there is an arrangement among several genetics laboratories over who is going to focus on which regions of interest. I'm not sure whether there is that type of coordination in the autism research efforts. If there are five or six areas that might harbour a susceptibility gene, should different laboratories be focusing on genes in specific areas?

Pericak-Vance: In general they are. Not every lab is focusing on every area. I agree with Tony Monaco: I know that there is a lot of overlap in the genes that we have both evaluated. I would still like to see more families, too, in terms of trying to differentiate some of these different genes. It takes a long time to evaluate these genes, and it is expensive.

Folstein: Another thing holding up the field is that very dense single nucleotide polymorphism (SNP) maps are not available yet. These will make some types of studies a little easier.

Bauman: Recently, there was an article suggesting that IVF babies had higher risk for neural problems of various types. There are some observations in Boston where we think we are seeing higher numbers of autistic children among the IVF population. I would be curious to know whether the genetics people are seeing this as part of their multiples?

Bailey: We have read similar reports, so we made a strategy decision to exclude IVF children from the IMGSAC multiplex sample.

Sigman: We heard the same thing, so we put it into the questionnaire that we are using in our regression study. We have also included it in an epidemiological study we are doing at UC Davis.

Bailey: I have seen about 3 or 4 cases clinically, but I don't know what the rate of IVF is in the general population.

Rutter: It is quite high, and going up.

Folstein: Why would this be?

Charman: My understanding, from talking to people who do the procedure, is that they do expect more medical problems, because of the nature of the couples and their age, and because of the physical effects that the procedures have. This might include autism.

Lipkin: There are many different reasons why couples pursue IVF. One may be maternal immunity to sperm. This sort of information could provide important clues to the pathogenesis of some cases of autism, particularly if there is a role for altered immunity in the mother or the child.

Rutter: Does anyone want to comment on the claim that the rate of twinning is associated with autism?

Folstein: All I can say is that we made a definite attempt in the original twin study of autism to estimate it versus what we would expect in the population. It wasn't different. At one point in the original twin study I kept being referred the same cases by a second source. Later some statisticians began to use this as a measure of the completeness of the ascertainment. If we keep getting the same families again and again, there aren't many new ones out there to be had. I think it is ascertainment bias. If you look at Ritvo's twin study, he was ascertaining in a very peculiar way. He got enormous numbers of monozygotic (MZ) twins.

Bailey: For the 1995 paper (Bailey et al 1995) I calculated the expected number of twins very carefully, and there was a small increase in the number of identical twin pairs. There has been a recent report by Greenberg et al (2001) suggesting that in the AGRE multiplex sample there is a large increase in the rate of twinning. We have looked at the rate of twins in the IMGSAC (unpublished data), and there is a very small increase, entirely compatible with a bias in terms of ascertainment. It is nowhere near the level of increase reported in the AGRE sample. My understanding is that Dan Geschwind, who is a central figure in the AGRE consortium, has a particular interest in MZ twins, and of course most groups studying molecular genetics of autism are not interested in MZ twins; so this combination may have given rise to this large number.

Folstein: If you ask the question of the population, send me all your autism families where there are two affected children, what proportion of families who have two affected children are twin families?

Bailey: In effect we have done this. It turns out there is only a slight increase in the number of dizygotic (DZ) twins over what one would expect from the population rate of twinning. We have effectively asked to be referred all UK multiplex autism families, and we have ascertained a twin sample simultaneously.

Rutter: The net effect, as I read it, would be that although we can't rule out a minor increase in twinning, it is extremely unlikely that there is the large increase that has been claimed. Given the source of the data, there is bound to be some ascertainment bias. The finding that makes me inclined to be sceptical about the postulated association between twinning and autism is that when we undertook a second British twin study (Bailey et al 1995), we checked whether the new ascertainment produced cases in the relevant age group missed first time. Very few such cases were found. It was about as complete a sample as you can get, but we found no significant excess of twins.

References

Bailey A, Le Couteur A, Gottesman I et al 1995 Autism as a strongly genetic disorder: evidence from a British twin study. Psychol Med 25:63–77

Greenberg DA, Hodge SE, Sowinski J, Nicoll D 2001 Excess of twins among affected sibling pairs with autism: implications for the etiology of autism. Am J Hum Genet 69:1062–1067

The neuropathology of the autism spectrum disorders: what have we learned?

Margaret L. Bauman and Thomas L. Kemper*

*Children's Neurology Service, Massachusetts General Hospital, 55 Fruit Street, Boston, MA 02114, and *Department of Anatomy and Neurobiology, Boston University School of Medicine, Boston, MA 02118, USA*

Abstract. Autism is a behaviourally defined disorder, initially described by Kanner in 1943. By definition, symptoms are manifested by 36 months of age and are characterized by delayed and disordered language, impaired social interaction, abnormal responses to sensory stimuli, events and objects, poor eye contact, an insistence on sameness, an unusual capacity for rote memory, repetitive and stereotypic behaviour and a normal physical appearance. Relatively few neuropathological studies have been performed on the brains of autistic subjects. Of those reported, abnormalities have been described in the cerebral cortex, the brainstem, the limbic system and the cerebellum. Although those with the disorder present with a specific set of core characteristics, each individual patient is somewhat different from another. Thus, it should not be surprising that the brains of these subjects should show a wide range of abnormalities. However, it is important to delineate the anatomic features, which are common to all cases, regardless of age, sex and IQ, in order to begin to understand the central neurobiological profile of this disorder. The results of our systematic studies indicate that the anatomic features that are consistently abnormal in all cases include reduced numbers of Purkinje cells in the cerebellum, and small tightly packed neurons in the entorhinal cortex and in the medially placed nuclei of the amygdala. It is known that the limbic system is important for learning and memory, and that the amygdala plays a role in emotion and behaviour. Research in the cerebellum indicates that this structure is important as a modulator of a variety of brain functions and impacts on language processing, anticipatory and motor planning, mental imagery and timed sequencing. Defining the differences and similarities in brain anatomy in autism and correlating these observations with detailed clinical descriptions of the patient may allow us greater insight into the underlying neurobiology of this disorder.

2003 Autism: neural basis and treatment possibilities. Wiley, Chichester (Novartis Foundation Symposium 251) p 112–128

Infantile autism is a behaviourally defined disorder, first described by Kanner in 1943. Symptoms become apparent by three years of age and include delayed and disordered language, impaired social interaction, isolated areas of interest, poor eye contact, abnormal responses to sensory stimuli, events and objects, an insistence on sameness and an unusual capacity for rote memorization. Physical appearance is normal and motor findings when present are subtle. Motor milestones are usually achieved on time.

Given the clinical features of the disorder, various anatomic sites have been hypothesized as being important to our understanding of autism including the limbic system (Darby 1976), medial temporal lobe (Damasio & Maurer 1978), thalamus (Coleman 1979), basal ganglia (Vilensky et al 1981), and vestibular system (Ornitz & Ritvo 1986). Efforts to define the neuropathology of this disorder began in 1968 with the publication of the results of a frontal lobe biopsy, which showed non-specific findings of arterioles, leptomeninges and cortex (Aarkrog 1968). Since that time, a number of anatomic studies have been reported with varying results. It is likely that differences in findings are, at least in part, related to the techniques used, and to differing clinical features of the subjects studied. However, some common microscopic themes are beginning to emerge.

Grossly, the autistic brain appears to be structurally normal, both by direct inspection and neuroimaging techniques (Fig. 1). Although most autistic children are born with normal head circumferences, the trajectory of head

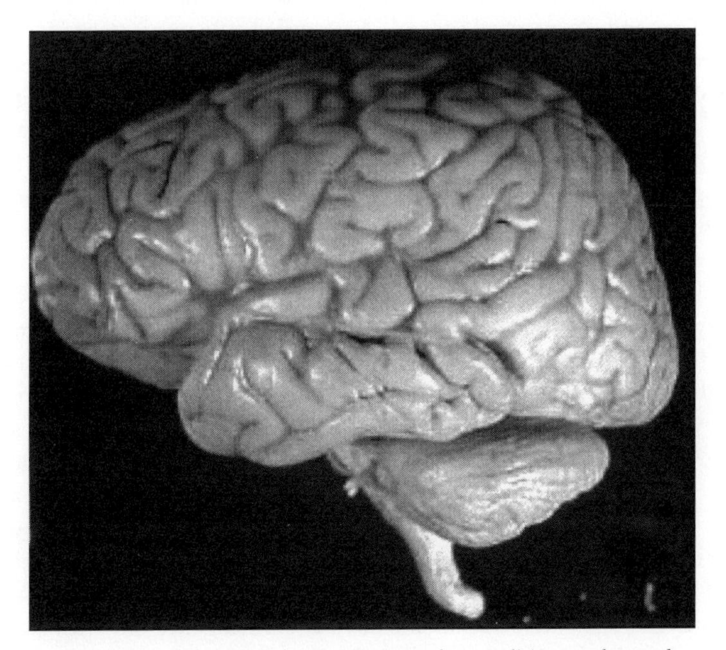

FIG. 1. Example of an adult autistic brain. No gross abnormalities are observed.

growth in these children has been found to increase during the preschool years (Lainhart et al 1997). This observation has recently been supported by magnetic resonance imaging (MRI) studies in which brain volume, primarily involving white matter, was observed to increase most markedly between 2–4.5 years of age, followed by a deceleration of brain growth in older autistic children (Courchesne et al 2001). Similar observations have been made in a series of autopsied autistic brains in which the fresh brain weight of children, ages 5–13 years, was increased by an average of 100–200 g when compared with expected brain weight for age and sex (Dekaban 1978). In contrast, brain weights for adults, ages 18–54 years, showed a decrease in weight by 100–200 g (Bauman & Kemper 1997). Thus, in terms of gross anatomy, there is a suggestion of a changing process with age.

Although some abnormalities have been reported in the cerebral cortex, these findings have been inconsistent in frequency and location. Bailey et al (1998) identified cortical abnormalities in the brains of four out of six mentally handicapped autistic subjects studied. Findings included irregular laminar pattern in the frontal lobe, ectopic neurons in the white matter, thickened areas in the parietal, temporal, frontal and cingulate cortices, regions of increased neuronal density and subpial gliosis in the right cerebral hemisphere. In contrast, histoanatomic observations made in our own material of nine autistic subjects and systematically studied in whole brain serial section in comparison with identically processed age and sex matched controls, have noted only two instances of cerebral cortical malformation (Kemper & Bauman 1998). In 8 of the 9 autistic brains, the anterior cingulate gyrus showed small neuronal cell size and increased cell packing density. A small cortical heterotopic lesion was observed on the infra-orbital region of one hemisphere in one child with a history of severe seizures, and heterotopic lesions were noted in the cerebellar molecular layer in a second child.. Several repeated surveys of all serial sections failed to identify any further evidence of cortical malformation in these brains.

The most consistent abnormalities in our material have been in the forebrain limbic system and in the cerebellum and related inferior olive. In the limbic system, the neurons of the hippocampal complex, subiculum, entorhinal cortex, mammillary body, amygdala and medial septal nucleus have been found to be unusually small in size and more densely packed when compared with age and sex-matched controls (Fig. 2). Golgi analysis of CA1 and CA4 neurons have shown reduced complexity and extent of dendritic arbors (Raymond et al 1996) (Fig. 3). In the amygdala, the finding of small neuronal size and increased cell packing density was most pronounced medially in the cortical, medial and central nuclei. The single exception to this pattern was found in the amygdala of a 12 year old autistic boy with well-documented average intelligence but with significant behavioural problems. In this brain, the entire amygdala was diffusely involved

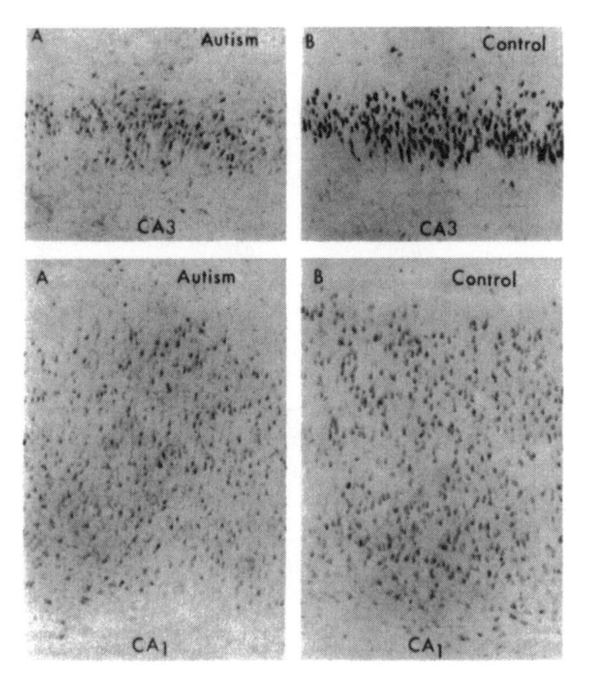

FIG. 2. Photomicrograph of Nissl-stained section of the hippocampus. Note the abnormally small, tightly packed cells in the CA3 and CA1 hippocampal subfields in the autistic brain as compared with the control.

with relative sparing of the hippocampus (Bauman & Kemper 1994, Kemper & Bauman 1998). All areas found to be abnormal in these autistic brains are known to be connected by closely interrelated circuits and comprise a major portion of the forebrain limbic system.

In the septum, a different pathology was found in the nucleus of the diagonal band of Broca (NDB). In all of the brains less than 12 years of age, the neurons of the NDB were unusually large and present in adequate numbers. In contrast, the NDB neurons in all of the adult brains were small and pale and markedly reduced in number.

Outside of the limbic system, the most apparent and consistent abnormalities were confined to the cerebellum and related inferior olive. All of our autistic brains, regardless of age, sex and intelligence showed a significant decrease in the number of Purkinje cells, primarily involving the posterolateral neocerebellar cortex and adjacent archicerebellar cortex of the cerebellar hemispheres (Arin et al 1991) (Fig. 4). We have found no abnormalities either in size or number of Purkinje cells in the vermis, despite some reports of hypo- and hyperplasia of this

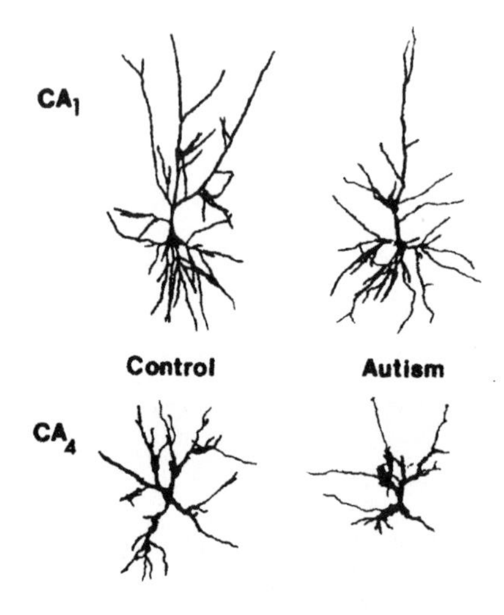

FIG. 3. Camera lucida drawings of Golgi-stained neurons from the CA4 and CA1 subfields of the hippocampus. Note the stunting of some of the dendritic arbors in the cells from the autistic brain, and the limited amount of secondary and tertiary branching in these neurons (from Raymond et al 1996 with permission).

cerebellar region (Courchesne et al 1994). Similar anatomic findings have been reported by Ritvo et al (1986) and more recently by Bailey et al (1998), thus making reduced numbers of Purkinje cells the most consistently reported pathological observation in the autopsied autistic brain.

In addition, abnormalities were noted in the fastigeal, globose and emboliform nuclei in the roof of the cerebellum that, like the NDB, appeared to differ with age. All of the adult brains showed small pale neurons that were significantly decreased in number, whereas those in the childhood brains, in addition to the neurons of the dentate nucleus, were found to be enlarged and plentiful in number.

Areas in the inferior olivary nucleus of the brainstem in our autistic brains, which are known to be related to the abnormal cerebellar cortex (Holmes & Stewart 1908), failed to show the expected retrograde cell loss and atrophy, which is invariably seen following perinatal or postnatal Purkinje cell loss in human pathology (Norman 1940, Greenfield 1954). The olivary neurons of the adult brains were present in adequate numbers but were small and pale (Fig. 5). In all of the childhood brains, these cells were markedly enlarged, but were otherwise normal in appearance and size (Fig. 6). In all of the autistic cases, some of the olivary neurons tended to cluster at the periphery of the nuclear convolutions, a pattern that has been reported in some syndromes of prenatal onset associated with mental retardation (Sumi 1980, DeBassio et al 1985).

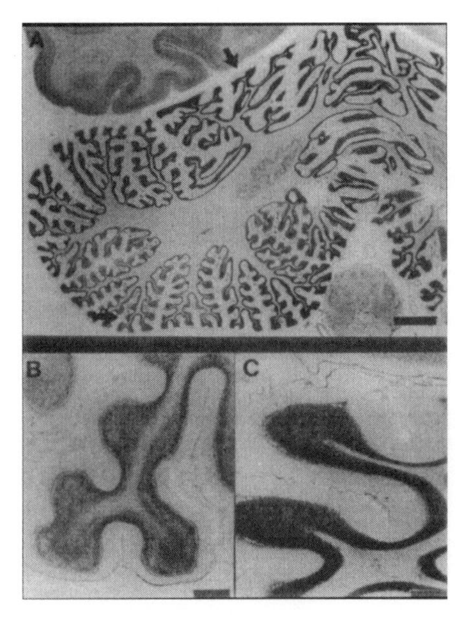

FIG. 4. Nissl-stained section of the cerebellum from the brain of an adult autistic male. Note the atrophy of the cerebellar cortex in the lateral and inferior portions of the hemispheres, and the relatively normal appearing anterior lobe and vermis. In B, the markedly reduced number of Purkinje cells and, to a lesser extent, granule cells, can be appreciated as compared with the more normal appearing anterior cerebellum (C). Systematic counting of the Purkinje cells, however, indicates a decreased number in the anterior lobe as well, but this reduction is less pronounced than in the posterior portions of the hemispheres.

Additional brainstem abnormalities have been reported by other investigators. Bailey et al (1998) have observed prominent arcuate nuclei and malformations of the inferior olive in some cases. Further, Rodier at al (1996) have reported decreased numbers of neurons in the facial nerve nucleus and superior olive, and shortening of the distance between the trapezoid body and the inferior olive in a patient with autism and Möbius syndrome. These findings appear to provide further supportive evidence for a prenatal onset for this disorder.

While a limited number of neuropathologic studies have been reported in autism, the information available raises a number of important questions. The findings in the NDB of the septum are puzzling. In the younger autistic subjects, the neurons of this nucleus are unusually large, while these same cells are small, pale and decreased in number in the older patients. It is possible that this variability in cell size and number with age may represent unstable circuitry involving the NDB. It is known that, in adult monkeys, this nucleus provides a strong highly focused cholinergic projection to the amygdala and hippocampus (Rosene & Van Hoesen

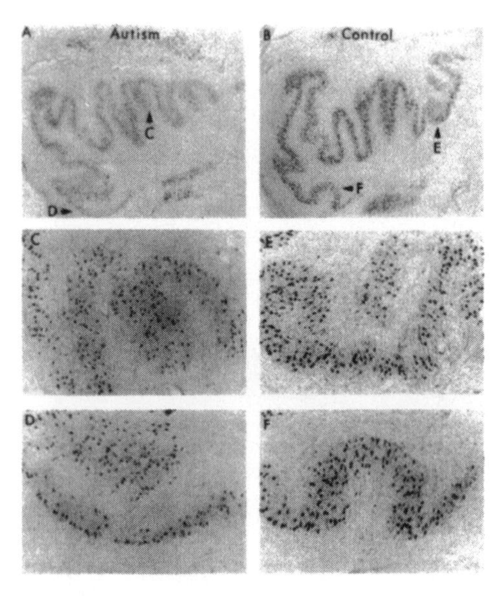

FIG. 5. High and low-power photomicrographs of the neurons of the inferior olive from the brain of an adult autistic male. Note that the neurons are small in size but present in adequate numbers. Note also the presence of a peripheral distribution of the neurons along the edge of the lower loop of the nucleus (D).

1987), The extent of its projection in fetal life is not known. Although, in the past, little attention had been paid to the role of acetylcholine in autism, the observation of abnormalities in the NDB have led to studies of cholinergic transmitter activity in several regions of the autistic brain. These investigations have documented decreased cortical muscarinic M1 receptor binding, most significantly in the parietal cortex, as well as decreased nicotinic receptor binding in both the frontal and parietal cortices (Perry et al 2001). Nicotinic receptor abnormalities have also been found in the granule cell, Purkinje and molecular layers of the cerebellar cortex in autistic brains when compared with controls (Lee et al 2002). How these observations may relate to the histoanatomic abnormalities reported in autism, and what role these findings may play in the neurodevelopmental and clinical features of this disorder remain to be determined.

Histoanatomic findings in the autistic brain suggest that these abnormalities had their onset before birth. In the cerebellar cortex, there is a bilateral symmetrical reduction in the number of Purkinje cells and to a lesser extent, granule cells without evidence of significant gliosis, primarily in the posterior inferior neocerebellar cortex and adjacent archicerebellar cortex. In animals, Brodal (1940) noted a progressively decreasing glial response following cerebellar

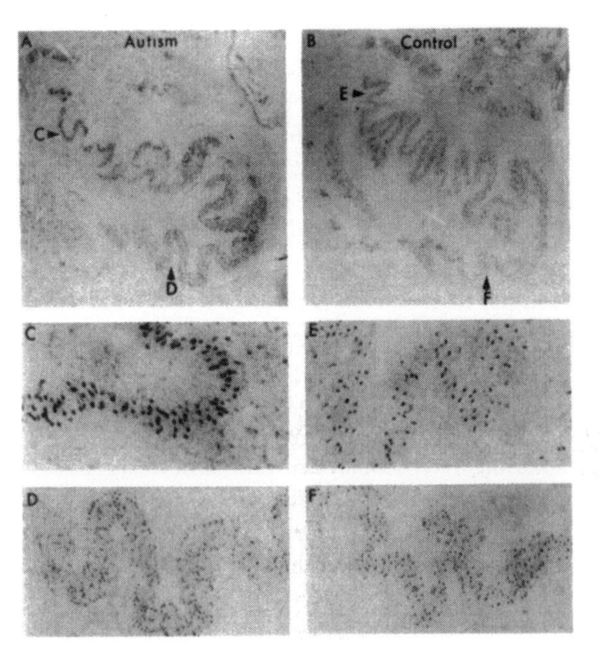

FIG. 6. High and low-power Nissl-stained photomicrographs of the inferior olivary nucleus from an autistic child and a control subject. In contrast to the findings in the autistic adult patient, the olivary neurons in this brain are significantly larger than are those in the control; otherwise they are normal in appearances. Note the peripheral distribution of neurons along the edges of the inferior loop of the nucleus (D).

lesions at increasingly early ages. Thus, the absence of glial hyperplasia appears to indicate that this lesion may have been acquired early in development. Further evidence for the early acquisition of the cerebellar abnormalities relates to the preservation of the neurons of the principal inferior olive. Retrograde loss of olivary neurons invariably occurs following cerebellar lesions occurring in the immature postnatal and adult animal (Brodal 1940) and in neonatal and adult humans (Holmes & Stewart 1908, Norman 1940, Greenfield 1954). The occurrence of retrograde olivary cell loss after cerebellar lesions is believed to be due to the close relationship of the olivary climbing fibre axons to the Purkinje cell dendrites (Eccles et al 1967). It has been shown in the fetal monkey that the olivary climbing fibres synapse with the Purkinje cell dendrites in a transitory zone beneath the Purkinje cells called a lamina dessicans, prior to establishing their definitive relationship with the Purkinje cells (Rakic 1971). Since this zone is no longer present in the human fetus after 30 weeks' gestation (Rakic & Sidman 1970), it is likely that the cerebellar cortical lesions noted in the autistic brains occurred at or before this time. In an analogous situation, expected retrograde cell loss of neurons of the medial dorsal nucleus of the thalamus failed to occur following prefrontal lesions in the rhesus monkey, prior to but not after, 106 days of gestation (Goldman & Galkin 1978). Thus, these findings, in addition to those reported by Rodier et al (1996) in the brainstem, appear to provide good evidence of a prenatal onset for autism. Whether there are also postnatal factors which later impact upon an already atypical neuroanatomic circuitry in these brains is unknown at this time. However, there is very little evidence to support this possibility in the autopsy cases studied to date.

There appears to be growing evidence that the underlying neurobiological processes involved in autism may be progressive. This suggestion has been stimulated by the observations of increased brain weight (Bauman & Kemper 1997) and brain volume (Courchesne et al 2001) in autistic children which is not seen in adults. In addition, microscopic observations of enlarged cells in some brain areas in autistic children, and small pale cells which are reduced in number in these same areas in adults, strongly indicates change with age. Clinically and pathologically this process does not appear to be degenerative in nature and may reflect the brain's attempt to compensate for its atypical circuitry over time. Research should continue to seek answers to the cause for and significance of these changes.

Additional questions stem from the fact that although multiple areas of the brain have been reported to show abnormalities. The most consistent and obvious areas of involvement appear to be the limbic system, cerebellum and related olivary nuclei. One might therefore ask what is common to these two brain regions. Both areas contain granule cells and it might be abnormalities in this cell population that results in involvement of both circuitries. Alternatively, it may be that significant cell populations in both systems are being generated in the fetal brain at roughly the same time during gestation. Clearly, more research is needed to understand these observations.

Given the fact that the clinical phenotype in autism can vary substantially between individuals, it is perhaps not surprising that a good deal of anatomic variability has also been reported. However, despite these sometimes seemingly discrepant reports, there is a suggestion of a unifying underlying substrate, which is common to all subjects regardless of age, sex and cognitive abilities. All of our subjects, including an adult with Asperger syndrome and a 12 year old boy with documented normal intelligence, have shown abnormalities in the entorhinal cortex, the medially placed nuclei of the amygdala and the presence of reduced numbers of Purkinje cells in the cerebellar cortex. This observation suggests that there may well be a common underlying pathology among the autism spectrum disorders and that the morphological differences between brains may be consistent with variations in clinical presentation. Future studies will need to devote more attention to careful clinicopathological correlations in order to better define the significance of these anatomic differences.

It is unlikely that defining the neuroanatomical abnormalities in the brain of autistic individuals will lead to a full understanding of this disorder. However, it should provide a guide for future research questions, particularly those related to the fields of genetics and neurochemistry which are beginning to yield important insights into the neurobiology of this disorder. In addition, however, it will be important to insure that whatever may be hypothesized in regard to the cause or causes of the autism spectrum disorders, is consistent with the neuroanatomic findings noted in the brains of affected individuals. Thus, the neuroanatomy of the autistic brain should provide a defining yardstick against which the results of future research can be measured.

References

Aarkrog T 1968 Organic factors in infantile psychoses and borderline psychoses: retrospective study of 45 cases subjected to pneumoencephalography. Dan Med Bull 15:283–288

Arin DM, Bauman ML, Kemper TL 1991 The distribution of Purkinje cell loss in the cerebellum in autism. Neurology 41(Suppl):307

Bailey A, Luthert P, Dean A et al 1998 A clinicopathological study of autism. Brain 121:889–905

Bauman ML, Kemper TL 1994 Neuroanatomic observations of the brain in autism. In: Bauman ML, Kemper TL (eds) The neurobiology of autism. Johns Hopkins University Press, Baltimore, p 119–145

Bauman ML, Kemper TL 1997 Is autism a progressive process? Neurology 48:285

Brodal A 1940 Modification of Gudden method for study of cerebral localization. Arch Neurol Psychiatry 43:46–58

Coleman M 1979 Studies of the autistic syndromes. In: Katzman R (ed) Congenital and acquired cognitive disorders. Raven Press, New York, p 265–303

Courchesne E, Saitoh O, Yeung-Courchesne R et al 1994 Abnormalities of cerebellar vermian lobules VI and VII in patients with infantile autism: identification of hypoplastic and hyperplastic subgroups by MR imaging. AJR 162:123–130

Courchesne E, Karns CM, Davis HR et al 2001 Unusual brain growth patterns in early life in patients with autistic disorder. Neurology 57:245–254

Dekaban AS, Sadowsky BS 1978 Changes in brain weights during the span of human life: Relation of brain weights to body heights and body weights. Ann Neurol 4:345–356

Damasio AR, Maurer RG 1978 A neurological model for childhood autism. Arch Neurol 35:777–786

Darby JH 1976 Neuropathological aspects of psychosis in childhood. J Autism Childhood Schizophrenia 6:339–352

DeBassio WA, Kemper TL, Knoefel JE 1985 Coffin–Siris syndrome: Neuropathological findings. Arch Neurol 42:350–353

Eccles JC, Ito M, Szentagothai J 1967 The cerebellum as a neuronal machine. Springer, New York

Goldman PS, Galkin TW 1978 Prenatal removal of frontal association cortex in the fetal rhesus monkey: anatomic and functional consequences in postnatal life. Brain Res 152:451–485

Greenfield JG 1954 The spino-cerebellar degenerations. CC Thomas, Springfield, IL

Holmes G, Stewart TG 1908 On the connection of the inferior olives with the cerebellum in man. Brain 31:125–137
Kanner L 1943 Autistic disturbances of affective contact. Nervous Child 2:217–250
Kemper TL, Bauman ML 1998 Neuropathology of infantile autism. J Neuropath Exp Neurol 57:645–652
Lainhart JE, Piven J, Wzorek M et al 1997 Macrocephaly in children and adults with autism. J Am Acad Child Adolesc Psychiatry 36:282
Lee M, Martin-Ruiz C, Graham A et al 2002 Nicotinic receptor abnormalities in the cerebellar cortex in autism. Brain 125:1483–1495
Norman RM 1940 Cerebellar atrophy associated with etat marbre of the basal ganglia. J Neurol Psychiatry 3:311–318
Ornitz EM, Ritvo ER 1986 Neurophysiologic mechanisms underlying perceptual inconsistency in autistic and schizophrenic children. Arch Gen Psychiatry 19:22–27
Perry EK, Lee MLW, Martin-Ruiz CM et al 2001 Cholinergic actvity in autism: abnormalities in the cerebral cortex and basal forebrain. Am J Psychiatry 158:1058–1066
Rakic P 1971 Neuron–glia relationship during granule cell migration in developing cerebellar cortex. A Golgi and electron microscopic study in macacus rhesus. J Comp Neurol 141:282–312
Rakic P, Sidman RL 1970 Histogenesis of the cortical layers in human cerebellum particularly the lamina dessicans. J Comp Neurol 139:7473–7500
Raymond GV, Bauman ML, Kemper TL 1996 Hippocampus in autism: a Golgi analysis. Acta Neuropathol 91:117–119
Ritvo ER, Freeman BJ, Scheibel AB et al 1986 Lower Purkinje cell counts in the cerebella of four autistic subjects: Initial findings of the UCLA-NSAC autopsy research report. Am J Psychiatry 146:862–866
Rodier PM, Ingram JL, Tisdale B, Nelson S, Roman J 1996 Embryological origins for autism: developmental anomalies of the cranial nerve nuclei. J Comp Neurol 146:862–866
Rosene DL, Van Hoesen GW 1987 The hippocampal formation of the primate brain. In: Jones EG, Peters A (eds) Cerebral cortex, vol 6. Plenum Press, New York, p 345–450
Sumi SM 1980 Brain malformation in the trisomy 18 syndrome. Brain 93:821–830
Vilensky JA, Damasio AR, Maurer RG 1981 Gait disturbances in patients with autistic behavior. Arch Neurol 38:646–649

DISCUSSION

Bishop: I have read about cerebellar involvement in ADHD and dyslexia (e.g. Livingstone et al 1991, Berquin et al 1998, Mostofsky et al 1998, Castellanos et al 2001, Leonard et al 2001, Rae et al 2002, Finch et al 2002). Are these different bits of the cerebellum that people are talking about?

Bauman: I don't know what the cerebellum looks like in ADHD or dyslexia.

Lipkin: What does peripheral nerve myelin look like in these subjects?

Bauman: I don't know. The children have good reflexes, but they are hypotonic. I don't think anyone has looked at peripheral myelin. Some of the children are having muscle biopsies for other reasons, so it could be studied easily.

Lipkin: Have the people you are examining had seizure disorders?

Bauman: Most of them are children and most of them have not had seizures.

Lipkin: The reason I am asking is that it is a white matter/grey matter difference: you would expect to see seizures in one and not the other. I was wondering whether there is selection there in the autopsy cases.

Bauman: Most of the children did not have seizures. There was one child who had infantile spasms. Most of the adults did have seizures.

Lipkin: Have you done evoked potential studies to assess central conduction?

Bauman: No. That's a good idea, though.

Rutter: One of the problems of determining whether children with autism have epilepsy is that follow-up studies have shown that many develop seizures in late adolescence or early adulthood (Rutter 1970, Volkmar & Nelson 1990). The finding that young children haven't had seizures doesn't mean that they won't have seizures when older.

Buitelaar: I have a question about your ideas of the onset of the abnormalities in the limbic system. For the cerebellum, it is likely to be prenatal. What about the changes in the amygdala?

Bauman: I don't know how to even judge that. All we can do is make a comment on the basis of the hook-ups of the cerebellum. It certainly looks as if the limbic system is developmentally immature relative to the rest of the brain. It doesn't look like it is damaged, it just appears to be immature. At one stage I went through sections from normal 2, 3 and 4 year olds, and these seemed very similar to those from a 29 year old man who was functioning at a three year old level. It correlated at a gross level in terms of the size and number of neurons.

Buitelaar: Is it similar to abnormalities found in other disorders?

Bauman: That's a tough question to answer because very few people have done whole brain serial sections in pathology. There has been some work done on Rubinstein–Taybi syndrome, congenital rubella syndrome and phenylketonuria (PKU) by my research colleague Tom Kemper. But these are diffusely abnormal brains. Next to those, the autistic brain seems to be more unusual in that it looks more selective.

Bolton: It is very early days in our understanding about the way in which the macrocephaly develops in children with autism spectrum disorders. Certainly from the data that we recently published looking at young infants with macrocephaly, it appeared as if they were at an increased risk of developing an autism spectrum disorder. The brain differences might therefore emerge earlier than other work suggests. Also, we have been doing some research on individuals with triplications and duplications of chromosome 15. It is interesting that although the phenotypic manifestations do seem to be correlated with whether or not the duplication is paternally or maternally derived, we have now identified a case with a duplication of chromosome 15 that is paternally derived. This raises the question that perhaps paternally derived duplications in that region do have phenotypic effects. Obviously, that would link with your thoughts about the GABAergic system.

Bauman: I agree.

Skuse: You mentioned that you thought there were many neurotransmitters that were possibly abnormal. Have you thought about the role played by trace amines? There was an interesting paper published recently by Borowsky et al (2001) on G protein-coupled receptors (GPCRs) that appeared to be activated by these amines, but which were not thought previously to have any particular function in humans. What intrigued me was that they then discovered three of the four classes of these receptors that are found in humans are expressed almost exclusively in the amygdala.

Bauman: Is that something you are studying?

Skuse: Not yet, but it is something that we are interested in following up. The reason why we were particularly interested in that finding was that MAOB (monoamine oxidase B) is involved in deamination of these trace amines. MAOB happens to lie within the critical interval in which we mapped an X-linked gene that is involved in the structural development of the amygdala (Good et al 2003). These two things seem to tie up logically.

Bauman: We haven't looked at this, but we would welcome the involvement of others.

Bishop: In terms of the differences between children and adults, might there be some sort of degenerative process? How far could this be experience dependent? I have read a little on the cerebellum, and it seems to be particularly sensitive to experience. There are accounts of it being big in musicians (Schlaug 2001), and enhanced in rats subjects to motor exercise (Black et al 1990). Might it be the child's experiences, rather than a neuropathological process, that is causing cerebellar abnormalities in autism?

Bauman: Potentially, yes. We had a hypothesis that because of the circuitry that was involved in the cerebellum, this might be an abnormal expression of an early normal situation. For example, most adults are wired so that the olive talks to the Purkinje cells, which then talk to the deep cerebellar nuclei which then communicates with the rest of the brain. In fetus there aren't any Purkinje cells there. The olive talks directly to the deep cerebellar nuclei. During later fetal development the Purkinje cells start hooking up to create the more elaborate circuitry found in the adult. If the autistic individuals lack sufficient Purkinje cells to make that hook up, are they forced to hang on to the more primitive circuit? If so, do the cells within that primitive circuit hypertrophy to compensate? At that point you have to make a bit of a leap and say that because it wasn't designed for the long haul, it can't continue, so there comes a point where it fades back. We don't know whether this is true, but it is the best hypothesis we have so far developed. An equally intriguing question is does this hypothesis have any kind of clinical significance? Are there differences that we could see clinically? Unfortunately, no one has any longitudinal data in the same group of subjects. You need to see the same group of children over time to answer this.

Bishop: In your adults and children the differences might not be related to age but to the interventions they have been exposed to.

Bauman: Most of these people didn't have a lot of intervention. This is long before early intervention was practised, even in children.

Rutter: You drew parallels with the Nelson findings (Nelson et al 2001). They have received a lot of publicity, but I am left puzzled as to what can be concluded from them because of the similarities in findings between autism and mental retardation, which are otherwise so different. I couldn't quite understand why people got excited about these findings.

Bauman: I'm not sure I can answer your question. I'm interested in this work because it points to the fact that something has gone awry prenatally, whether or not this is specific to autism or mental retardation. My hope would be that they would get more specific. The investigators cast a net out, selected some neurotrophins and neuropeptides that they thought might be involved, and this is what fell out. My hope would be that they would go back and re-think this, and ask whether there are other substances that might be more specific. It is an interesting observation that someone can pick up some abnormalities in a 24–48 h old baby that already suggest there is something wrong with brain development.

Rutter: That is interesting, but what is so striking to me is that most mental retardation is prenatal in origin, as it seems to be with autism, yet the brains of children with severe mental retardation are grossly abnormal in a way that is not the case in autism. This is such a marked difference that it suggests that a radically different mechanism must be responsible.

Bauman: I agree.

Lipkin: Several people have tried to repeat the neonatal bloodspot studies without success. I'm not saying the data aren't true, but it would be nice to see the work replicated.

Bauman: I was led to believe that this was a hugely complex process.

Lipkin: It's biochemistry. There's nothing mysterious to it; it's just a question of building the appropriate assays.

Howlin: When you showed the comparison between the average brain sizes and weights of the autistic children, most were bigger and heavier than normals but quite a few were substantially lower. Is there any difference between these two groups?

Bauman: That's a good question, but I haven't looked.

Rogers: Several people have suggested that the large brains in childhood may be because of the lack of a normal pruning process. Your suggestion that this is a myelin difference suggests that a different process is responsible.

Bauman: It could be both. It's certainly possible that there is a pruning problem here. What throws me a little is that I am having a hard time correlating that with weight. I can correlate that with volume, but extra processes don't weigh that much. When brain weight increases it is usually myelin that is putting on the weight.

Folstein: Joe Piven has published a paper looking at the relative increases of brain size in the grey versus white matter. I believe that the increased size reflects mainly increased white matter. This is consistent with what you say. If you have a failure of pruning, might you get more myelinated fibres?

Bauman: Then you have to think about how elaborate the dendritic tree gets. Normally in immature neurons the dendritic arbors branch up, but they are not as elaborate as in the adult. I wonder if it just doesn't get out that far. The neurons that we have seen with Golgi staining look pretty stunted. It looks like there has been an over-pruning problem, if anything.

Rutter: Tony Bailey, would you like to comment on the somewhat different pattern of findings in your study from those Margaret Bauman has described?

Bailey: The general point is that yesterday we discussed the molecular genetic findings, where there is quite a surprising amount of agreement across many different studies. The key aspect is the sheer number of cases that have been genotyped. What Margaret and I are having difficulty with is that we are making inferences from just a few cases from our two sites. There has to be a major effort to systematically recruit more post-mortem material. Everyone expects that the first autism susceptibility loci will be identified in the next few years and of course, there will be many problems interpreting their roles and significance if we don't have more detailed knowledge of the neuropathology. Given that there is limited evidence, it is clear that we do have some exceptionally large brains in adults. We know that relatives show increased head circumference, and this is a finding in adults as well as children. So we have to be somewhat cautious: is this a developmental phenomenon in which, as Eric Courchesne argues, the brains grow unusually rapidly and then, finally achieve an unremarkable size? My guess is that this isn't true, and that in a proportion of cases there is a persistent increased brain size. There is then a second question about the pathological basis of this increase in size. In mega-encephaly not associated with autism, there is usually an increased number of neurons and glia, which is associated with an increase in neuropil and white matter. We are about to start counting neurons in our sample, but our impression from subjective examination of these cortices is that one can see areas where the cortex is unusually thick, and there appear to be too many neurons, at least in these regions. Our starting hypothesis is that this is not primarily a pruning problem, but an increase in total neuron number. The structural MRI data are still a little contradictory. Several groups have reported an increase in grey matter volume. Additionally if there is an increased number of neurons then one would expect an increase in white matter volume. The paradox is that despite most MRI studies finding an increase in brain volume, most studies also find decreases in the size of the corpus callosum. To me this suggests that although there may be an increase in the number of neurons, these may not be normally connected. We are left with the point that I was making yesterday: do we tend to focus on the localized abnormalities, or do we see autism as a disorder in which there are quite widespread neurodevelopmental abnormalities?

Then the next challenges are to uncover the mechanisms that give rise to diverse developmental abnormalities and to establish the relationship between the observed pathology and the symptomatology. There has been quite a lot of discussion this morning about the cerebellum and Purkinje cells. There is no doubt that there is a decrease in Purkinje cell number in virtually every brain examined. It is possible that epilepsy may be a contributory factor. But the fact that there are developmental abnormalities of other sorts in the brain stem and cerebellum suggest that there is a partly developmental contribution. I think there has been an assumption that the decreased number of Purkinje cells is directly related to symptomatology. Given our relative lack of knowledge about the role of the cerebellum in cognition, this certainly is an assumption. There have been only a couple of studies looking at cerebellar physiology. The published study is from Nancy Minshew's group who have not been able to find any evidence of cerebellar motor impairments (Minshew et al 1999), and neither have we (unpublished data). As Patrick mentioned yesterday, genes have pleiotropic effects and it is possible that there are many neurodevelopmental abnormalities, not all of which may give rise to cognitive consequences relevant to the symptomatology. However, the main difference between Margaret's data and our own is in terms of obvious abnormalities in cortical structure. It is too early to know whether these are regionally localized or not, but I am very struck by the fMRI data. If one looks at studies of theory of mind, language and face processing, the findings suggest that cortical activity is not localized in exactly the same region as in control individuals, but equally it is not in completely the wrong spot, but localized in contiguous areas. Given the tentative evidence that we have already for abnormalities in connectivity, this raises the possibility that one underlying problem may be difficulties in the final stages of axonal path finding, particularly between cortical and sub-cortical structures. The considerable challenge is to establish whether there are only some functions that are abnormally localized, or whether if we were to systematically assess functions that we have not yet thought of examining in autism, we would find more general cortical disorganization. To end with my original point: these thoughts are based on a very small number of post-mortem cases combined with data from other sources to aid interpretation of these limited data. There has to be a concerted effort to recruit more post-mortem material.

References

Berquin PC, Giedd JN, Jacobsen LK et al 1998 The cerebellum in attention-deficit hyperactivity disorder: a morphometric MRI study. Neurology 50:1087–1093

Black JE, Isaacs KR, Anderson BJ, Alcantara AA, Greenough WT 1990 Learning causes synaptogenesis, whereas motor activity causes angiogenesis, in cerebellar cortex of adult rates. Proc Natl Acad Sci USA 87:5568–5572

Borowsky B, Adham N, Jones KA et al 2001 Trace amines: identification of a family of mammalian G protein-coupled receptors. Proc Natl Acad Sci USA 98:8966–8971

Castellanos FX, Giedd JN, Berquin PC et al 2001 Quantitative brain magnetic resonance imaging in girls with attention-deficit/hyperactivity disorder. Arch Gen Psychiat 58:289–295

Finch AJ, Nicolson RI, Fawcett AJ 2002 Evidence for a neuroanatomical difference within the olivo-cerebellar pathway of adults with dyslexia. Cortex 38:529–539

Good CD, Elgar K, Thomas NS et al 2003 Dosage sensitive X-linked locus influences the development of amygdala and orbito-frontal cortex, and fear recognition in humans. Brain, in press

Leonard CM, Eckert MA, Lombardino LJ et al 2001 Anatomical risk factors for phonological dyslexia. Cereb Cortex 11:148–157

Livingstone MS, Rosen GD, Drislane FW, Galaburda AM 1991 Physiological and anatomical evidence for a magnocellular defect in developmental dyslexia. Proc Natl Acad Sci USA 88:7943–7947

Minshew NJ, Luna B, Sweeney JA 1999 Oculomotor evidence for neocortical systems but not cerebellar dysfunction in autism. Neurology 52:917–922

Mostofsky SH, Reiss AL, Lockhart P, Denckla MB 1998 Evaluation of cerebellar size in attention-deficit hyperactivity disorder. J Child Neurol 13:434–439

Nelson KB, Grether JK, Croen LA et al 2001 Neuropeptides and neurotrophins in neonatal blood of children with autism or mental retardation. Ann Neurol 49:597–606

Rae C, Harasty JA, Dzendrowskyj TE et al 2002 Cerebellar morphology in developmental dyslexia. Neuropsychologia 40:1285–1292

Rutter M 1970 Autistic children: infancy to adulthood. Semin Psychiatry 2:435–450

Schlaug G 2001 The brain of musicians: a model for functional and structural adaptation. In: Zatorre RJ, Peretz I (eds) The biological foundations of music. Annals of the New York Academy of Sciences. Vol 930. New York Academy of Sciences, New York, p 281–299

Volkmar FR, Nelson DS 1990 Seizure disorders in autism. J Am Acad Child Adolesc Psychiatry 29:127–129

Microbiology and immunology of autism spectrum disorders

W. Ian Lipkin and Mady Hornig

Center for Immunopathogenesis and Infectious Diseases, Mailman School of Public Health, Columbia University, 722 West 168th Street, New York, NY 10032, USA

Abstract. Both genetic and environmental factors are likely to contribute to the pathogenesis of neurodevelopmental disorders. Even in heritable disorders of high penetrance, variability in timing of onset or severity of disease indicate a role for modifying principles. Investigation in animal models of the consequences of interactions between host response genes and microbes, toxins, and other environmental agents in a temporal context may elucidate the pathophysiology of a wide spectrum of chronic diseases. Here we review the evidence that infectious and immune factors may contribute to the pathogenesis of neurodevelopmental disorders, describe an animal model of neurodevelopmental disorders based upon viral infection, identify processes by which neural circuitry may be compromised, and outline plans for translational research in animal models and prospective human birth cohorts.

2003 Autism: neural basis and treatment possibilities. Wiley, Chichester (Novartis Foundation Symposium 251) p 129–148

Establishing a causal relationship between infection with a microbial agent and a specific brain disease can be complex. In some instances, for example, herpes simplex encephalitis, the agent is readily implicated: the virus is present in brain and destroys infected tissue through replication. Alternatively, tissue damage and disease may be the indirect result of a host immune response to microbial gene products present in neural cells. Immune responses to microbial agents can also lead to breakdown of tolerance to host antigens and result in tissue damage (Zhao et al 1998). The agent responsible for induction of autoimmunity need not be present in central nervous system (CNS) at the time of clinical presentation. Furthermore, the original infection may have been peripheral, as is the case in Sydenham's chorea, or as is proposed for tics and obsessions following streptococcal infection. Yet another mechanism for brain disease is persistent non-cytopathic viral infection. Such infections can profoundly impact neurotransmitter function or brain development, yet remain cryptic unless specific reagents are used for detecting viral gene products (Lipkin et al 1988a, 1988b, Oldstone 1989a, 1989b, 1989c).

An additional wrinkle to consider is that developmental differences in host inflammatory and neuroendocrine capacities and in rates of maturation of nervous and immune system elements (Rubin et al 1999) could contribute to the differential susceptibility of neuronal and glial populations to pre- or postnatal inflammatory stressors (infectious, immune) and impact the phenotypic expression of autism and other neurodevelopmental disorders (Briese et al 1999, Hornig et al 1999). Thus, expression of complex neuropsychiatric diseases such as autism may require the presence of specific genes, an environmental trigger, and exposure at a particular time during brain development.

Perinatal CNS infection (Chess 1971, Gillberg & Gillberg 1983, Hoon & Reiss 1992, Barak et al 1999) and disturbed neuroimmune networks (Warren et al 1986, 1987, 1990, 1991, 1994, 1995, Singh et al 1991, 1993, 1997a, 1997b, 1998, Singh 1996, 1997, Burger & Warren 1998, Warren 1998) are proposed as factors in the pathogenesis of autism spectrum disorders. An immune or infectious basis for autism is supported by studies suggesting an increased rate of autism in some geographic regions (Gillberg et al 1991, Baron-Cohen et al 1999, California Department of Developmental Services 1999 *http://www.dds.ca.gov/Autism/pdf/Autism_Report_1999.PDF*, Gillberg & Wing 1999), season-of-birth effects (Bartlik 1981, Kostantareas et al 1986, Burd 1988, Tanoue et al 1988, Atlas 1989, Gillberg 1990, Bolton et al 1992, Mouridsen et al 1994, Barak et al 1995, 1999, Ticher et al 1996, Torrey et al 1997, Stevens et al 2000), and linkage to viral and/or immune factors (Chess 1971, Warren et al 1986, 1987, 1990, 1991, 1994, 1995, Singh et al 1991, 1993, 1997a, 1997b, 1998, Singh 1996, 1997, Burger & Warren 1998, Warren 1998, Stevens et al 2000). Heritable predisposing factors may include immunologically relevant influences such as linkages to major histocompatibility complex (MHC) genes (Warren et al 1992, 1996a, 1996b, Warren & Singh 1996, Warren 1998, Torres et al 2002); increased frequency of the null allele of the complement component 4b locus, located in the MHC (Warren et al 1991); and increased frequency of a family history of autoimmune disorders (Comi et al 1999).

Despite a finding of season-of-birth effects in several studies, (Bartlik 1981, Burd 1988, Gillberg 1990, Barak et al 1995, 1999, Ticher et al 1996, Stevens et al 2000) many suggesting a preponderance of March and/or August births or a second trimester pathogen exposure amongst children with autism, other studies have not confirmed this link (Landau et al 1999). Even if a microbial link is confirmed for a subset of children with autism, other factors may also vary seasonally (e.g. nutritional differences) and influence expression of disease (Gillberg & Coleman 1992).

Studies of immunological function in children with autism reveal a wide array of abnormalities, including decreased cellular immune capacity (Warren et al 1986, 1990, Wright et al 1990, Yonk et al 1990, Denney et al 1996); decreased plasma complement component C4b (Warren et al 1994, 1995); and increased humoral immune and autoantibody responses (Weizman et al 1982, Singh et al 1993). These abnormalities provide general support for the hypothesis that children with autism may be predisposed to respond abnormally to viral infections either through the establishment of persistent infections or a virally triggered autoimmune diathesis.

Several reports are consistent with diminished Th1 and increased Th2 responses in autism: (1) mitogen-stimulated T cell proliferation is decreased in some (Stubbs & Crawford 1977, Warren et al 1986), but not all (Ferrari et al 1988) studies; (2) Th1 (interferon [IFN]γ^+ $CD4^+$ and interleukin [IL2]$^+$ $CD4^+$ cells) and Tc1 (IFNγ^+ $CD8^+$ and $IL2^+$ $CD8^+$ cells) T cells are reduced, and Th2 ($IL4^+$ $CD4^+$ cells) and Tc2 ($IL4^+$ $CD8^+$ cells) T cells are reportedly increased (Gupta et al 1998); (3) reduced natural killer cell activity (Warren et al 1987) despite unchanged numbers of natural killer cells (Warren et al 1990); and (4) increased levels of serum IgE (Gupta et al 1996, Trottier et al 1999). Furthermore, as Th2 cells are implicated in systemic (non-organ specific) autoimmune disorders (De Carli et al 1994, Singh et al 1999), the increased autoantibody production (Todd & Ciaranello 1985, Todd et al 1988, Plioplys et al 1989a, 1989b, Yuwiler et al 1992, Singh et al 1993, 1997a, 1997b, 1998, Connolly et al 1999) and family history of autoimmune disorders (Money et al 1971, Raiten & Massaro 1986, Gillberg et al 1992, Comi et al 1999) reported for children with autism lends further support to the idea that a Th2 predominant immune response may play a role in autism pathogenesis. In contrast, Singh reported increased plasma IL12 and IFNγ levels in autism (1996), findings more consistent with Th1 than Th2-weighted responses. Lastly, a preliminary report by Nelson and colleagues indicates a *lower* level of autoantibodies to MBP, GFAP and NAFP at birth in children with autism or mental retardation than in normal controls (Nelson et al 2000). These apparent inconsistencies in levels of autoantibodies in children with autism might be explained by the difference in time of sampling, with later time-points reflecting a break in immune tolerance.

It is conceivable that there is a link between susceptibility to infection with measles virus and HHV6, and an autoimmune diathesis. The observation that the CD46 receptor binds complement proteins C3b and C4b is interesting in light of reports of decreased plasma levels of C4b (Warren et al 1994, 1995) and increases in the null allele of the C4b gene (Warren et al 1991) in autism. The C4b gene is located in the MHC on chromosome 6; partial C4 deficiency and the C4b null allele are associated with an increased susceptibility to a variety of autoimmune diseases (Brai et al 1994, Ulgiati & Abraham 1996, Naves et al 1998, Kawano et al 1999). CD46 receptor, in addition to serving as the entry site for vaccine strains of measles (Tatsuo et al 2000), is also the entry site for HHV6 (Santoro et al 1999, Clark 2000). Increased levels of antibodies to HHV6 and measles are positively associated with peripheral autoantibodies to CNS antigens in children with autism (Singh et al

1998); studies have not yet been reported regarding such associations with autoantibodies in cerebrospinal fluid.

Alterations in T cell subsets may also fit with the hypothesis that increased susceptibility to specific types of viral infections may be mediated by regulation of virus-specific receptors and of Th1 vs. Th2 immunity, and may provide a link to development of CNS-directed autoimmune responses in autism. In conjunction with reports of an increase in T cells expressing a 'late activated' pattern (i.e. positive for DR or class II MHC molecules, but negative for IL2 receptor or CD25) in autism (Plioplys et al 1994), consistent with the pattern seen in several autoimmune disorders (Burmester et al 1984, Hafler et al 1985, Bergroth et al 1988), levels of DR^+ $IL2^-$ T cells in children with autism are inversely correlated with plasma levels of C4b (Warren et al 1995). The *DRB1* gene is located in close proximity to the C4b gene on the HLA region of chromosome 6, and is also near to genes encoding IgA and 21-hydroxylase (class III region) (Wilton et al 1985, Brai et al 1994, Fiore et al 1995, Reil et al 1997, Schroeder et al 1998). IgA deficiency, also noted in autism (Warren et al 1997), is associated with the presence of autoantibodies (Brai et al 1994, Fiore et al 1995) and an increased incidence of overt autoimmune disease (Barka et al 1995). Some DRB1 alleles reportedly have a very strong association with autism (Warren et al 1992, Daniels et al 1995, Warren et al 1996a, 1996b, Warren & Singh 1996, Torres et al 2002), although one larger study of multiplex sibships with autism did not confirm this result (Rogers et al 1999). Given the presumed heterogeneity of the disorder, and the possibility that genetic loading in families with multiple affected members may be less likely to require exposure to an environmental factor (e.g. virus, bacteria, toxin, or other agent) in order to lead to the autistic phenotype, the absence of linkage in this one study does not rule out a role for an HLA-linked immunogenetic vulnerability in a subset of children with autism. Immunogenetic studies that compare subpopulations with and without evidence of immune dysfunction will be required to address this possibility.

Neonatal bornavirus infection of Lewis rats as a model of neurodevelopmental damage based on host–environment interactions

Borna disease virus (BDV) is a neurotropic, negative-strand RNA virus that causes a spectrum of behavioural deficits depending on the age, immune status, CNS maturity and genetics of the host. In adult Lewis rats, infection results in humoral and cellular immunity to the virus, meningoencephalitis, diffuse CNS damage, dopamine neurotransmitter disturbances, and disorders of movement and behaviour (Solbrig et al 1994). In contrast, infection of Lewis rat neonates causes a behavioural syndrome (Bautista et al 1994, Hornig et al 1999) without robust immunity that includes subtle disturbances of sensorimotor development and activity (Hornig et al 1999), play (Pletnikov et al 1999) emotional reactivity (Hornig et al 1999), social communication (M. Hornig and W. I. Lipkin, unpublished results), and spatial and aversive learning (Dittrich et al 1989), as well as anatomical and neurochemical pathology consistent with some reports of autism spectrum disorders. Although our data do not indicate a role for BDV in pathogenesis of autism (Hornig et al 1999) this animal model, which roughly corresponds to viral infection during the third trimester of primate gestation, provides insights into mechanisms by which pre- and perinatal infection can cause neurodevelopmental damage and supports efforts to pursue an infectious/immune basis for autism in prospective birth cohorts using high throughput methods for genetic and microbial epidemiology.

Neonatal bornavirus infection results in loss of neurons in specific regions by apoptosis (Hornig et al 1999, Weissenböck et al 2000). Despite spread of virus throughout limbic circuitry and cerebellum by two weeks postinfection (pi), programmed cell death reaches a maximum at 4 weeks pi and is largely restricted to granule cells of dentate gyrus (DG), granule and Purkinje cells of cerebellum, and pyramidal neurons of layers V and VI of retrosplenial and cingulate cortex (Hornig et al 1999, Weissenböck et al 2000). Infection alone is therefore an insufficient signal for cell loss, as many neuronal populations remain persistently infected without evident reduction of cell numbers. In addition, modest levels of apoptosis are seen in granule cells of cerebellum, yet these cells are spared from infection.

One means by which a virus might disrupt neural function and development in the absence of inflammation is through the induction of neuronotrophic cytokines. Neuronotrophic cytokines comprise a burgeoning set of immunoregulatory molecules, including the haematolymphoietic factors (e.g. interleukins, tumour necrosis factor family, interferons), the transforming growth factor (TGF)β superfamily factors (including TGFβ1, 2, 3; glial-derived neurotrophic factor [GDNF]), and the classic neurotrophic factors (nerve growth factor [NGF], brain-derived neurotrophic factor [BDNF], NT3, NT4/5). A large subset of the neuronotrophic, haematolymphoietic cytokines may be roughly categorized according to their origin from one of two types of Th cells: Th1 (cell-mediated immunity and stimulation of antigen-presenting cells) or Th2 (humoral or B cell-mediated immunity). The potential mechanisms of cytokine-mediated damage in the context of the developing brain include: direct effects on neuronal elements; activation or suppression of second messenger/intracellular signalling pathways; induction of shifts in excitotoxic elements such as quinolinic acid or acute phase proteins such as neopterin or β2 microglobulin; direct alterations of neuronal function (e.g. inhibition of long-term potentiation in hippocampus); activation or suppression of glial cells; or alteration of glial cell proliferation or differentiation (including expression of adhesion molecules such as the integrins)

(Benveniste 1997, Mehler et al 1996). Given that the postnatal expression of neurotrophic cytokine and cytokine receptor mRNAs in brain differs for each cytokine (Benveniste 1997), and that the sensitivity of neuronal populations to the trophic or apoptosis-inducing effects of cytokine changes during development, wide variation in the patterns of virus-induced, cytokine-related damage would be expected, depending on the relative maturity of the evolving nervous system at the time of infection. In addition, cell loss induced by either BDV or developmentally programmed changes may alter the capacity of resident CNS cells to both produce and respond to neurotrophic cytokines.

One of the primary mechanisms of host defence following viral infection begins with the induction of IFNγ and other cytokines, which in turn initiate a cascade of host responses in a wide variety of cell types. In the CNS, IFNγ modulates oligodendrocyte, neuronal and glial cell functions, and is important in activating glial cells to produce mediators of cell damage or death, including toxic intermediates of nitrogen and oxygen, and complement components (St. Pierre et al 1996). Damage to neurodevelopmental circuitry may thus parallel the production of these downstream mediators following IFNγ induction, and provide a means by which BDV, or other factors that result in elevated levels of proinflammatory cytokines, might disrupt brain cell differentiation and function without inflammatory cell infiltration.

Recent studies concerning cytokine expression during neonatal infection provide a converging view of the potential importance of cytokines as mediators of BDV-related CNS injury in neonatally infected rats (Hornig et al 1999, Plata-Salamán et al 1999, Sauder & de la Torre 1999). Cytokine expression changes over time in different brain regions, with maximal shift occurring at 4 weeks. Higher levels of mRNAs for cytokine products of CNS macrophages/microglia (IL1α, IL1β, IL6, TNFα) are noted in hippocampus, amygdala, cerebellum, prefrontal cortex, and nucleus accumbens (Hornig et al 1999). Elevated levels of these proinflammatory cytokines were first apparent at 2 weeks, peaked at 4 weeks, and then declined at 6 and 12 weeks. No alterations in other proinflammatory cytokines, including IL2, IL3, TNFβ and IFNγ, were observed. Given that production of several of these latter proinflammatory cytokines is unique to T cells, B cells, mast cells, and bone marrow stromal cells, and not to macrophages or microglia, these data suggest that BDV may exert a selective effect on cells of microglial or macrophage lineage.

Interestingly, in post-mortem cerebellum and parietal cortex of individuals with autism, the anti-apoptotic factor, bcl2, is reduced, and in parietal cortex, the pro-apoptotic factor, p53, is increased, suggesting a biochemical basis for enhanced neuronal losses by apoptosis in autism (Fatemi & Halt 2001). However, changes in neurotrophic factor, apoptosis-related product, and cytokine gene expression in the neonatal rat model fail to fully explain the distribution and timing of cell death.

Changes in neurotrophic factor gene expression are only observed in hippocampus at 4 weeks pi, possibly representing reduced synthesis following dropout of specific neuronal subsets, and cytokine and apoptosis-related product mRNA alterations wane after apoptosis peaks at 4 weeks but are still significant to 12 weeks pi, long after apoptosis has ceased. Furthermore, increases in serotonin levels are noted in hippocampus at days 21 (Pletnikov et al 2000) and 28 pi (unpublished data) — changes which should serve to promote cell survival and prevent apoptosis. Serotonin is critical in establishment and maintenance of synaptic connections, yet persistent elevations of serotonin into the early post-weaning period (beginning approximately postnatal day 21) is associated with arrest of spine development (Norrholm & Ouimet 2000). Alterations in serotonin may also modulate regional patterns of α-amino-3-hydroxy-5-methyl-4-isoxazolpropionate (AMPA) receptors, thereby contributing to plasticity and localized glutamatergic neurotransmission (Okado et al 2001). Events leading to designation of specific subsets of infected neurons for death during this early period of brain plasticity remain unclear. AMPA and *N*-methyl-D-aspartate (NMDA) receptors appear to have contrasting, age-dependent effects on programmed cell death. Excitotoxic, glutamatergic injuries of newborn rat brain result in morphologic evidence of apoptotic neuronal death, approximating that which occurs during normal brain development (Portera-Cailliau et al 1997); NMDA receptor blockade increases developmental elimination of neurons by apoptosis (Ikonomidou et al 1999). The contribution of different glutamate receptor subtypes appears to be age-sensitive: in adult models of excitotoxic damage, blockade of NMDA receptors increases cell survival following glutamate exposure. Furthermore, transient AMPA receptor blockade protects developing chick brainstem auditory neurons from programmed cell death (Solum et al 1997). Studies undertaken to determine whether viral infection affected excitatory synaptic connections and distribution of developmentally-regulated glutamate receptors revealed losses in dendritic complexity and spines, and increases in expression of the calcium-dependent AMPA receptor, GluR1. The latter were closely correlated in time and space to the peak of apoptosis of specific neuronal subsets following neonatal infection. These results suggest that neonatal viral infection interferes with the age-dependent establishment of synaptic connections necessary for cell survival, and with normal development-related shifts in glutamate receptor expression in key brain regions.

Several factors could contribute to the loss of dendritic complexity that heralds apoptotic losses in this infection-based neurodevelopmental model. Excess glutamate exposure can also lead to dendritic pruning; we (unpublished data) and others (Billaud et al 2000) find that persistent BDV infection of glial cells *in vitro* is associated with reduced presynaptic glutamate uptake, possibly contributing to early, localized increases in glutamate levels and decreased dendritic branching.

Arrest of spine development due to inappropriate persistence of serotonin during the early post-weaning period, as described in neonatal infections of Lewis rats (unpublished data) is another possible mechanism underlying the changes in neuronal morphology preceding cell death.

The relevance of GluR1 receptor clustering on cell populations targeted for death by apoptosis is unclear. Relative reductions of GluR2 receptors may increase susceptibility to excitotoxic damage (Pellegrini-Giampietro et al 1997). An alternate possibility is that an increased concentration of GluR1 receptors in the flip isoform, with their enhanced resistance to desensitization following chronic glutamate exposure, might underlie the enhanced vulnerability of certain neurons to excitotoxic death. During normal AMPA receptor ontogeny in regions such as hippocampus and cerebellum, a developmental switch from the alternatively spliced flip to the flop isoform, corresponding to a declining need for neuronal pruning, occurs at the same time-point at which GluR1 receptors are increased in neonatally infected animals. The reliance of BDV on alternative splicing (Schneider et al 1994) raises the intriguing possibility that competition for splicing machinery may impede progression of normal developmental programs, contributing to abnormal persistence of the excitotoxicity-sensitive flip isoform of GluR1 at a critical juncture of CNS development.

The epidemiology of BDV and its role in human disease remain controversial (Hatalski et al 1997, Staeheli et al 2000). Similarities between some behaviours observed in neonatally infected rats and in autistic children led to the hypothesis that the virus might be implicated in pathogenesis of autism. Although neither serologic nor molecular data support a role for BDV as an aetiologic agent in autism (Hornig et al 1999), multicentre studies are underway to assess whether it may be implicated in other neuropsychiatric disorders.

MMR vaccine and autism

MMR was introduced in 1971, and added to the recommended childhood immunization schedule by 1979. It is administered between the ages of 12 and 18 months, an interval when many children with autism spectrum disorders (ASDs) are first noted to have abnormal behaviour. As in other neurodevelopmental disorders, children with ASDs may have gastrointestinal disturbances (GID) (Chong 2001); in some instances parents report coincidence of behavioural disturbances and GID (Horvath et al 1999). A cohort of British children with developmental disability and GID were referred for evaluation to a research group investigating a potential relationship between MMR and inflammatory bowel disease (IBD). Endoscopic assessment indicated ileocolonic lymphonodular hyperplasia in a high proportion of children with ASD and GID (Wakefield et al 1998, 2000, Quigley & Hurley 2000). Thereafter, molecular analysis revealed the presence of MMR vaccine gene products in PBMC of 33% (Kawashima et al 2000) and in enteric biopsies of 77% (Uhlmann et al 2002) of affected children. Cases and controls were not well-matched for age in some groups in the study by Kawashima et al (2000), as children with developmental disorders were younger (3–10 years of age) than individuals with IBD (15–34 years); Uhlmann et al (2002) provide insufficient detail to determine whether age may be significantly different among groups. The recent MMR vaccine history of subjects participating in either of these studies is also unknown.

Differences in methods and assay sensitivity may also contribute to disparities among study results. Whereas O'Leary and colleagues at the Royal Free and University College Hospital (London, UK) and Coombe Women's Hospital (Dublin, Ireland) used real time polymerase chain reaction (PCR) to analyse bowel lesions (Uhlmann et al 2002), Afzal and coworkers at the National Bureau of Standards (UK) used nested PCR for their studies of the potential association of MV with IBD (2000). The two groups determined sensitivity of their assays using different standards (titred amounts of infectious virus [Afzal et al 2000] vs. synthetic transcripts [Uhlmann et al 2002]); thus, a direct comparison of the sensitivity of their assays is not feasible. An additional difference that could explain discordance in results is that whereas one group (Uhlmann et al 2002) extracted only bowel lesions, the other (Afzal et al 2000) extracted template from full thickness bowel.

A final complication in analysing results is that the presence of MV in subjects with specific, immune-mediated disorders might also be non-specifically related to persistence of MV in localized inflammatory cells or tissue, without MV playing a role in the disorder's pathogenesis. In this regard, MV sequences are reported to persist in circulating inflammatory cells in a large proportion of individuals with certain autoimmune (Andjaparidze et al 1989) or inflammatory disorders, such as otosclerosis (Niedermeyer & Arnold 1995), and in brain, kidney, spleen, liver, and lung tissues of up to 45.1% (23/51) of autopsy subjects (Katayama et al 1998). Some investigators hypothesize that persistence results from altered immune responses; for example, serum levels of anti-measles IgG are reduced in subjects with otosclerosis compared to healthy individuals (Lolov et al 2001), whereas increased levels of antibodies to measles are positively associated with the presence of peripheral autoantibodies to CNS antigens in children with autism (Singh et al 1998). Given the potential impact of these findings for autism research and treatment and public confidence in vaccine safety, it is imperative that the prevalence of MMR sequences in gastrointestinal tracts and peripheral blood of children with autism be assessed independently by other investigators using sampling methods and real time PCR methods equivalent to those employed by Uhlmann, Kawashima and colleagues. Furthermore, to ensure that differences in results do not reflect variability in assay sensitivity, groups reporting MMR sequences in children with autism should test coded samples submitted by independent laboratories.

Direction for future research in the microbiology and immunology of neurodevelopmental disorders

A comprehensive approach to investigating the pathogenesis of neurodevelopmental disorders must consider the interaction of both host and environmental factors. Evidence from epidemiology and animal models suggest that early infection, possibly with a variety of agents, may trigger complex behavioural disorders by impacting viability of specific neural cells and circuits. Future work should focus on dissecting the mechanisms of developmental neuropathology in animal models and using clues derived from these more simple systems to target infectious disease investigation. Toward this end we and others are establishing high throughput genomic and proteomic methods for characterizing the microbial and immune environment during gestation and in the prenatal period, and employing these tools in prospective birth cohorts.

Acknowledgements

Research in the Center for Immunopathogenesis and Infectious Diseases is supported by grants from the National Institutes of Health (NINDS, NIAID, NIMH, NICHD), the Ellison Medical Foundation (Pandora's Box Project), the CDC, and the MIND Institute of UC Davis.

References

Afzal MA, Armitage E, Ghosh S, Williams LC, Minor PD 2000 Further evidence of the absence of measles virus genome sequence in full thickness intestinal specimens from patients with Crohn's disease. J Med Virol 62:377–382

Andjaparidze OG, Chaplygina NM, Bogomolova N et al 1989 Detection of measles virus genome in blood leukocytes of patients with certain autoimmune diseases. Arch Virol 105:287–291

Barak Y, Kimhi R, Stein D, Gutman J, Weizman A 1999 Autistic subjects with comorbid epilepsy: a possible association with viral infections. Child Psychiatry Human Dev 29:245–251

Barka N, Shen GQ, Shoenfeld Y et al 1995 Multireactive pattern of serum autoantibodies in asymptomatic individuals with immunoglobulin A deficiency. Clin Diagn Lab Immunol 2:469–472

Baron-Cohen S, Saunders K, Chakrabarti S 1999 Does autism cluster geographically? A research note. Autism 3:39–43

Bartlik BD 1981 Monthly variation in births of autistic children in North Carolina. J Amer Med Women's Assoc 36:363–368

Bautista JR, Schwartz GJ, de la Torre JC, Moran TH, Carbone KM 1994 Early and persistent abnormalities in rats with neonatally acquired Borna disease virus infection. Brain Res Bull 34:31–40

Benveniste EN 1997 Cytokine expression in the nervous system. In: Keane RW, Hickey WF (eds) Immunology of the nervous system. Oxford University Press, New York, p 419–459
Bergroth V, Konttinen YT, Pelkonen P et al 1988 Synovial fluid lymphocytes in different subtypes of juvenile rheumatoid arthritis. Arthritis Rheum 31:780–783
Billaud JN, Ly C, Phillips TR, de la Torre JC 2000 Borna disease virus persistence causes inhibition of glutamate uptake by feline primary cortical astrocytes. J Virol 74:10438–10446
Brai M, Accardo P, Bellavia D 1994 Polymorphism of the complement components in human pathology. Ann Ital Med Int 9:167–172
Briese T, Hornig M, Lipkin WI 1999 Bornavirus immunopathogenesis in rodents: models for human neurological diseases. J Neurovirol 5:604–612
Burd L 1988 Month of birth of non-speaking children. Dev Med Child Neurol 30:685–686
Burger RA, Warren RP 1998 Possible immunogenetic basis for autism. Mental Retardation Dev Disabil Res Rev 4:137–141
Burmester GR, Jahn B, Gramatzki M, Zacher J, Kalden JR 1984 Activated T cells in vivo and in vitro: divergence in expression of Tac and Ia antigens in the nonblastoid small T cells of inflammation and normal T cells activated in vitro. J Immunol 133:1230–1234
California Department of Developmental Services 1999 Changes in the population of persons with autism and pervasive developmental disorders in California's Developmental Services System: 1987 through 1998. A report to the legislature. Sacremento, March 1999. *http://www.dds.ca.gov/Autism/pdf/Austism_Report_1999.PDF*
Chess S 1971 Autism in children with congenital rubella. J Autism Child Schizophr 1:33–47
Chong SK 2001 Gastrointestinal problems in the handicapped child. Curr Opin Pediatr 13: 441–446
Clark DA 2000 Human herpesvirus 6. Rev Med Virol 10:155–173
Comi AM, Zimmerman AW, Frye VH, Law PA, Peeden JN 1999 Familial clustering of autoimmune disorders and evaluation of medical risk factors in autism. J Child Neurol 14:388–394
Connolly AM, Chez MG, Pestronk A, Arnold ST, Mehta S, Deuel RK 1999 Serum autoantibodies to brain in Landau-Kleffner variant, autism, and other neurologic disorders. J Pediatrics 134:607–613
De Carli M, D'Elios MM, Zancuoghi G, Romagnani S, Del Prete G 1994 Human Th1 and Th2 cells: functional properties, regulation of development and role in autoimmunity. Autoimmunity 18:301–308
Denney DR, Frei BW, Gaffney GR 1996 Lymphocyte subsets and interleukin-2 receptors in autistic children. J Autism Dev Disord 26:87–97
Desmond MM, Wilson GS, Melnick JL et al 1967 Congenital rubella encephalitis. Course and early sequelae. J Pediatrics 71:311–331
Dittrich W, Bode L, Ludwig H, Kao M, Schneider K 1989 Learning deficiencies in Borna disease virus-infected but clinically healthy rats. Biol Psychiatry 26:818–828
Fatemi SH, Halt AR 2001 Altered levels of Bcl2 and p53 proteins in parietal cortex reflect deranged apoptotic regulation in autism. Synapse 42:281–284
Ferrari P, Marescot MR, Moulias R et al 1988 [Immune status in infantile autism. Correlation between the immune status, autistic symptoms and levels of serotonin] (French). Encephale 14:339–344
Fiore M, Pera C, Delfino L, Scotese I, Ferrara GB, Pignata C 1995 DNA typing of DQ and DR alleles in IgA-deficient subjects. Eur J Immunogenet 22:403–411
Gillberg C 1990 Do children with autism have March birthdays? Acta Psychiatr Scand 82:152–156
Gillberg C, Gillberg IC 1983 Infantile autism: a total population study of reduced optimality in pre-, peri-, and neonatal periods. J Autism Dev Disord 13:153–166

Gillberg C, Coleman M 1992 Infectious diseases. In: Gillberg C, Coleman M (eds) The biology of the autistic syndromes. 2nd edn, Mac Keith Press, London p 218–225

Gillberg C, Wing L 1999 Autism: not an extremely rare disorder. Acta Psychiatr Scand 99: 399–406

Gillberg C, Steffenburg S, Schaumann H 1991 Is autism more common now than ten years ago? Brit J Psychiatry 158:403–409

Gillberg IC, Gillberg C, Kopp S 1992 Hypothyroidism and autism spectrum disorders. J Child Psychol Psychiatry 33:531–542

Gupta S, Aggarwal S, Heads C 1996 Dysregulated immune system in children with autism: beneficial effects of intravenous immune globulin on autistic characteristics. J Autism Dev Disord 26:439–452

Gupta S, Aggarwal S, Rashanravan B, Lee T 1998 Th1- and Th2-like cytokines in CD4+ and CD8+ T cells in autism. J Neuroimmunol 85:106–109

Hafler DA, Hemler ME, Christenson L et al. 1985 Investigation of in vivo activated T cells in multiple sclerosis and inflammatory central nervous system diseases. Clin Immunol Immunopathol 37:163–171

Hatalski CG, Lewis AJ, Lipkin WI 1997 Borna disease. Emerg Infect Dis 3:129–135

Hoon AH Jr, Reiss AL 1992 The mesial-temporal lobe and autism: case report and review. Dev Med Child Neurol 34:252–259

Hornig M, Weissenböck H, Horscroft N, Lipkin WI 1999 An infection-based model of neurodevelopmental damage. Proc Natl Acad Sci USA 96:12102–12107

Horvath K, Papadimitriou JC, Rabsztyn A, Drachenberg C, Tildon JT 1999 Gastrointestinal abnormalities in children with autistic disorder. J Pediatr 135:559–563

Ikonomidou C, Bosch F, Miksa M et al 1999 Blockade of NMDA receptors and apoptotic neurodegeneration in the developing brain. Science 283:70–74

Katayama Y, Kohso K, Nishimura A, Tatsuno Y, Homma M, Hotta H 1998 Detection of measles virus mRNA from autopsied human tissues. J Clin Microbiol 36:299–301

Kawashima H, Mori T, Kashiwagi Y, Takekuma K, Hoshika A, Wakefield A 2000 Detection and sequencing of measles virus from peripheral mononuclear cells from patients with inflammatory bowel disease and autism. Dig Dis Sci 45:723–729

Kostantareas MW, Hauser P, Lennox C, Homatidis S 1986 Season of birth in infantile autism. Child Psychiatry Hum Dev 17:53–65

Landau EC, Cicchetti DV, Klin A, Volkmar FR 1999 Season of birth in autism: a fiction revisited. J Autism Dev Disord 29:385–393

Lipkin WI, Battenberg ELF, Bloom FE, Oldstone MBA 1988a Viral infection of neurons can depress neurotransmitter mRNA levels without histologic injury. Brain Res 451:333–339

Lipkin W, Carbone K, Wilson M, Duchala C, Narayan O, Oldstone M 1988b Neurotransmitter abnormalities in Borna disease. Brain Res 475:366–70

Lolov SR, Encheva VI, Kyurkchiev SD, Edrev GE, Kehayov IR 2001 Antimeasles immunoglobulin G in sera of patients with otosclerosis is lower than that in healthy people. Otol Neurotol 22:766–770

Luckett WP 1980 The suggested evolutionary relationships and classification of tree shrews. In: Luckett WP (ed) Comparative biology and evolutionary relationships of tree shrews. Plenum Press, New York p 3–31

Mehler MF, Goldstein H, Kessler JA 1996 Effects of cytokines on CNS cells: neurons. In: Ransohoff RM, Benveniste EN (eds), Cytokines and the CNS. CRC Press, Boca Raton, p 115–150

Money J, Bobrow NA, Clarke FC 1971 Autism and autoimmune disease: a family study. J Autism Child Schizophr 1:146–160

Naves M, Hajeer AH, Teh LS et al 1998 Complement C4B null allele status confers risk for systemic lupus erythematosus in a Spanish population. Eur J Immunogenet 25:317–320

Niedermeyer HP, Arnold W 1995 Otosclerosis: a measles virus associated inflammatory disease. Acta Otolaryngol 115:300–303

Norrholm SD, Ouimet CC 2000 Chronic fluoxetine administration to juvenile rats prevents age-associated dendritic spine proliferation in hippocampus. Brain Res 883:205–215

Okado N, Narita M, Narita N 2001 A biogenic amine-synapse mechanism for mental retardation and developmental disabilities. Brain Dev 23:S11–S15

Oldstone MB 1989a Molecular mimicry as a mechanism for the cause and a probe uncovering etiologic agent(s) of autoimmune disease. Curr Top Microbiol Immunol 145:127–135

Oldstone MB 1989b Viral alteration of cell function. Sci Am 261:42–48

Oldstone MB 1989c Viruses can cause disease in the absence of morphologic evidence of cell injury: implication for uncovering new diseases in the future. J Infect Dis 159:384–389

Pellegrini-Giampietro DE, Gorter JA, Bennett MV, Zukin RS 1997 The GluR2 (GluR-B) hypothesis: Ca^{2+}-permeable AMPA receptors in neurological disorders. Trends Neurosci 20:464–470

Plata-Salaman CR, Ilyin SE, Gayle D, Romanovitch A, Carbone KM 1999 Persistent Borna disease virus infection of neonatal rats causes brain regional changes of mRNAs for cytokines, cytokine receptor components and neuropeptides. Brain Res Bull 49:441–451

Pletnikov MV, Rubin SA, Vasudevan K, Moran TH, Carbone KM 1999 Developmental brain injury associated with abnormal play behavior in neonatally Borna disease virus-infected Lewis rats: a model of autism. Behav Brain Res 100:43–50

Pletnikov MV, Rubin SA, Schwartz GJ, Carbone KM, Moran TH 2000 Effects of neonatal rat Borna disease virus (BDV) infection on the postnatal development of the brain monoaminergic systems. Brain Res Dev Brain Res 119:179–185

Plioplys AV, Greaves A, Kazemi K, Silverman E 1989a Autism: anti-210K neurofilament immunoglobulin reactivity. Neurology 39:187

Plioplys AV, Greaves A, Yoshida W 1989b Anti-CNS antibodies in childhood neurologic diseases. Neuropediatrics 20:93–102

Plioplys AV, Greaves A, Kazemi K, Silverman E 1994 Lymphocyte function in autism and Rett syndrome. Neuropsychobiol 29:12–16

Portera-Cailliau C, Price DL, Martin LJ 1997 Excitotoxic neuronal death in the immature brain is an apoptotic-necrosis morphological continuum. J Compar Neurol 378:70–87

Quigley EM, Hurley D 2000 Autism and the gastrointestinal tract. Am J Gastroenterol 95: 2154–2156

Raiten DJ, Massaro T 1986 Perspectives on the nutritional ecology of autistic children. J Autism Dev Disord 16:133–144

Reil A, Bein G, Machulla HK, Sternberg B, Seyfarth M 1997 High-resolution DNA typing in immunoglobulin A deficiency confirms a positive association with DRB1*0301, DQB1*02 haplotypes. Tissue Antigens 50:501–506

Rogers T, Kalaydjieva L, Hallmayer J et al 1999 Exclusion of linkage to the HLA region in ninety multiplex sibships with autism. J Autism Dev Disord 29:195–201

Rubin SA, Bautista JR, Moran TH, Schwartz GJ, Carbone KM 1999 Viral teratogenesis: brain developmental damage associated with maturation state at time of infection. Brain Res Dev Brain Res 112:237–244

Santoro F, Kennedy PE, Locatelli G, Malnati MS, Berger EA, Lusso P 1999 CD46 is a cellular receptor for human herpesvirus 6. Cell 99:817–827

Sauder C, de la Torre JC 1999 Cytokine expression in the rat central nervous system following perinatal Borna disease virus infection. J Neuroimmunol 96:29–45

Schneider PA, Schneemann A, Lipkin WI 1994 RNA splicing in Borna disease virus, a nonsegmented, negative-strand RNA virus. J Virol 68:5007–5012

Schroeder HWJ, Zhu ZB, March RE et al 1998 Susceptibility locus for IgA deficiency and common variable immunodeficiency in the HLA-DR3, -B8, -A1 haplotypes. Mol Med 4:72–86

Singh VK 1996 Plasma increase of interleukin-12 and interferon-gamma. Pathological significance in autism. J Neuroimmunol 66:143–145

Singh VK 1997 Immunotherapy for brain diseases and mental illnesses. Prog Drug Res 48: 129–146

Singh VK, Warren RP, Odell JD, Cole P 1991 Changes of soluble interleukin-2, interleukin-2 receptor, T8 antigen, and interleukin-1 in the serum of autistic children. Clin Immunol Immunopathol 61:448–455

Singh VK, Warren RP, Odell JD, Warren WL, Cole P 1993 Antibodies to myelin basic protein in children with autistic behavior. Brain Behav Immunity 7:97–103

Singh VK, Singh EA, Warren RP 1997a Hyperserotoninemia and serotonin receptor antibodies in children with autism but not mental retardation. Biol Psychiatry 41:753–755

Singh VK, Warren R, Averett R, Ghaziuddin M 1997b Circulating autoantibodies to neuronal and glial filament proteins in autism. Pediatr Neurol 17:88–90

Singh VK, Lin SX, Yang VC 1998 Serological association of measles virus and human herpesvirus-6 with brain autoantibodies in autism. Clin Immunol Immunopathol 89:105–108

Singh VK, Mehrotra S, Agarwal SS 1999 The paradigm of Th1 and Th2 cytokines: its relevance to autoimmunity and allergy. Immunol Res 20:147–161

Solbrig MV, Koob GF, Fallon JH, Lipkin WI 1994 Tardive dyskinetic syndrome in rats infected with Borna disease virus. Neurobiol Dis 1:111–119

Solum D, Hughes D, Major MS, Parks TN 1997 Prevention of normally occurring and deafferentation-induced neuronal death in chick brainstem auditory neurons by periodic blockade of AMPA/kainate receptors. J Neurosci 17:4744–4751

Staeheli P, Sauder C, Hausmann J, Ehrensperger F, Schwemmle M 2000 Epidemiology of borna disease virus. J Gen Virol 81:2123–2135

St. Pierre BA, Merrill JE, Dopp JM 1996 Effects of cytokines on CNS cells: glia. In: Ransohoff RM, Benveniste EN (eds) Cytokines and the CNS. CRC Press, Boca Raton, p 151–168

Stevens MC, Fein DH, Waterhouse LH 2000 Season of birth effects in autism. J Clin Exp Neuropsychol 22:399–407

Stubbs EG, Crawford ML 1977 Depressed lymphocyte responsiveness in autistic children. J Autism Child Schizophr 7:49–55

Tanoue Y, Oda S, Asano F, Kawashima K 1988 Epidemiology of infantile autism in southern Ibaraki, Japan: differences in prevalence in birth cohorts. J Autism Dev Disord 18:155–166

Tatsuo H, Ono N, Tanaka K, Yanagi Y 2000 SLAM (CDw150) is a cellular receptor for measles virus. Nature 406:893–897

Ticher A, Ring A, Barak Y, Elizur A, Weizman A 1996 Circannual pattern of autistic births: reanalysis in three ethnic groups. Human Biol 68:585–592

Todd R, Ciaranello R 1985 Demonstration of inter- and intraspecies differences in serotonin binding sites by antibodies from an autistic child. Proc Natl Acad Sci USA 82: 612–616

Todd RD, Hickok M, Anderson GM, Cohen DJ 1988 Antibrain antibodies in infantile autism. Biol Psychiatry 23:644–647

Torres AR, Maciulis A, Stubbs EG, Cutler A, Odell D 2002 The transmission disequilibrium test suggests that HLA-DR4 and DR13 are linked to autism spectrum disorder. Hum Immunol 63:311–316

Torrey EF, Miller J, Rawlings R, Yolken RH 1997 Seasonality of births in schizophrenia and bipolar disorder: a review of the literature. Schizophr Res 28:1–38

Trottier G, Srivastava L, Walker CD 1999 Etiology of infantile autism: a review of recent advances in genetic and neurobiological research. J Psychiatry Neurosci 24:103–115
Uhlmann V, Martin CM, Sheils O et al 2002 Potential viral pathogenic mechanism for new variant inflammatory bowel disease. Mol Pathol 55:84–90
Ulgiati D, Abraham LJ 1996 Comparative analysis of the disease-associated complement C4 gene from the HLA-A1, B8, DR3 haplotype. Exp Clin Immunogenet 13:43–54
Wakefield AJ, Murch SH, Anthony A et al 1998 Ileal-lymphoid-nodular hyperplasia, non-specific colitis, and pervasive developmental disorder in children. Lancet 351:637–641
Wakefield AJ, Anthony A, Murch SH et al 2000 Enterocolitis in children with developmental disorders. Am J Gastroenterol 95:2285–2295
Warren RP 1998 An immunologic theory for the development of some cases of autism. CNS Spect 3:71–79
Warren RP, Singh VK 1996 Elevated serotonin levels in autism: association with the major histocompatibility complex. Neuropsychobiology 34:72–75
Warren RP, Foster A, Margaretten NC, Pace NC 1986 Immune abnormalities in patients with autism. J Autism Dev Disord 16:189–197
Warren RP, Foster A, Margaretten NC 1987 Reduced natural killer cell activity in autism. J Am Acad Adolesc Psychiatry 26:333–335
Warren RP, Yonk LJ, Burger RA et al 1990 Deficiency of suppressor-inducer (CD4+CD45RA+) T cells in autism. Immunol Invest 19:245–251
Warren RP, Singh VK, Cole P et al 1991 Increased frequency of the null allele at the complement C4B locus in autism. Clin Exp Immunol 83:438–440
Warren RP, Singh VK, Cole P et al 1992 Possible association of the extended MHC haplotype B44-SC30-DR4 with autism. Immunogenetics 36:203–207
Warren RP, Burger RA, Odell D, Torres AR, Warren WL 1994 Decreased plasma concentrations of the C4B complement protein in autism. Arch Pediatr Adolescent Med 148:180–183
Warren RP, Yonk J, Burger RW, Odell D, Warren WL 1995 DR-positive T cells in autism: association with decreased plasma levels of the complement C4B protein. Neuropsychobiol 31:53–57
Warren RP, Singh VK, Averett RE et al 1996a Immunogenetic studies in autism and related disorders. Mol Chem Neuropathol 28:77–81
Warren RP, Odell JD, Warren WL et al 1996b Strong association of the third hypervariable region of HLA-DR beta 1 with autism. J Neuroimmunol 67:97–102
Warren RP, Odell JD, Warren WL et al 1997 Brief report: immunoglobulin A deficiency in a subset of autistic subjects. J Autism Dev Disord 27:187–192
Weissenböck H, Hornig M, Hickey WF, Lipkin WI 2000 Microglial activation and neuronal apoptosis in Bornavirus infected neonatal Lewis rats. Brain Pathol 10:260–272
Weizman A, Weizman R, Szekely GA, Wijsenbeek H, Livini E 1982 Abnormal immune response to brain tissue antigen in the syndrome of autism. Am J Psychiatry 139:1462–1465
Wilton AN, Cobain TJ, Dawkins RL 1985 Family studies of IgA deficiency. Immunogenetics 21:333–342
Wright HH, Abramson RK, Self S, Genco P, Cuccaro M 1990 Serotonin may affect lymphocyte cell surface markers in autistic probands. American Academy of Child and Adolescent Psychiatry. San Francisco, CA, NR (abstr 12)
Yonk LJ, Warren RP, Burger RA et al 1990 CD4+ helper T cell depression in autism. Immunol Lett 25:341–345
Yuwiler A, Shih JC, Chen CH et al 1992 Hyperserotoninemia and antiserotonin antibodies in autism and other disorders. J Autism Dev Disord 22:33–45
Zhao Z-S, Granucci F, Yeh L, Schaffer PA, Cantor H 1998 Molecular mimicry by herpes simplex virus-type 1: autoimmune disease after viral infection. Science 279:1344–1347

DISCUSSION

Bolton: Your data indicate that following neonatal Borna virus infection there is no evidence of inflammatory response in the brain that might signal an infectious disease. In terms of the brain's response to pathogens, at what stage would you expect an inflammatory reaction to become evident in humans? How does this relate to the timing of the MMR vaccination programme?

Lipkin: Early exposure of infectious agents to an immature lymphoid system can result in failure to perceive those agents as foreign. This state of tolerance facilitates persistent infection. Brain damage can nonetheless occur in the absence of an inflammatory response. Borna disease virus doesn't replicate as well as many other viruses; thus, it takes more time for the titre to grow high enough in the brain to spill over into the lymphatics. In our hands, infection must occur within the first 12 hours of life to achieve tolerance. In contrast, if you use lymphocytic choriomeningitis virus, an agent which replicates much more rapidly, tolerance can be seen even with infections as far out as 24 or 48 h postnatal. One can construct a variety of models by using different viruses, or changing the timing, titre, or route of inoculation. In the Borna disease virus model we are approximating 30 weeks gestation in humans.

Charman: I have a general question. In the chronic diseases where there is evidence for a role for viral infections, is it possible in a population or an individual to determine a proportion of cases in whom the viral infection will have definitely played a role versus cases where it might not have been involved?

Lipkin: There are many different mechanisms by which infectious agents can cause disease. I tried to illustrate a few of them. In hepatitis B and C for example, the virus persists and causes chronic inflammation. In poliomyelitis, the virus causes damage and then it is gone. You may not be able to recover polio virus sequence in anterior horn cells, however, the victim is nonetheless left with a paralytic disease. A third example would be brain or cardiac disease due to indirect effects of streptococcal infection. Here a peripheral infection causes an immune response which, through cross reactivity or other mechanisms, results in recognition of host components as foreign. Many rheumatological conditions are thought to involve this type of process. When an investigator reports having looked for infection with a specific microbe but not finding it, the questions to ask are whether the search was done at the right time, the reagents were appropriate, or the assays were sufficiently sensitive. In some types of cancer, the virus that initiated the neoplastic process may have long gone. Thus, we may need extremely sensitive methods to detect the few footprints that remain. Another confounding problem may be that the virus persists but in only a small fraction of cells. This has been described in some herpes virus infections where only 1 in 100 000 white blood cells contains nucleic acid. Unless you survey enough white blood cells you are not going to find the agent. Nonetheless, addressing these problems is just a question of will and resources. We already have tools available to sort out whether MMR can be associated with autism. The kinds of data that have come out of Wakefield and O'Leary's work indicate that we can resolve the issue very easily. First, the number of cells that they report are infected and the number of individuals infected is high. Second, the real-time PCR method that they use is relatively straightforward. The tools required have been developed in our lab and at CDC. My prediction is that we are not going to replicate their findings. What we are likely to find is that the proportion of individuals who have these sequences will be far less or perhaps none at all. If we find measles sequences it is important to remember that infiltrating inflammatory cells present in bowel may just harbour these sequences and not be involved with pathogenesis. The argument has been raised by a number of my colleagues that wild-type measles virus does not cause autism, therefore vaccine strain measles virus can't either. One problem with that argument is that the cellular receptors used by wild-type measles virus and vaccine virus are different. CD46, the cellular receptor for measles vaccine virus, has other interactions potentially consistent with models of autoimmunity. CD46 binds some complement proteins as well as another virus, HHV6. Antibodies to HHV6 have been associated with increased levels of antibodies to brain antigens. Therefore, it is conceivable that susceptibility to infection with vaccine virus might be a marker for susceptibility to autoimmune diseases. Having said this, however, there are good reasons to think that MMR is not likely to be a significant factor in autism. Most convincing, as Eric Fombonne has shown, is that the introduction of the vaccine is not historically correlated with the proposed increase in the frequency of autism.

In 1998 Wakefield described finding ileocolonic lymphonodular hyperplasia in 7 out of 12 children with autism and GI symptoms (Wakefield et al 1998). This prompted him to pursue a measles virus hypothesis as this had guided his earlier work in inflammatory bowel disease. With Kawashima, in 2000, he found measles vaccine virus RNA in white blood cells of a third of children with autism (Kawashima et al 2000), and then with O'Leary he found these sequences in bowel samples from approximately three-quarters of children with autism (Uhlmann et al 2002). Although these sequences were infrequent in controls, I find it difficult to interpret the significance of the differences between normal children and autistic children because age matching was either poor or not described. We would not be surprised if viral sequences persist for up to several months after vaccination; thus, it is critical to know how recently children with autism or normal controls were vaccinated. In the Kawashima report children with autism were younger (3–10 years of age) than individuals with inflammatory bowel disease (15–34 years). This does not mean that I believe or disbelieve the reports of measles vaccine virus sequences. I am simply noting that if present, these sequences might simply reflect proximity to vaccination.

There is only one published report from an independent group that examined the frequency of measles virus sequences in gut biopsies; however, this was in inflammatory bowel disease rather than autism (Afzal et al 2000). The problem with comparing this work from Afzal and Minor head to head with that from the Wakefield and O'Leary group (Uhlmann et al 2002) is that they used different methods. O'Leary and Wakefield used real-time PCR and examined gut lesions; Afzal used nested PCR on random biopsies. Primers were different. My estimate on the basis of talking to members of both of these groups is that the sensitivity of O'Leary and Wakefield's assays is likely to be better than Afzal's. This does not mean that O'Leary and Wakefield are correct and Afzal and Minor are incorrect. It does mean, however, that we need to revisit the data using appropriately sensitive methods in well-matched subject populations.

If O'Leary and Wakefield's findings are confirmed we will still need to determine whether the presence of viral sequences is relevant to pathogenesis of autism. Measles virus may simply be related to the accumulation of white blood cell subsets. There have been a number of reports that indicate that patients with a variety of autoimmune disorders may have circulating white blood cells harbouring measle virus sequences. Additionally, measles virus sequences have been in tissues obtained post-mortem from diseases clearly unrelated to measles virus infection.

Folstein: Why haven't you done the study yet? Is the problem funding or time?

Lipkin: We are still waiting for funding.

Bishop: How strong is the evidence that there is no increase in autism after a measles outbreak? Have people looked for this?

Lipkin: There are no reports on this. Last week, a senior measles virologist in Europe retired. Several of the most prominent measles virus investigators from all over the world sat around for an hour discussing this issue. No one could come up with any sort of correlation between wild-type measles virus and autism. There aren't a lot of measles outbreaks any more because of the vaccine. One of our major concerns is that the acceptance of the vaccine has fallen dramatically because of concerns relating to autism. In Germany it is 80% and in Italy it is less than 60%.

Rutter: Andy Wakefield has been very helpful here: this problem of measles going away has been reversed! We have had epidemics in South East London that might soon provide the data!

Fombonne: There was a study in the late 1970s investigating the risk associated with measles, mumps, rubella and flu in prenatal and postnatal periods in children with autism and controls. The associations were very weak and going in all sorts of directions, casting doubts on the validity of the findings. We recently have tested in a UK investigation whether or not prior to the introduction of the first measles vaccinations in 1968, and at a time when there were high peaks of measles epidemics following a typical biannual cycle, rates of autism births were linked with the incidence of measles. In a very large sample we found no evidence of an association.

Lipkin: I am not convinced that data from measles outbreaks tells us whether or not measles vaccine can be implicated. Wild-type virus and vaccine strains have different properties and routes of infection are different.

Dawson: Help me understand Wakefield's recent study in terms of its potential methodological problems. He had two groups, both of which had inflammatory bowel disease, but one of which had developmental disorders. There could be a confounder relating to age which would be important. But the methods were the same and they both had inflammatory bowel disease. The two things I can figure out that differed between those groups are age and time since vaccine.

Lipkin: We also don't have neurological or cognitive data on the controls in the Uhlmann study.

Dawson: You have to accept that they say the children in the comparison group are without any developmental disorders. You would want a tight study there. But how could there be such a dramatic difference between the groups in the presence of that virus?

Lipkin: There are similar examples of discordance between laboratories with many infectious agents. What we have learned is that unless specific precautions are taken you can have problems like these. There has to be a separate site that collects the samples, cuts them and aliquots them. The samples don't need to go to more than one laboratory. The critical point is that an independent group needs to send samples to the laboratory and the lab should have consistent results with independent aliquots of the same clinical materials.

Baird: I wanted to pursue the issue of molecular mimicry and the antibodies in the streptococcus story. Is it reasonable when you are trying to make an association between disorder and measuring auto-antibodies to expect that one might have a level of antibodies that would relate to the level of disorder? Might finding them be coincident with the onset of the illness; you might expect fluctuation of any disorder to go along with fluctuation in antibody levels? My understanding is that this is not really what is happening in the OCD story. Also, we are finding these antibodies in a very wide range of neurological conditions.

Lipkin: These are excellent questions. Unfortunately, we don't yet have the appropriate assay for PANDAS. We have been unable to identify anything in sera of these patients that will cause disease in the animals, despite use of a variety of experimental approaches. One of the reasons we shifted to the animal model was that we figured if we could reproduce PANDAS in an animal model, we might then be able to clone the target of immunity from the animal, and back-translate into the humans to see whether we could identify the appropriate epitopes. Another marker that many people will have heard of is the D8/17 marker described by John Zabriski of the Rockefeller University. He has been working in this field for a long time but has been unable to do any interesting biochemistry or immunology because he only has an IgM antibody. John tells me that his monoclonal antibody recently switched from IgM to IgG. Thus, he now has the tools to do some very interesting work in Sydenham's chorea patients. Sydenham's chorea has features similar to PANDAS.

Folstein: There is a probability of encephalitis with measles. If you just looked at the cases with measles encephalitis there might be a relationship.

Lipkin: This would represent just a few in a thousand measles cases.

Folstein: I wanted also to get back to the point of Patrick Bolton's initial question: if you had a measles virus-related regression at 18–24 months, would you expect to see an inflammatory response at that age in the brain?

Lipkin: We need not restrict ourselves to brain inflammation. Perhaps we have measles in the GI tract. This is in line with what has been proposed about alterations in the gut permeability and opioid peptides. If measles entered into the brain at 18–24 months we might expect to see inflammation there. However, virus need not get into the brain to cause a behavioural disorder.

Folstein: So you could look peripherally for the inflammation.

Lipkin: We have proposed to look at gut and white blood cells for measles virus transcripts, and also use a multiplex approach to examining antibodies and other biomarkers in serum.

References

Afzal MA, Armitage E, Ghosh S, Williams LC, Minor PD 2000 Further evidence of the absence of measles virus genome sequence in full thickness intestinal specimens from patients with Crohn's disease. J Med Virol 62:377–382

Kawashima H, Mori T, Kashiwagi Y, Takekuma K, Hoshika A, Wakefield A 2000 Detection and sequencing of measles virus from peripheral mononuclear cells from patients with inflammatory bowel disease and autism. Dig Dis Sci 45:723–729

Uhlmann V, Martin CM, Sheils O et al 2002 Potential viral pathogenic mechanism for new variant inflammatory bowel disease. Mol Pathol 55:84–90

Wakefield AJ, Murch SH, Anthony A et al 1998 Ileal-lymphoid-nodular hyperplasia, non-specific colitis, and pervasive developmental disorder in children. Lancet 351:637–641

What do imaging studies tell us about the neural basis of autism?

Chris Frith

Wellcome Department of Imaging Neuroscience, Institute of Neurology, 12 Queen Square, London WC1N 3BG, UK

Abstract. There is no clear evidence from imaging studies for specific structural abnormalities in the brains of people with autism. The most robust observation is of greater total brain volume. There is evidence that this greater volume is not present at birth, but appears during the first few years. This brain enlargement might be a marker of abnormal connectivity due to lack of pruning. While abnormalities have often been reported in the cerebellum and the amygdala, these are difficult to interpret since both increases and decreases in the size of these structures have been observed. Another way of identifying the neural basis of autism is to investigate brain systems underlying cognitive functions compromised in this disorder such as face perception and 'theory of mind'. Autistic people fail to activate the 'fusiform face area' during face perception tasks and show weak activation of medial frontal cortex and superior temporal gyrus when performing theory of mind tasks. These problems stem from a lack of integration of sensory processing with cognitive evaluation. I speculate that this problem reflects a failure of top-down modulation of early sensory processing. The problem could result from abnormal connectivity and lack of pruning.

2003 Autism: neural basis and treatment possibilities. Wiley, Chichester (Novartis Foundation Symposium 251) p 149–176

What do imaging studies tell us about the neural basis of autism?

Autism is a particularly exciting and challenging disorder for a neuropsychologist to study. On the one hand this is a disorder with a biological basis. Even though the details remain unknown there is substantial evidence that autism is a disorder of brain development with a genetic cause. On the other hand this is a disorder in which the diagnostic signs reflect impairments of the highest human mental faculties — the ability to communicate, to imagine and to make social contact. If we are to understand autism then we will have to make links between physical activity in the brain and these high level mental processes.

The development of brain imaging in the last decades of the 20th century provided the opportunity to explore such links much more directly than ever before, and yet, so far, the results from studies of autism have been disappointing. There is no fact about brain structure or function in autism on which everyone agrees. In the first part of this review I shall present what I consider to be the most reliable findings from imaging studies of brain structure and brain function. In the second part I shall speculate on what these findings could mean for our understanding of autism.

MRI studies of brain structure

Magnetic resonance imaging (MRI) can provide detailed pictures of brain structure with a resolution in the millimetre range. This exquisite detail creates great problems for quantitative statistical analysis. The traditional approach has been to specify regions of interest (ROIs) such as the amygdala and then measure their volume. There are some disadvantages to this approach. First, there may be difficulties and disagreements about how to define the region of interest. Second, subtle differences in brain structure will be missed if they do not fall into a definable region of interest. For example, while the volume of frontal cortex could be measured as a whole it is difficult to define clear subdivisions within this large region. More recently the technique of voxel based morphometry (VBM) has been developed (Ashburner & Friston 2000) which overcomes these particular problems. VBM detects structural differences anywhere in the brain without the need for defining prior regions of interest. However, the method requires that the brains being studied are first normalized into a standard template. This procedure preserves differences in structure at the scale of 5–15 mm, but eliminates the large-scale differences that might be seen in a regions-of-interest approach.

A problem specific for the study of brain structure in autism arises from the frequent co-occurrence of mental retardation and epilepsy. Both these disorders are associated with structural brain abnormalities. In order to identify abnormalities that are specific to autism, the non-autistic comparison group must be matched in all other respects. Fortunately there are a number of such well-conducted studies of autism from which three themes emerge. These concern brain size, the cerebellum and the amygdala.

Brain size (see Table 1)

A number of MRI studies of autism have observed autistic people to have greater brain volumes than controls (Piven et al 1996, Hardan et al 2001b). This is consistent with observations of greater head circumference (Lainhart et al 1997, Miles et al 2000) and greater size and weight in post-mortem brains (see Bailey et al 1998). However, autistic individuals with abnormally small brains can also

TABLE 1 Studies showing increased brain size in autism

Reference	*Numbers*	*Diagnosis*	*Characteristics*	*Matching/ covariates*	*Measures*	*Qualifications*
Piven et al 1996	35 autistic 36 controls	ADI DSM-III-R	Age 18 (12–29) IQ 91 (52–136)	Age Height IQ	Brain volume, lobe volumes (MRI)	Effect only in males No difference in frontal lobes
Hardan et al 2001b	16 autistic 19 controls	ADI	Age 22 (12–52) IQ 100 (83–136)	Age IQ	Brain volume (MRI, adjusted for intra-cranial volume)	
Lainhart et al 1997	91 autistic	ADI DSM-III-R	Age 14 (3–38) Low IQ	Age, height norms	Occipitofrontal circumference	Effect not present at birth
Fombonne et al 1999	126 autistic	ICD-10	Age 8 (2–16) IQ low-normal	Age norms	Occipitofrontal circumference	Microcephaly associated with medical disorders Macrocephaly increased with age
Miles et al 2000	137 autistic	DSM-IV	Age: 1–41 yrs IQ: 20–131	Age, height norms	Occipitofrontal circumference	Microcephaly associated with low IQ and seizures
Courchesne et al 2001	60 autistic 52 controls	ADI-R DSM-IV	Age 6 (2–16) IQ: 36–122	Age	Head circumference Brain volume (MRI)	Effect not present at birth Maximum at 2–4 yrs
Bailey et al 1998	6 autistic (post mortem)	ADI	Age 20 (4–27) Low IQ	Age norms	Brain weight	4/6 macrocephalic

be found. In the study by Miles et al (2000), microcephaly, but not macrocephaly, was associated with low IQ and seizures suggesting that microcephaly is a feature of mental retardation (Cole et al 1994) and is not specific to autism. This complication may explain why a few studies have not observed an increase in brain size in autism.

At least two studies have reported that the brains of autistic individuals were not abnormally large at birth (Lainhart et al 1997, Courchesne et al 2001). Longitudinal data from both these studies suggests that in autism there is an abnormally rapid rate of brain growth during infancy, which then slows again during adolescence. More longitudinal studies of brain development in autism are needed in addition to cross-sectional studies in which cases are stratified in terms of age and IQ.

The period of rapid brain growth, which occurs during the preschool years and appears to be abnormal in autism, coincides with a major reorganisation of connectivity in the human brain during which new synapses are formed and there is dendritic growth and myelination. At the same time large numbers of synapses are eliminated (Huttenlocher 1999). Does something go wrong with these processes in autism? I shall return to this question in my speculations at the end of this paper.

Cerebellum (see Table 2)

Since the original report (Courchesne et al 1988) there has been considerable controversy (Piven et al 1997, Courchesne 1999, Piven et al 1999) as to whether regions of the cerebellum are abnormally small in autism. The situation is complicated by observations that, in some cases, the same regions can be abnormally large (Courchesne et al 1994). More recent studies have not clarified the situation. Levitt et al (1999) observed small cerebellar regions in a group of autistic individuals who scored, on average, 33 IQ points lower than their control group. The significance of the difference in cerebellar size disappeared when IQ was entered as a covariate in the group comparison. Hardan et al (2001a) compared non-retarded adolescent autistic individuals with controls matched for age and IQ and found that the cerebellum was larger in the autistic individuals possibly commensurate with their greater overall brain size. No differences were found in sub regions of the cerebellum.

Two independent studies using VBM (Abell et al 1999, Salmond et al 2002) both found evidence of increased grey matter density bilaterally in the posterior lobes of the cerebellum.

TABLE 2 The cerebellum in autism

Reference	*Numbers*	*Diagnosis*	*Characteristics*	*Matching*	*Measures*	*Results*
Courchesne et al 1988	18 autistic 12 controls	DSM-III	Age 21 (6–30)	Age	Regions of interest (MRI)	Vermal lobes VI & VII smaller
Courchesne et al 1994	50 autistic	DSM-III?	Age (2–39)	Age	Regions of interest (MRI)	Smaller in 43 Larger in 6
Piven et al 1997	35 autistics 36 controls	ADI DSM-III	Age 18 (12–29) IQ 91 (52–136)	Age IQ	Regions of interest (MRI)	Cerebellum enlarged
Levitt et al 1999	8 autistic 21 controls	ADI DSM-IV	Age 12 (8–17) IQ 83	Age	Regions of interest (MRI)	Vermis smaller Effect removed when IQ covaried
Hardan et al 2001a	22 autistic 22 controls	ADI	Age 22 (12–52) IQ 100 (83–136)	Age IQ	Regions of interest (MRI)	Cerebellum enlarged
Abell et al 1999	15 autistic 15 control	DSM-IV	Age 28 IQ 110	Age IQ	Voxel-based morphometry	Increased grey matter density, posterior lobes +46 −54 −33 −52 −66 −23
Salmond et al 2002	14 autistic 18 controls	?	Age 13 IQ 104	Age IQ	Voxel-based morphometry	Increased grey matter density, posterior lobes 22 −74 −45 28 −66 −22

Amygdala (see Table 3)

The amygdala has been implicated in many theoretical accounts of autism (e.g. Hetzler & Griffin 1981, Baron-Cohen et al 2000) largely on the grounds that lesions in this region of the temporal lobe in animals can lead to abnormalities of behaviour that may resemble those seen in autism. Unfortunately studies of the amygdala using MRI are as inconsistent as for the cerebellum.

Two studies (Aylward et al 1999, Pierce et al 2001) reported reduced amygdala volume, one (Haznedar et al 2000) reported no difference, and one (Howard et al 2000) reported an increase in volume. The two VBM studies (Abell et al 1999, Salmond et al 2002) both reported increased grey matter density in the general region of the amygdala.

The large brain volume observed in many autistic individuals probably reflects abnormal development of connectivity during infancy and early childhood. This abnormal connectivity is likely to be manifest in structural abnormalities in many brain areas, including the cerebellum and the amygdala, but it is not yet clear precisely what effect these abnormalities will have on volumes of structures measured with a regions-of-interest approach or on grey matter density measured with VBM. In this context it is interesting to note that neither of the VBM studies detected differences in white matter density anywhere in the brain. However, it must be remembered that the normalization that is a necessary component of this technique will eliminate any large volume differences in a single type of tissue such as white matter.

Functional studies of resting blood flow *(see Table 4)*

A number of studies have been published in which positron tomography (PET) or single photon computed tomography (SPECT) has been used to measure blood flow in the brains of autistic individuals 'at rest'. Such studies probably tells us more about brain structure than brain function, since they reveal areas of persistently abnormal perfusion as well as patterns of blood flow specific to the state in which the subjects were in at the time of the measurement. Few consistent differences in blood flow were observed in the early studies (see Boddaert & Zilbovicius 2002). However, two recent, well-controlled studies have revealed consistent abnormalities in the temporal cortex (Ohnishi et al 2000, Zilbovicius et al 2000). Both these studies investigated school-age children who were mentally retarded as well as autistic and used control groups matched for age and IQ. The children were sedated during the scanning procedure in order to minimize head movements. The data were analysed using a whole-brain voxel-based approach (statistical parametric mapping) rather than ROIs. In both studies bilateral reductions of blood flow were observed in the temporal lobes

TABLE 3 The amygdala in autism

Reference	*Numbers*	*Diagnosis*	*Characteristics*	*Matching*	*Measures*	*Results*
Aylward et al 1999	14 autistic 14 control	ADI	Age 21 (11–37) IQ 106	Age IQ Height	Regions of interest	Amygdala smaller
Pierce et al 2001	7 autistic 8 control	ADI DSM-IV	Age 30 (21–41) IQ 84 (73–102)	Age	Regions of interest	Amygdala smaller
Haznedar et al 2000	17 autistic 17 control	ADI DSM-IV	Age 28 IQ (55-125)	Age	Regions of interest	No difference
Howard et al 2000	10 autistic 10 control	DSM-IV Wing check list	Age (16–40) IQ?	Age IQ	Regions of interest	Amygdala larger
Abell et al 1999	15 autistic 15 control	DSM-IV	Age 28 IQ 110	Age IQ	Voxel-based-morphometry	Increased grey matter density, amygdala -14 -05 -28
Salmond et al 2002	14 autistic 18 control	?	Age 13 IQ 104	Age IQ	Voxel-based-morphometry	Increased grey matter density, amygdala ± 20 -14 -24

TABLE 4 Resting cerebral blood flow in the temporal lobes in autism

Reference	*Numbers*	*Diagnosis*	*Characteristics*	*Matching*	*Measures*	*Results*
Zilbovicius et al 2000	33 autistic 10 mentally retarded	DSM-IV	Age 8 (5–13) IQ 45	Age IQ	Voxel-based (rCBF)	Reduced rCBF superior temporal gyrus (BA22) +40 −16 +4 −40 −14 +4
Ohnishi et al 2000	23 autistic 26 mentally retarded	DSM-IV	Age 7 (3–13) IQ 48	Age IQ	Voxel-based (rCBF)	Reduced rCBF superior temporal gyrus (BA22) +52 0 −2 −50 −16 0

in auditory association cortex (superior temporal gyrus). It remains to be seen whether the same pattern of temporal lobe hypo-perfusion occurs in autistic individuals with IQs in the normal range.

Cognitive activations

The aim of most functional imaging studies is to identify patterns of brain activity associated with particular cognitive processes. This is achieved by comparing activity associated with the performance of at least two tasks only one of which engages the cognitive process of interest. The major difficulty for such studies is to design experimental paradigms that successfully isolate a cognitive process. When this approach is used to study autism what we hope to find is an interaction (in the statistical sense) between task and group. In other words the pattern of activity associated with some cognitive process (the difference between the experimental task and the control task) will be different in the autistic group compared to the control group. There are essentially two approaches for choosing tasks relevant for functional imaging studies of autism. One approach is to identify a brain region of interest (e.g. the amygdala) and then select a task that is known to elicit activity in this region in normal volunteers (e.g. looking at fearful faces, see LeDoux 2000). The other approach is to identify a cognitive process of interest (e.g. mentalizing) and then select tasks (e.g. 'theory of mind' tasks) that have successfully related this process with activity in circumscribed brain regions (see Frith 2001). Since the evidence from the studies of structural brain abnormalities reviewed so far provides little evidence for abnormalities in specific brain regions, I shall organize my review of functional imaging studies in terms of cognitive processes. There is general agreement that three broad classes of cognitive process are implicated in autism. These are executive functions, central coherence and social cognition.

Cognitive systems implicated in autism

Executive function. Human behaviour is organised by a hierarchy of systems with simple reflexes at the bottom and a 'central executive' (Baddeley 1986) at the top. Many aspects of the functioning of this central executive are captured by Shallice's (1988, chapter 14) concept of a Supervisory Attentional System (SAS). This system selects and controls lower level automatic processing routines, a function that is especially important in novel situations. Executive functions of this kind are required for successful performance of working memory tasks, for tasks involving planning and for tasks in which pre-potent responses have to be suppressed. Neurological patients who show a lack of flexibility and an inability to control their processing resources would be said to suffer from a disorder of executive function. Such patients typically have large lesions in prefrontal cortex. Disorders of certain executive functions are considered to be responsible for the repetitive behaviour and narrow interests of autistic individuals. Such individuals typically perform badly on tasks that depend upon executive functions.

There have been many brain imaging studies of normal volunteers while they perform executive tasks. These studies confirm the involvement of prefrontal cortex. However, it has proved much more difficult to relate particular components of executive function to particular brain regions (Duncan & Owen 2000). The tasks most frequently used in these studies of executive function involve selective attention. Such studies show that selective attention depends upon top-down modulation of activity in relevant sensory processing areas. The response to faces in the face-processing region of the fusiform gyrus is enhanced when the volunteer has been instructed to attend to faces (Wojciulik et al 1998). This increased activity in the relevant sensory processing area can be detected even before any stimuli have been presented. The ultimate source of the signals that induces this modulation is believed to lie in a system of frontal and parietal areas (Kastner & Ungerleider 2000). While there is no evidence for gross structural abnormalities in the frontal cortex of autistic individuals there could be abnormal connectivity between frontal cortex and the posterior regions that are modulated during performance of executive tasks.

I am not aware of any study, as yet, in which autistic individuals have been scanned while performing executive function tasks.

Central coherence. Weak central coherence is believed to be the source of the superior performance of autistic individuals on certain perceptual tasks such as finding hidden figures (Happé 1999). Central coherence defines a style in cognitive processing. People who adopt the style of weak central coherence pay attention to local details at the expense of the whole. Autistic individuals typically adopt this style. As a result they can find hidden figures more easily and are less susceptible to visual illusions. On the other hand they find it more difficult to extract the gist from the details of a scene or a narrative. People who adopt the style of strong central coherence cannot find the hidden figures, but are good at extracting the gist.

Fink and his colleagues (e.g. Fink et al 1997) have performed a series of imaging studies in which normal volunteers had to attend to the global or the local aspect of complex visual figures. As in the studies of selective attention mentioned in the last section, attention to global or local aspects was associated with modulation in specific regions of extra-striate visual cortex. Medial extra-striate cortex was more activated when attention was directed towards global features while lateral extra-striate cortex was more activated during attention to local features. Again the source of the signals that sustained attention to either the local or the global level seemed to lie in a fronto-parietal network of regions.

Ring and his colleagues (Ring et al 1999) scanned autistic individuals and control subjects while they performed a task in which figures have to be found which were hidden in a background (an embedded figures task). The control task required subjects to a view passively a blank screen. From such a comparison it is difficult to isolate the cognitive processes that are specifically involved in finding hidden figures since the tasks differ in so many respects. Nevertheless interesting differences were observed between the autistic individuals and the controls. The autistic people showed relatively greater activation in lateral extra-striate cortex, while the controls showed relatively greater activation in dorso-lateral prefrontal cortex. This observation would be consistent with the idea that the early stages of sensory processing (emphasizing local features) are intact in autism while the top-down modulation of these early processing stages (which would be required to extract global features) is not functioning properly.

Social cognition. The term social cognition covers a variety of processes only some of which are impaired in autism. In this discussion of functional imaging studies I shall consider just two problems associated with autism; face recognition and mind blindness.

Face recognition. Faces are very important stimuli in social interactions since they provide clues as to who people are and to what they are thinking and feeling. Autistic individuals have problems with many aspects of face perception including face recognition (see Blair et al 2002). In normal volunteers the presentation of faces, in contrast to objects such as houses or furniture, robustly activates a region of the fusiform gyrus in the inferior temporal lobe that has become known as the fusiform face area (FFA) (Puce et al 1996, Kanwisher et al 1997). Reading the emotional expressions on faces activates a variety of areas determined by the nature of the expression. In particular, fearful faces activate the amygdala (e.g. Morris et al 1996).

There have been three studies in which autistic individuals were scanned while looking at faces. (Schultz et al 2000) and (Pierce et al 2001) used face recognitions tasks, while (Critchley et al 2000) studied recognition of emotional expressions in faces (see Table 5). In all three studies the fusiform face area (medial inferior temporal cortex in the fusiform gyrus) was robustly activated in the control subjects, but not in the autistic individuals. In the studies of Pierce and Critchley the autistic individuals also showed reduced activation of the amygdala. In the study by Schultz et al the autistic individuals showed greater activity in an area of

TABLE 5 Functional imaging in autism: activity associated with face perception

Reference	*Numbers*	*Diagnosis*	*Characteristics*	*Matching*	*Measures*	*Tasks*	*Results*
Schultz et al 2000	14 autistic 28 control	ADI ICD-10	Age 24 IQ 109	Age IQ	Regions of interest (BOLD)	Same/different judgements: faces, objects or patterns	Reduced in FFA (37 −54 −10) Increased in ITG (48 −48 −14)
Pierce et al 2001	7 autistic 8 control	ADI DSM-IV	Age 30 (21–41) IQ 84 (73–102)	Age	Regions of interest (BOLD)	Target detection: faces or shapes	Reduced in FFA (30 −59 −13) Normal in ITG
Critchely et al 2000	9 autistic 9 control	ICD-10	Age 37 IQ 102	Age IQ	Voxel-based (BOLD)	Target detection: faces (expression or gender)	Reduced in FFA (23 −58 −13)

inferior temporal cortex when processing faces which was lateral to the fusiform face area. This area was activated by the controls when processing objects. Schultz et al interpret this as evidence that the autistic individuals were using low-level feature based strategies for processing faces.

Mind blindness. This term is synonymous with 'impaired mentalizing' or 'lack of theory of mind'. People with mind blindness have difficulty understanding the behaviour of others in terms of mental states such as beliefs, intentions and desires. Mind blindness has been studied intensively in autism and is considered to be a core deficit since it can explain many of the problems with communication and social interaction. There have been several studies in which normal volunteers have been scanned while solving problems which require thinking about the mental states of others. Many different paradigms have been used varying from reading stories to watching animated films. Three regions have been consistently activated by the requirement to consider the mental states of others (Frith 2001). These are the medial prefrontal cortex, the posterior superior temporal sulcus (STS) and the temporal pole in the vicinity of the amygdala. In most studies this activation is bilateral. The precise function of these different regions is not yet known although STS is also consistently activated by perception of the movement of living things (biological motion).

There are three studies in which autistic individuals were scanned while thinking about the mental states of others. In the study by Happé et al (1996) the volunteers read stories that could only be understood in terms of the mental states of the characters. In the study by Baron-Cohen et al (2000) volunteers had to match the expression of a pair of eyes with a complex mental state term (e.g. suspicious). In the study by Castelli et al (2002) volunteers watched animated cartoons in which two triangles moved around in such a way as to elicit attributions of deceiving, teasing and so on. In all three studies regions of the 'mind reading' network were less active in the autistic individuals. Of particular relevance to the functional imaging studies mentioned in previous sections are the detailed results of the study using animations. In addition to the mind reading network the animations that elicited mentalistic descriptions also activated extra-striate visual areas (probably V3 and the lateral occipital complex; areas concerned with form and object perception), probably because the movements in these animations were more complex. These early processing regions were activated normally in the autistic individuals. In addition correlational measures showed significantly reduced connectivity in the autistic group between the extra-striate regions and STS. It was as if relevant information was being successfully extracted in the early stages of processing, but was failing to get through to the mind reading system.

Speculations

These functional imaging studies reveal a consistent abnormality in the pattern of activity observed in individuals with autism across a wide range of tasks. This consistency does not lie in the observation that some regions are consistently under or over-active. It lies in the observation that early sensory processing areas are activated normally or may even be over-active, while later processing areas are under-active. The location of these areas depends upon the task. In the face recognition studies there was normal activity in early general object processing areas, but reduced activity in the more specific face processing area. In the mind reading studies there was normal activity in early sensory processing areas, but reduced activity in areas that seem to have a more specific role in mind reading such as STS.

This general abnormality of sensory processing would lead to weak central coherence. Local processing would be intact since this depends upon neurons in the early processing stages with small receptive fields. Global processing would be impaired since this depends upon bringing together information from different locations. This can only occur at a later stage of processing in neurons with large receptive fields.

Why does information fail to be transmitted from the early to the later processing areas in the autistic brain? In early accounts, information processing in the normal brain was strictly bottom-up (or feed forward). A visual scene was first analysed in terms of picture elements such as orientation, these were then combined into units of shape such as corners, these in turn were combined into objects, and so on (e.g. Hubel & Wiesel 1962). Recent evidence from anatomy and neurophysiology suggests that this is not how the brain works. There are horizontal and feedback connections between neurons in addition to the feed forward connections, and the feed forward connections are in the minority (Douglas et al 1995). The way that neurons handle incoming information is modified by the horizontal and feedback connections (Lamme & Roelfsema 2000). For example, Sugase et al (1999) recorded from neurons in the inferior temporal cortex of monkeys (including both banks of the STS) looking at faces and geometric figures. There were two distinct phases in the activity elicited by the presentation of a face. In the first phase ($\sim$90 ms) the neural activity only distinguished between faces and geometric figures. In the second phase ($\sim$150 ms) the activity distinguished between faces with different expressions or different identities. The authors concluded that, during the second phase of firing the sensory properties of the neurons had become more finely tuned due to feedback from regions such as the amygdala. Friston & Buchel (2000) studied the changes in functional connectivity that occur when volunteers selectively attend to visual motion. They showed that, during attention, connectivity is enhanced between early visual cortex (V2) and visual motion processing areas (V5). This enhanced connectivity is generated by top-down signals from an attentional system located in parietal and prefrontal cortex.

The functional imaging studies suggest that in autism these top-down signals fail to modulate connectivity so that, for example, when autistic individuals are looking at faces, information about shape is not appropriately routed to specialist face processing areas. This hypothesis could be tested at the behavioural and at the physiological level by using tasks that depend upon feedback control (e.g. cross-modal effects in attention) and measuring changes in functional connectivity.

Plausible computational models of how the brain perceives and categorizes the causes of its sensory inputs use feedback connections in order to predict sensory input (see Friston 2002 for a review). These connections form a generative or forward model of sensory experience that is driven by high-level representations. Neurodevelopmental abnormalities of these connections would (i) render sensory-evoked responses unpredictable, producing large physiological activations in sensory areas (a correlate of prediction error) and (ii) impair the formation of high-level representations and corresponding physiological responses in parietal, temporal and prefrontal association cortices. From a cognitive perspective these effects might lead to (i) perceptual strategies that treat all sensory information as novel and (ii) a lack of central coherence, in which high-level representations that would normally bind together different sensory attributes, would be missing.

These speculations are consistent with the structural brain abnormalities associated with autism if we assume that the abnormally rapid growth during early childhood reflects something going wrong with the reorganization of anatomical connectivity within the brain that occurs at this time. In the visual cortex feed-forward connections are mature a few months after birth, while feedback connections take much longer to develop (Burkhalter 1993). During childhood an early phase of synaptogenesis is followed by a later phase of synapse elimination that continues up to the age of 12 in auditory cortex and even later in prefrontal cortex (Huttenlocher & Dabholkar 1997). During this stage in the monkey there is 'massive' pruning of feed back projections (Batardiere et al 2002). It is assumed that this pruning eliminates faulty connections and optimises the functioning of the feed back control system. Lack of pruning in autism might therefore lead to an increase in brain size and be associated with poor functioning of the feedback control system.

References

Abell F, Krams M, Ashburner J et al 1999 The neuroanatomy of autism: a voxel-based whole brain analysis of structural scans. Neuroreport 10:1647–1651

Ashburner J, Friston KJ 2000 Voxel-based morphometry—the methods. Neuroimage 11:805–821

Aylward EH, Minshew NJ, Goldstein G et al 1999 MRI volumes of amygdala and hippocampus in non-mentally retarded autistic adolescents and adults. Neurology 53:2145–2150
Baddeley AD 1986 Working memory. Clarendon Press, Oxford
Bailey A, Luthert P, Dean A et al 1998 A clinicopathological study of autism. Brain 121:889–905
Baron-Cohen S, Ring HA, Bullmore ET, Wheelwright S, Ashwin C, Williams SCR 2000 The amygdala theory of autism. Neurosci Biobehav Rev 24:355–364
Batardiere A, Barone P, Knoblauch K et al 2002 Early specification of the hierarchical organization of visual cortical areas in the macaque monkey. Cereb Cortex 12:453–465
Blair RJR, Frith U, Smith N, Abell F, Cipolotti L 2002 Fractionation of visual memory: agency detection and its impairment in autism. Neuropsychologia 40:108–118
Boddaert N, Zilbovicius M 2002 Functional neuroimaging and childhood autism. Pediatr Radiol 32:1–7
Burkhalter A 1993 Development of forward and feedback connections between areas V1 and V2 of human visual-cortex. Cereb Cortex 3:476–487
Castelli F, Frith C, Happé F, Frith U 2002 Autism and brain mechanisms for the attribution of mental states to animated shapes. Brain 125:1–11
Cole G, Neal JW, Fraser WI, Cowie VA 1994 Autopsy findings in patients with mental handicap. J Intellect Disabil Res 38:9–26
Courchesne E 1999 An MRI study of autism: the cerebellum revisited. Neurology 52:1106–1107
Courchesne E, Yeung-Courchesne R, Press GA, Hesselink JR, Jernigan TL 1988 Hypoplasia of cerebellar vermal lobule-VI and lobule-VII in autism. New Engl J Med 318:1349–1354
Courchesne E, Saitoh O, Townsend JP et al 1994 Cerebellar hypoplasia and hyperplasia in infantile-autism. Lancet 343:63–64
Courchesne E, Karns CM, Davis HR et al 2001 Unusual brain growth patterns in early life in patients with autistic disorder: an MRI study. Neurology 57:245–254
Critchley HD, Daly EM, Bullmore ET al 2000 The functional neuroanatomy of social behaviour: changes in cerebral blood flow when people with autistic disorder process facial expressions. Brain 123:2203–2212
Douglas RJ, Koch C, Mahowald M, Martin KAC, Suarez HH 1995 Recurrent excitation in neocortical circuits. Science 269:981–985
Duncan J, Owen AM 2000 Common regions of the human frontal lobe recruited by diverse cognitive demands. Trends Neurosci 23:475–483
Fink GR, Halligan PW, Marshall JC, Frith CD, Frackowiak RSJ, Dolan RJ 1997 Neural mechanisms involved in the processing of global and local aspects of hierarchically organized visual stimuli. Brain 120:1779–1791
Friston KJ 2002 Beyond phrenology: what can neuroimaging tell us about distributed circuitry. Annu Rev Neurosci 25:221–250
Friston KJ, Buchel C 2000 Attentional modulation of effective connectivity from V2 to V5/MT in humans. Proc Natl Acad Sci USA 97:7591–7596
Frith U 2001 Mind blindness and the brain in autism. Neuron 32:969–979
Happé F 1999 Autism: cognitive deficit or cognitive style? Trends Cogn Sci 3:216–222
Happé F, Ehlers S, Fletcher P et al 1996 'Theory of mind' in the brain. Evidence from a PET scan study of Asperger syndrome. Neuroreport 8:197–201
Hardan AY, Minshew NJ, Harenski K, Keshavan MS 2001a Posterior fossa magnetic resonance imaging in autism. J Am Acad Child Adolesc Psychiatry 40:666–672
Hardan AY, Minshew NJ, Mallikarjuhn M, Keshavan MS 2001b Brain volume in autism. J Child Neurol 16:421–424
Haznedar MM, Buchsbaum MS, Wei TC et al 2000 Limbic circuitry in patients with autism spectrum disorders studied with positron emission tomography and magnetic resonance imaging. Am J Psychiatry 157:1994–2001

Hetzler BE, Griffin JL 1981 Infantile autism and the temporal lobe of the brain. J Autism Dev Disord 11:317–330
Howard MA, Cowell PE, Boucher J et al 2000 Convergent neuroanatomical and behavioural evidence of an amygdala hypothesis of autism. Neuroreport 11:2931–2935
Hubel DH, Wiesel TN 1962 Receptive fields, binocular interaction and functional architecture in the cat's visual cortex. J Physiol 160:106–154
Huttenlocher PR 1999 Dendritic synaptic development in human cerebral cortex: time course and critical periods. Dev Neuropsychol 16:347–349
Huttenlocher PR, Dabholkar AS 1997 Regional differences in synaptogenesis in human cerebral cortex. J Comp Neurol 387:167–178
Kanwisher N, McDermott J, Chun MM 1997 The fusiform face area: a module in human extrastriate cortex specialized for face perception. J Neurosci 17:4302–4311
Kastner S, Ungerleider LG 2000 Mechanisms of visual attention in the human cortex. Annu Rev Neurosci 23:315–341
Lainhart JE, Piven J, Wzorek M et al 1997 Macrocephaly in children and adults with autism. J Am Acad Child Adolesc Psychiatry 36:282–290
Lamme VAF, Roelfsema PR 2000 The distinct modes of vision offered by feedforward and recurrent processing. Trends Neurosci 23:571–579
LeDoux JE 2000 Emotion circuits in the brain. Annu Rev Neurosci 23:155–184
Levitt JG, Blanton R, Capetillo-Cunliffe L, Guthrie D, Toga A, McCracken JT 1999 Cerebellar vermis lobules VIII-X in autism. Prog Neuropsychopharmacol Biol Psychiatry 23:625–633
Miles JH, Hadden LL, Takahashi TN, Hillman RE 2000 Head circumference is an independent clinical finding associated with autism. Am J Med Genet 95:339–350
Morris JS, Frith CD, Perrett DI et al 1996 A differential neural response in the human amygdala to fearful and happy facial expressions. Nature 383:812–815
Ohnishi T, Matsuda H, Hashimoto T et al 2000 Abnormal regional cerebral blood flow in childhood autism. Brain 123:1838–1844
Pierce K, Muller RA, Ambrose J, Allen G, Courchesne E 2001 Face processing occurs outside the fusiform 'face area' in autism: evidence from functional MRI. Brain 124:2059–2073
Piven J, Arndt S, Bailey J, Andreasen N 1996 Regional brain enlargement in autism: a magnetic resonance imaging study. J Am Acad Child Adolesc Psychiatry 35:530–536
Piven J, Saliba K, Bailey J, Arndt S 1997 An MRI study of autism: the cerebellum revisited. Neurology 49:546–551
Piven J, Saliba K, Bailey J, Arndt S 1999 An MRI study of autism: the cerebellum revisited — reply. Neurology 52:1106–1107
Puce A, Allison T, Asgari M, Gore JC, McCarthy G 1996 Differential sensitivity of human visual cortex to faces, letterstrings, and textures: a functional magnetic resonance imaging study. J Neurosci 16:5205–5215
Ring HA, Baron-Cohen S, Wheelwright S et al 1999 Cerebral correlates of preserved cognitive skills in autism — a functional MRI study of embedded figures task performance. Brain 122:1305–1315
Salmond CH, Ashburner J, Friston KJ, Gadian DG, Vargha-Khadem F 2003 Convergent evidence for the neural basis of autism. Phil Trans Roy Soc Lond B 358:405–413
Schultz RT, Gauthier I, Klin A et al 2000 Abnormal ventral temporal cortical activity during face discrimination among individuals with autism and Asperger syndrome. Arch Gen Psychiatry 57:331–340
Shallice T 1988 From neuropsychology to mental structure. Cambridge University Press, Cambridge
Sugase Y, Yamane S, Ueno S, Kawano K 1999 Global and fine information coded by single neurons in the temporal visual cortex. Nature 400:869–873

Wojciulik E, Kanwisher N, Driver J 1998 Covert visual attention modulates face-specific activity in the human fusiform gyrus: fMRI study. J Neurophysiol 79:1574–1578
Zilbovicius M, Boddaert N, Belin P et al 2000 Temporal lobe dysfunction in childhood autism: A PET study. Am J Psychiatry 157:1988–1993

DISCUSSION

Bishop: You made a strong case that one should control for IQ in studies of structural imaging. But this is not as easy as it sounds with low IQ: there are many different conditions that can result in low IQ, most of which might have rather specific effects on the brain. Are you suggesting that we should only look at children with IQ in the normal range? Are you talking about verbal IQ or performance IQ?

C. Frith: This problem has come up before. The concept of IQ is difficult, particularly in relation to autism. But there are a number of studies already in the literature that involved people with low IQs

Bishop: You wouldn't want people with Down syndrome or Williams syndrome. They have very distinctive patterns of brain pathology (Jernigan & Bellugi 1994).

C. Frith: In that case you could control by having people with Down syndrome with and without autism.

Bishop: I think it's a real logical problem. There are now findings with people with specific language impairment that suggest specific cerebellar abnormalities. There are an awful lot of things we could be controlling for. Finding the appropriate control group in cases like this is a problem.

Amaral: We are in the process of looking at the amygdala using structural MRI. We have done about 82 subjects, both in low IQ, high IQ, Asperger and control. The age range is 8–18, and we haven't seen a significant difference over this age range. As we did this, we realized that the protocols vary from one study to the other. Most of the protocols outline the entorhinal cortex as well as the amygdala. It may or may not be a confound, but since Margaret Bauman has found pathological changes in the entorhinal cortex it is worthwhile controlling for this. We have our protocol for outlining the amygdala on our ftp site (*ftp://brainshop.ucdavis.edu*), and we would love people to use this. Permission to obtain these materials is provided by Dr Jane Pickett, Director of the Autism Tissue Program (*ATP@brainbank.org*). This consists of a text description plus a series of MRI sections where we show what we do.

C. Frith: This is one of the problems with the regions of interest approach: you will get different definitions of what the region of interest is. This is overcome by VBM, where you don't need any preconceptions at all (although the method does have other problems). I am sure that there will be many more studies in the future using VBM.

Schultz: We have measured more than 40 amygdalas by MRI, mostly in high functioning individuals, and contrasted these with a similar sized group of healthy controls. Like David Amaral, we are not seeing any differences in volume.

C. Frith: Was there an effect of low IQ on the size of the amygdala?

Amaral: There is a tendency for autistic subjects with low IQ to have a smaller amygdala, although it hasn't reached significance in our sample.

Charman: Is this in the context of a small brain?

Amaral: This is controlled for overall brain size.

Schultz: We see some correlation between IQ and brain size in healthy controls, but less so among persons with autism. Among controls, our data and those in the extant literature suggest a brain size correlation with general intelligence of about $r=0.40$. Here, extra mass provides a processing advantage. However, in autism, the relationship seems much less strong, such that increases in brain size don't confer the same cognitive advantage. This could be because of bottlenecks in information processing ability. In other words, extra neurons usually can be used to process information more efficiently or more deeply. However, in autism, organizational differences in their brains may disallow this general principal. There may be bottlenecks at critical junctures of the flow of information processing from raw perceptual material to organized, abstracted output that are of primary importance in determining the whole system's processing capacity.

I wanted to ask Chris Frith about the mentalizing task. You see hypo-activation in several nodes. We are using a similar test and find comparable results. One of the key questions for us as we do this is how do we assure ourselves that the patients are really working on the same material; that they are really processing this as deeply and are attending as much as the comparison group of healthy controls? We would like some sort of task to prove that they are attending.

C. Frith: We didn't have an online test of attention, but after each scan the volunteers had to tell the experimenter what they thought was going on. To this extent they were required to attend.

Schultz: We have a similar phenomenon even with the fusiform face area. Changes in one's level of attention can change how much the fusiform reacts. We think that this issue of attention or depth of processing is emerging as key in understanding how hypo-activation to task contrasts among persons with autism.

C. Frith: It would be interesting to do a study in which you directly manipulated attention. The face could be irrelevant in a comparison task and you could see whether there was modulation of activity in the face area by attention.

Hollis: I picked up the difference described in the primary visual and mentalizing tasks. You suggested that the differences in the activation of visual areas might be due to a 'bottom–up' process driven by the complexity of the visual stimulus. Is this actually the case, or could it be related to some 'top-down' regulation that generates more interest and attention to the stimulus and hence greater activation of visual cortical areas?

C. Frith: It could be either way. What you would need to do is have identical movements which are interpreted in different ways.

Hollis: Could you do this by giving a prompt?

C. Frith: We did try prompting them in the first experiment, but this didn't have any detectable effect.

Happé: Our guess is that the processes are obligatory: you can't switch off your mentalizing. Asking a subject not to attend to that aspect wouldn't work.

Schultz: In our version of this task we are not finding any visual cortical activation. It may be that the complexity of the movements is matched.

Rogers: I want to underlie the importance of Dorothy Bishop's point. We have data gathered on several groups of two-year olds. They are all matched on language level and age. When we do this kind of matching we find that the children in all of the developmental delay groups have more delayed motor milestones, they have more CNS dysfunction, motor impairment and many more medical problems than the children with autism. When you match on language you end up feeling that your group with mental retardation is much more impaired than the group with autism.

Charman: In some ways, it is surprising that in the results of functional imaging of mentalizing tasks, given how different the tasks are, the same regions are coming up. As an outsider, to me this is very striking.

Schultz: In an fMRI study that we have just finished, where we are studying the impact of person familiarity on a task involving pictures of faces, we find activation in the precuneus region, what some have called the 'retro splenial' region. This would encompass portions of the posterior cingulate gyrus, immediately posterior to the splenium of the corpus callosum.

Sigman: What about the precuneus region?

Schultz: We get activation of the precuneus region for familiarity effects.

Sigman: We did a study of children with autism and found the same differences in activation as Bob found with adults, but we also found differences in activation in the precuneus region.

Schultz: This also has been shown in a study by Shah et al (2001) published in *Brain* last year. They found that the retrosplenial cortex was engaged by both familiar faces and by familiar voices. This is interesting and it suggests that this area is modality independent, and perhaps part of a higher order network that could be important in understanding autism.

C. Frith: The precuneus was certainly active in some of our studies, but not all.

Dawson: I wanted to discuss your top-down hypothesis, and think about face processing from the developmental point of view. One of the things that we have argued is that a child with autism isn't looking at faces naturally. To explain this, it is helpful to think about how face processing and attention develop in the first year of life. In the first six months, attention is drawn by stimulus features: novelty and unpredictability are key. Then there is a shift in the development of face processing, and infants' attention is not just influenced by the particular stimulus features, such as configuration, but rather their attention is more intentional and directed by representation. This kind of attention is frontally mediated in terms of the brain. We think this is where things go awry in autism. It is in the second half of the first year, for the child with autism, attention isn't getting directed by representations of anticipating social reward. In normal infants, one of the things that directs attention to faces is the anticipation of social reward or other important information that is associated with the face. Their attention is directed with that representation in mind. The prediction from this is that you wouldn't really be able to develop an infant screening test in the first six months on the basis of some sort of perceptual input abnormality. In one infant, where we were able to get detailed records of the infant's behaviour from birth to three years of life. We conducted a comprehensive evaluation of him at 13 months. The physician recorded the infants' behaviour during the first year in much detail. For the first six months of life the infant was very interested in faces and very socially responsive. It wasn't until around 7–8 months of life that they started to see the abnormalities arise in social interaction. This all fits with the idea that these higher cortical processes are disrupted in autism. This is important for thinking about the impact of intervention.

Bailey: Would you accept that there are some parents who will give a report of the child being disinterested in faces very early in life?

Dawson: My idea is speculation at this point. We could discover that there is something very early on. This is a model of what might be going on that would fit with Chris Frith's model. My prediction is that is going to be very hard to find a problem in the first six months, and easier when higher cortical systems coming online.

Bailey: What is going through my mind is that there is quite a lot of variability when symptoms first become apparent, or indeed when children lose some skills. The evidence from genetic studies is that in families where there are two affected individuals one can see a remarkable discordance in age of onset. Yet they presumably have the same underlying disorder, so there must be a lot of variability from individual to individual.

Dawson: Regarding early-onset cases, we do have a home video tape study of 8–10 month old infants, in which we can distinguish infants with autism from typically developing infants. We know that some cases autism can be detected at 8–10 months. Earlier than this it is questionable.

Folstein: Chris Frith, I was interested in the Burkhalter (1993) paper that you quoted. What methods could we use to test that hypothesis in autism? Can it be done with an imaging method?

C. Frith: I'm not sure. The BOLD signal and bloodflow measure slightly different things, which have to do with synaptic activity. One very far-fetched idea is that if you have an abnormal number of synapses, will you see an abnormal relationship between the bold signal which you get in MRI and the bloodflow signal you get in PET. This is the sort of way that I thought you might want to get at this. You need to talk to someone who knows about what these signals really mean.

Bailey: One could use electrophysiological techniques. I am not aware of studies finding significant differences in the size of evoked potentials, so there is not evidence that more nerve cells are firing or more synapses are being recruited. With reference to the discussion we had earlier about myelin abnormalities, grossly most studies find that the timing of evoked potentials in autism is pretty similar to that in normals.

C. Frith: The other speculation that I had is that if there was all this proliferation and it wasn't properly pruned, would this make you more at risk for epilepsy because there would be too many interconnections?

Bailey: My speculation is that at least some of the epilepsy is due to cortical dysgenesis. We have a number of post-mortem cases who had epilepsy in whom we can see areas of cortical dysgenesis (Bailey et al 1998). One of the interesting things is that cortical dysgenesis in individuals without autism is a significant cause of late-onset epilepsy, and there has never been an adequate account about why there is a bimodal peak in the age of onset of epilepsy in autism.

Bishop: Mike Merzenich has claimed that when he does his training experiments on monkeys, in which he is getting them to change the representation on the brain, training a monkey to do these very difficult tasks incessantly can generate epilepsy. He has argued that this may be a model of what happens in autism. I haven't seen the evidence for this, but it is an interesting alternative to think that we should consider not just what the autistic brain brings to the world, but what experience then subsequently does.

Dawson: The question of how experience or intervention can affect brain development is an important one. We would all expect that it could affect cortical maps, organization of the brain and connectivity and circuitry. With respect to whether intervention would affect the next level of abnormalities in cells and brain size, it totally depends on what mechanism is underlying these abnormalities in the first place. If they are experientially caused, then maybe we will find intervention effects. I suspect that that the anatomical differences of brain size and cellular abnormalities detected on autopsy are genetically driven. I don't know how much effect we will be able to have on them through experience or intervention.

Buitelaar: In Utrecht, Herman van Engeland, Chantal Kemner and others have found on a number of occasions that in odd-ball paradigms children with autism have abnormal ERPs, particularly when using auditory stimuli. This suggests a disturbance in information processing.

Bailey: The point I was making was not that there are not abnormalities. It was that if you examine evoked potentials, their timing is not obviously different in people with autism. It does not look as though the fundamental problem is that activity is slow.

Buitelaar: It is not slow. It is differences in cortical activity.

Bailey: I'm not disagreeing, but that is a different question—whether or not there is a basic problem in myelin conductivity.

Schultz: I wanted to go back to the issue of what these fMRI activations mean, and what they have to do with experience. For the fusiform face area hypo-activation, our interpretation is that it is not representing some fundamental pathophysiology of autism, but it is actually a reflection of autism: the person with autism has less experience with people, so this area is wired and built differently. Even for these hypo-activations that we see, I think of them as extended phenomenology. I don't think of them in terms of mechanism and understanding the causes of autism, because they are so correlational. One of the things we are trying to do now is to train people with autism to become better at face recognition, to see whether we can change the function of the fusiform face area and whether this predicts any change in their behaviour. We need to move to these sorts of experimental designs and away from pure correlations between activations and diagnostic groups.

Charman: How do you train them to improve face recognition?

Schultz: We have created a computerized platform of games where the faces are all the stimuli in the games. In some games they have to throw water balloons at all the faces with a certain emotional expression and they get points for this. We do fMRI before and after this.

Bailey: We have wondered the same thing, as I suppose everyone who is doing face-processing studies has. Is this the chicken or the egg? I am struck that Isabelle Gautier can take someone who is grebal naïve and who doesn't activate fusiform gyrus, and two weeks later when they are grebal experts they do activate the fusiform gyrus. Grebals are plasticene-like objects with groups of features. These individuals change from not activating the fusiform gyrus to activating it with just two weeks' training. It seems implausible to me, given the amount of time that older people with autism do look at faces, that this failure of activation of fusiform is simply due to a lack of experience.

Schultz: Yes, she takes university students and trains them in 10 h to start activating the fusiform.

Bishop: Do the grebals do anything interesting to capture the participants' attention?

Schultz: No. She showed the same thing with bird watchers, for whom birds activate the fusiform face area. She considers this as to be an 'expertise' area, rather than just a face area. The interesting thing is that anyone who is an expert at grebals or birds or dogs becomes passionate about it. They can't actually become expert at grebals until they become interested in their differences and speculate about what it means.

Bishop: Do you get the same effect with completely inanimate objects, such as cars?

Schultz: Yes.

Bailey: We saw a child recently whose father is passionate about motorbikes, and has been taught about different makes of motorbikes from the age of five. He has a clear, specific motorbike-evoked response that we have seen with some adults.

Sigman: Bob, you told me you had an interesting finding with autistic people who are experts.

Schultz: Yes, we have a case study of a boy who is a digimon expert. He is passionate about digimon and he is very good at discriminating. He doesn't have a fusiform face activation to faces, but he does to digimon.

Bishop: How do you know that fusiform face area is not just activating to something you are interested in?

Schultz: We don't. I don't think you can become visually expert unless you are interested. I think it takes that kind of emotional input, and this is part of the process.

Rutter: Can I clarify? As I understand it, you are both making the same point, Tony and Bob. An experimental approach does help to sort out cause and effect. The fact that you can change the effect in a very dramatic way provides a leverage for determining whether the findings in autistic individuals can be similarly altered, and, if so, for what, in what circumstances.

Bailey: Presumably the acid test is whether autistic subjects can be trained to the same performance as other people as some index of whether they have been motivated to learn. We have some evidence (Wallace, unpublished PhD thesis, University of London, 2002) that even when affected individuals perform relatively well on face-processing tasks, they are still using different strategies. They might still achieve good performance, and one might still find they are activating another part of the brain.

Dawson: That was the point I wanted to make. As autistic people get older they start looking at faces: but what this means depends on how they are processing that information. They could be looking at the hair, or they could be analysing individual features. We have had anecdotal evidence that some people with autism eventually become interested in looking at faces and facial expressions: they are obviously looking at faces but the strategies and ways that they process the information are still atypical.

Rutter: Dorothy Bishop, can we come back to the point that you were raising about the difficulties in matching groups for cognitive level. The need for matching was highlighted in the 1960s when in a range of studies comparing children with autism and normal controls (mainly undertaken in Southern California) it became clear that the deficits that were supposedly specific to autism were actually a function of low mental age (Yule 1978). The basic point was that supposed pattern differences may actually be a consequence of generally low intellectual functioning. The problem that you are raising seems to be a different one, raising two further issues. The first is that the early studies using matching tended to make the naïve assumption that outside autism, all individuals with mental retardation were fairly similar. It has since become quite clear that that is not so. Accordingly, investigators have to ask whether any differences found between groups of children with autism and groups with non-autistic mental retardation reflect the peculiarities of autism or of particular mental retardation sample used. The basic logic was well outlined in a paper by Gillies (1965). Your second point, based on language findings focuses on the need to go on to identify the mediating mechanism. If it has been shown that a pattern difference is not simply an artefact of level, and that it appears to be specific to autism, there is then the further question of the cognitive process that is mediating the difference between the groups with and without autism? Amongst other things this means asking whether a particular skill (or deficit) is a secondary consequence of language level, or whether it is independent of language. That question would seem to require a matching approach, but it argues for multiple matching.

Bishop: I'm not sure I agree. It depends on the level of description. Certainly, if you are explaining some cognitive deficit in terms of an underlying cognitive process, then you are right. But if you are talking about neurobiology, the concept of matching for IQ seems to be a strange one, because there are so many different ways that someone can achieve a low IQ. One approach would be multiple matching of lots of different conditions with low IQ, but I wonder how meaningful that is. As a first step, a more sensible approach might be to look at a range of IQs within your autistic group and see whether any of the neurobiological variables correlate with IQ. If they don't, we needn't worry about matching. It might be sensible to deliberately go for a wide range and measure not just IQ, but a range of other things that might be driving any correlation that exists. This could include language and measures of attention.

Sigman: My research group became interested in this issue because of Eric Courchesne's findings of differences in the cerebellum between individuals with autism and normal individuals. Someone in my research group asked him whether there was any relationship between IQ and the size of the cerebellum, and he said there was a robust correlation. This raises the question of whether the differences in the cerebellum are specific to autism or are generally found with developmental disabilities. A matched control group is necessary to determine specificity.

Bishop: What I would then say is that perhaps we should focus solely on high-functioning children with autism, so we know where we are. Then we could be fairly confident that we have identified the correlate of the autism, rather than the correlate of the IQ. I am very uneasy about these low IQ control groups, which are typically a mish mash.

Dawson: The issue of control group is very complex. I don't like the solution of just focusing on the high-functioning children. 75% of autistic individuals function in the mentally retarded range. The variability is interesting; it is something we have to understand. If it is the case that lower-functioning individuals have more hippocampal involvement, that is of interest. We will get a lot of genetic information out of studying the variability. I am reluctant to neglect 75% of the people who have the syndrome.

Bishop: Who would you use as their control?

Dawson: The controls are difficult. I do like the idea of multiple matching. It really depends on the hypothesis that you are going to address.

Bishop: Suppose your hypothesis is just a structural one. For example, that cerebellar size is abnormal in autism. What would you do then?

Dawson: If the hypothesis concerns whether this brain abnormality is a general phenomenon of mental retardation versus specific to the autism syndrome, then I would want two or three different mental retardation syndromes. If whenever there is mental retardation present, I always saw the brain abnormality, I would start sorting it out. For example, we haven't talked about the fact that large brains occur in other syndromes. We need to think about this. There is no perfect control group that will give the final answer. It is converging evidence with multiple control groups.

Rogers: Along these lines there is an interesting paper by Miles & Hillman (2000) which looked at the different gender ratios in different IQ groups. The conclusion was that different gender ratios suggested different biology and genetics to high- and low-functioning autism. One doesn't necessarily represent the other.

Lord: Another difficulty in studying high functioning children is that if many of the participants with autism are relatively 'high functioning' but actually in the borderline range of IQ, it is difficult to find controls, and non-autistic individuals with borderline IQs who are identified by either health or educational systems have usually been so because of other problems. Some researchers have included only autistic groups with normal IQs, but this is a small proportion of the autistic population. Individuals with IQs of 80–90 may be ignored and this is a pitfall.

Rutter: We must not forget the dramatic progress made back in the 1960s by Beate Hermelin, Neil O'Connor and Uta Frith, showing that it is possible to use experimental approaches with very handicapped autistic individuals, and still obtain interesting answers (Hermelin & O'Connor 1970). Their research provided the essential basis for modern cognitive studies. There is a lot to be gained by studying individuals with autism and an IQ in the normal range, but it would be a mistake to assume that severely handicapped individuals cannot be investigated fruitfully.

Bishop: I wouldn't disagree. My proposal was in response to the idea that IQ could be a major factor driving the results. Then I would say that it might be sensible to shore up that result by focusing on the normal range. But my general strategy would be to first look at the correlates of IQ, and not just assume that IQ is such an important variable in every context. Even on the behavioural side, the tendency to use children with Down syndrome as a control group is a principled decision that people often don't realise they are making. As we do more research on Down syndrome it is quite clear that language is disproportionately impaired compared with non-verbal abilities (e.g. Miller 1988). We need to consider the aetiology of our control groups.

Dawson: People recognize this.

Charman: The practicality of going out and recruiting three other groups as controls with IQs under 70 is daunting. You'd need three times the money.

U. Frith: The question of the most appropriate control group produces continuing debate. It is worth pointing out that before the 1960s autism seemed to be part of the huge undifferentiated concept of mental retardation. Hardly anyone thought of looking for specific cognitive deficits. When we were studying autism in the '60s, we were hoping even then that we would take out a subgroup from this great mass that was basically thought to suffer from generally low cognitive abilities. To a large extent we succeeded in defining specific cognitive difficulties in autism, and more recently this has been done with some other syndromes. It is heartening when we can see disproportionate impairments over and above generally low cognitive abilities in anything at all. This is a triumph of the methodologies that have been used, in particular the mental age match design. This is of course quite crude when an identified subgroup is compared to a mixed and largely unidentified group. However, the solution is not rigid matching with ever more refined subgroups. There have to be other approaches. Correlational methods are a possibility. For instance, David Amaral mentioned that the amygdala is disproportionally larger in some cases of autism than others. What might be the behavioural correlates? From a historical point of view we are still only at the beginning in finding out the basis of mental retardation. To make more progress we need to understand the neurological basis of intelligence. If we had such a basis, then it would become much easier to study the specific impairments over and above general impairments.

Rutter: Dorothy Bishop, you were making the point about being able to rule this out if you failed to find a correlation within your autistic group. But this needs to be done in both samples. There are examples where a correlation exists within one group but not the other. This in itself becomes interesting in raising questions about the process.

Happé: We have already had an example of correlation between amygdala and whole brain size, and the fact that it holds in one group and not the other is immediately a clue.

Folstein: To come back to the heterogeneity issue, we know that autism is genetically and phenotypically heterogeneous. There is a lot to be learned from within-group comparisons. Rather than fussing over which control group to use, it might be more useful to consider control groups within the autism sample. For example, you might compare cases with a lot of compulsions and those without.

References

Bailey A, Luthert P, Dean A et al 1998 Clinicopathological study of autism. Brain 121:889–905

Burkhalter A 1993 Development of forward and feedback connections between areas V1 and V2 of human visual-cortex. Cereb Cortex 3:476–487

Gillies S 1965 Some abilities of psychotic children and subnormal controls. J Mental Deficiency Res 9:89–101

Hermelin B, O'Connor N 1970 Psychological experiments with autistic children. Pergamon Press, Oxford & New York

Jernigan T, Bellugi U 1994 Neuroanatomical distinctions between Williams and Down syndromes. In: Broman SH, Grafman J (eds) Atypical cognitive deficits in developmental disorders. Lawrence Erlbaum Associates, Hillsdale, NJ, p 57–66

Miles JH, Hillman RE 2000 Value of a clinical morphology examination in autism. Am J Med Genet 91:245–253

Miller JF 1988 The developmental asynchrony of language development in children with Down syndrome. In: Nadel L (ed) The psychobiology of Down syndrome. MIT Press, Cambridge, MA, p 167–198

Shah NJ, Marshall JC, Zafiris O et al 2001 The neural correlates of person familiarity. A functional magnetic resonance imaging study with clinical implications. Brain 124:804–815

Yule W 1978 Research methodology: what are the 'correct controls'? In: Rutter M, Schopler E (eds) Autism: a reappraisal of concepts and treatment. Plenum Press, New York, p 155–162

The amygdala, autism and anxiety

David G. Amaral*† and Blythe A. Corbett†

**Department of Psychiatry, Center for Neuroscience and California National Primate Research Center and †The MIND (Medical Investigation of Neurodevelopmental Disorders) Institute, University of California Davis, Davis, CA 95616, USA*

Abstract. Brothers has proposed that the amygdala is an important component of the neural network that underlies social cognition. And Bauman and Kemper observed signs of neuropathology in the amygdala of the post-mortem autistic brain. These findings, in addition to recent functional neuroimaging data, have led Baron-Cohen and colleagues to propose that dysfunction of the amygdala may be responsible, in part, for the impairment of social functioning that is a hallmark feature of autism. Recent data from studies in our laboratory on the effects of amygdala lesions in the macaque monkey are at variance with a fundamental role for the amygdala in social behaviour. If the amygdala is not essential for normal social behaviour, as seems to be the case in both non-human primates and selected patients with bilateral amygdala damage, then it is unlikely to be the substrate for the abnormal social behaviour of autism. However, damage to the amygdala does have an effect on a monkey's response to normally fear-inducing stimuli, such as snakes, and removes a natural reluctance to engage novel conspecifics in social interactions. These findings lead to the conclusion that an important role for the amygdala is in the detection of threats and mobilizing an appropriate behavioural response, part of which is fear. If the amygdala is pathological in subjects with autism, it may contribute to their abnormal fears and increased anxiety rather than their abnormal social behaviour.

2003 Autism: neural basis and treatment possibilities. Wiley, Chichester (Novartis Foundation Symposium 251) p 177–197

In the best of biomedical research endeavours, there is a natural symbiosis between basic research (and often basic animal research), and careful assessment of clinical populations. The issues raised in this paper draw from efforts to establish the neurobiological basis of primate social behaviour, on the one hand, and attempts to determine brain systems that are impacted in autism and lead to impairments of social behaviour, on the other. One effort has enormous potential to inform the other. If, for example, a neural system—let's call it the Social System—is established that underlies the various components of social interaction, and given that impairments of social interaction are a major deficit in autism spectrum disorders, then a reasonable hypothesis might be that a region of primary brain pathology might be in the Social System[1]. Conversely, if specific and reproducible areas of brain pathology were identified in autism spectrum disorder, this information might provide a useful heuristic as to which brain regions might be components of the Social System.

Life, of course, is rarely so simple and autism is certainly one of the most complex of neurological disorders. It is complex because it has many diverse symptoms, including social impairment, language problems and motor stereotypies. These symptoms are observed heterogeneously throughout the population that makes up the autism spectrum. There are also a number of co-morbid conditions, such as sleep disturbances, gastrointestinal distress and psychiatric symptoms including anxiety and obsessive–compulsive behaviour. The following is a short summary of the thought and experimental process that we have followed — starting with the notion that the amygdala is a fundamental component of the Social System and likely to be heavily involved in the pathophysiology of autism — to our current view that the amygdala is involved in detecting and reacting to environmental threats. And, if the amygdala is impaired in autism, it may be more responsible for alterations in fear and anxiety rather than social behaviour.

The amygdala

The primate amygdala is a relatively small brain region located in the temporal lobe, just anterior to the hippocampus. In the macaque monkey it is approximately 0.6 cm^3 in volume and in the human it is about 3.0 cm^3. The amygdala is comprised of at least 13 nuclei and cortical regions, many of which are partitioned into two or more subdivisions. The amygdala has widespread extrinsic connections including those with the neocortex, hippocampal formation, cholinergic basal forebrain, striatum, hypothalamus and brainstem. While neocortical inputs to the amygdala arise mainly from higher-order unimodal and polymodal association cortices, projections back to the neocortex extend monosynaptically even to primary sensory areas such as visual area V1. There is an extensive network of intrinsic connections within the amygdala that generally brings information from more laterally situated nuclei, such as the lateral nucleus, to more medially situated nuclei, such as the central nucleus. The amygdala contains a plethora of neuroactive substances and has some of the highest brain levels of benzodiazepine receptors and opiates. Detailed descriptions of the neuroanatomy of the amygdala can be found in Amaral et al (1992). One can

[1]Of course, this is not necessarily the case. It could well be that autism is due to brain dysfunction(s) at a much more fundamental level of sensory or motor processing. And this dysfunction only manifests itself in complex situations such as social encounters.

conclude from the neuroanatomy of the amygdala that it is privy to much of the sensory processing that occurs in the neocortex and, that through its widespread efferent connections, it has the ability to influence the activity of numerous functional systems that range from elemental physiological processes such as heart rate and respiration to the highest processes of perception, attention and memory.

The amygdala and social behaviour

Several lines of evidence have indicated that the amygdala plays an important role in socioemotional behaviour. Macaque monkeys with bilateral lesions that include the amygdala are typically more tame than normal animals, demonstrate abnormal food preferences and have alterations of sexual behaviour (Brown & Schafer 1888, Kluver & Bucy 1938, 1939). Rosvold et al (1954) designed studies explicitly to evaluate changes in social behaviour in macaque monkeys following amygdala damage. They established artificial social groups of male rhesus monkeys and studied the dominance hierarchy that emerged. They then carried out two-stage bilateral destructive lesions of the amygdala of the most dominant animal and studied the dominance hierarchy as the group reorganized. They found that the lesions led to a decrease in social dominance with the lesioned animal typically falling to the most subordinate position of the group.

A more extensive program of studies was carried out by Kling and colleagues using both captive and free-ranging non-human primates (Kling et al 1970, Kling & Cornell 1971, Kling & Steklis 1976). Dicks et al (1968), for example, retrieved rhesus monkeys from social troops on the island of Cayo Santiago. These animals were subjected to bilateral amygdalectomy and then returned to their social groups. While it was difficult to follow the minute-to-minute interactions of the lesioned animals, the typical finding was that they were invariably ostracized and would often perish without the support of the social group.

From the results of these and similar studies carried out by several laboratories, Brothers (1990) formalized the view that the amygdala is one of a small group of brain regions that form the neural substrate for social cognition. This view predicts that the amygdala is essential for certain aspects of the interpretation and production of normal social gestures such as facial expressions and body postures. It also predicts that damage to the amygdala would invariably lead to a decrease in the amount or quality of conspecific social interactions.

The amygdala and autism

In their seminal studies on the neuropathology of the autistic brain, Bauman & Kemper (1985) noted that the medially situated nuclei of the amygdaloid complex had clusters of small, tightly packed neurons that were not observed in control brains. The amygdala neuropathology was only one area among many that included alterations in the hippocampus, septum, cerebellum and other structures. Unfortunately, these observations have not yet been independently replicated. Neuroimaging studies have thus far produced conflicting results on whether there is a gross change in the volume of the amygdala. Abell et al (1999) reported an increased left amygdala volume in cases of autism and Asperger's syndrome. Howard et al (2000) also reported increased amygdala volumes in both hemispheres of the brain in subjects with autism. In contrast to these studies, Aylward et al (1999) reported the amygdala to be decreased in volume compared to age matched control cases. Pierce et al (2001) also reported amygdala volumes to be significantly smaller. Thus, these studies appear inconclusive as to whether there is a size difference in the autistic amygdala. Even if the size was significantly different, it is unclear whether this would imply better or worse function.

More suggestive evidence for a role of the amygdala in autism comes from a variety of functional imaging studies. Individuals with high functioning autism or Asperger syndrome showed significantly less amygdala activation than control subjects during a task that required them to judge what a person might be feeling or thinking from images of their eyes (Baron-Cohen et al 1999). A more recent fMRI study, comparing adult males with autism to control subjects, measured the neural activation in areas of the brain that are associated with a social perception task (Ashwin et al 2001). Subjects were shown images of real faces that varied in intensity of facial affect from neutral expressions to extreme fear expressions, as well as scrambled faces. The subject was simply required to press a button every time they saw a picture on the screen. During this social perception task, the subjects with autism showed less activation of the amygdala and orbitofrontal cortex. Moreover, the subjects with autism showed increased activity (implying greater reliance) on the superior temporal gyrus and anterior cingulate cortex. These data would appear to suggest that when normal subjects are carrying out tasks that require social evaluation, the amygdala is activated. And this activation is decreased in individuals with autism.

The amygdala theory of autism

Based on these converging lines of evidence, Baron-Cohen et al (2000) wrote a very compelling review that concluded, 'The amygdala is therefore proposed to be one of several neural regions that are abnormal in autism.' An implication of the paper is that pathology of the amygdala leads to an impairment in social intelligence, which is a hallmark feature of autism. That the amygdala might be at the heart of the pathophysiology of autism was also suggested somewhat earlier by Bachevalier (1994, 1996) based on observations of neonatal macaque monkeys who had been subjected to bilateral medial temporal lobe lesions. Bachevalier described these monkeys (at 6 months of age) as dramatically decreasing their social behaviour as compared to controls in dyadic social encounters with conspecifics. The lesioned animals actively avoided social contacts and had 'blank, inexpressive faces and poor body expression (i.e. lack of normal playful posturing) and they displayed little eye contact. Furthermore, animals with early medial temporal lobe lesions developed locomotor stereotypies and self-directed activities' (Bachevalier 1994). Since selective lesions of the hippocampus did not produce this pattern of behavioural alterations, Bachevalier attributed them to damage of the amygdala.

The literature that figured prominently in the generation of the amygdala theory of autism and the notion that the amygdala is essential for normal social behaviour was very influential on our own program of studies aimed at unravelling the neurobiology of primate social behaviour. While we would have been delighted to have generated data consistent with the hypothesis that the amygdala is central to social behaviour, the data we did generate has led us to a distinctly different conclusion.

The amygdala is not essential for social behaviour in the adult monkey

We have carried out a series of experimental studies to re-examine the role of the amygdala in conspecific social behaviour using the rhesus monkey as a model system (Emery et al 2001). Adult, male rhesus monkeys with bilateral ibotenic acid lesions of the amygdala, and age-, sex- and dominance-matched control monkeys were observed during dyadic interactions with 'stimulus monkeys' (two males and two females). This stereotaxic, neurotoxic lesion technique has the merit of removing the neurons of the amygdala while sparing fibres that pass through it. A variety of both affiliative (groom, present sex, etc.) and agonistic (aggression, displace, etc.) behaviours were quantitatively recorded while animals interacted in a large (18 ft×7 ft×6.5 ft) chain link enclosure. Each experimental animal interacted with each stimulus animal for four 20 minute periods in what we called the unconstrained dyad format. In what was initially a very surprising observation, the amygdala-lesioned monkeys generated significantly greater amounts of affiliative social behaviour towards the stimulus monkeys than the control monkeys. Control monkeys, when they first met the stimulus monkeys, demonstrated a typical and appropriate reluctance to engage in social interactions. They appeared to go though a period of evaluation to determine the intentions of the other animal. The lesioned monkeys, in contrast, appeared to be socially uninhibited since they did not go through the normal period of evaluation of the social partner before engaging in social interactions.

The inevitable conclusion from this study is that in dyadic social interactions, monkeys with extensive bilateral lesions of the amygdala can interpret and generate social gestures and initiate and receive more affiliative social interactions than normal controls. In short, they are clearly not critically impaired in carrying out social behaviour. We would suggest that the lesions have produced a socially uninhibited monkey since their normal reluctance to engage a novel animal appears to have been eliminated. This, as well as evidence that the amygdala-lesioned animals are not fearful of normally fear-inducing stimuli such as snakes, has led us to the hypothesis that a primary role of the amygdala is to evaluate the environment for potential threats or dangers. Without a functioning amygdala, macaque monkeys do not evaluate other novel conspecifics as potential adversaries and whatever system(s) are involved in mediating social interactions run in default mode of approach.

Early amygdala lesions do not eliminate social behaviour

One caveat of this conclusion that the amygdala is not essential for social behaviour is that these experiments were carried out in mature monkeys. One might argue that while the amygdala is not necessary for generating social behaviour, perhaps it is essential for gaining social knowledge. We have carried out a series of studies in which the amygdala is lesioned bilaterally in primates at two weeks of age (Prather et al 2001). This is at a point in time when infant macaque monkeys are mainly found in ventral contact with their mothers and there is virtually no play or other types of social interactions with other animals. We found that the interactions of the lesioned animals with their mothers was similar to that of control animals. Moreover, we found that, like adult animals with bilateral amygdala lesions, they showed little fear of normally fear-provoking objects such as rubber snakes. However, they showed increased fear, as indicated by more fear grimaces and more screams during novel dyadic social interactions. Most germane to the discussion, however, is the finding that the lesioned animals generated substantial social behaviour that was similar to that generated by age-matched controls. In a larger replication study that is currently under way (Prather et al, unpublished observations 2002) the quality and quantity of social interactions in a number of social formats is being investigated and there may be subtle differences in these parameters. However, the inescapable conclusion from observation of these animals is that there are none that are markedly impaired in generating species typical social behaviours such as grooming, play and facial expressions. All of the animals appear to be visually attentive of the other animals when they are involved in large 'play groups' comprised of two control animals, two animals with amygdala lesions and two animals with hippocampal lesions as well as male and female adult animals. And none appear to have developed motor stereotypies despite the fact that they have now reached one year of age.

The results from studies carried out both in adult and mature rhesus monkeys with complete bilateral lesions of the amygdala have forced us to consider the conclusion that the amygdala is not essential either for interpretation or expression of species-typical social behaviours or for gaining social knowledge. If the amygdala is not a central component of the Social System, it is unlikely that pathology of it would lead directly to the impairments of social behaviour that are observed in autism.

Subject S. M.

There are relatively few human subjects who have bilateral and discrete lesions of the amygdala. One outstanding exception is patient S. M. who has been extensively studied by Adolphs and colleagues (Adolphs et al 1994, 1995). Patient S. M. suffers from Urbach–Wiethe syndrome that has produced bilateral space occupying lesions of the amygdala. Interestingly, she is impaired in her ability to identify fearful faces despite the fact that she can reliably detect happiness and other emotions in faces. S. M. is also unable to determine which individuals would typically be considered untrustworthy based on their facial appearance (Adolphs et al 1998).

Despite these difficulties, patient S. M. leads a reasonably normal life. She is capable of holding a job, has been married and is raising children. One is impressed not so much with the deficits in this subject who has no amygdala, but rather by how intact is much of her everyday behaviour, including social behaviour. A similar conclusion can be drawn from patient H. M. who had bilateral temporal lobectomies for intractable seizures. His surgery has completely removed the amygdala and rostral half of the hippocampal formation (Corkin et al 1997). While H. M. is densely amnesic, he is nonetheless capable of normal social interactions. And neither he nor patient S. M. demonstrate typical autistic symptomatology. These patients would seem to support the contention that the amygdala is not essential for normal social behaviour and that damage to the amygdala does not necessarily lead to autistic behaviour.

Anxiety in autism

How does the concept of threat detection figure into the picture of autism? In Kanner's (1943) original report on autism, not only did he describe social and language impairments, but he also highlighted the anxious behaviour exhibited in his initial sample of children. Fear of threatening events is considered a common experience among primates and an adaptive response in humans (Reynolds & Richmond 1994). Anxiety, on the other hand, is an emotional response evoked when an individual perceives a situation as threatening even in the absence of direct danger. We would suggest that dysregulation of the amygdala might manifest itself in the individual with autism as alterations either of fear or anxiety. Although the presence of anxiety has been alluded to in descriptions (American Psychiatric Association 1994) and classifications of autism (Rescorla 1988, Wing & Gould 1979), the characteristics and pervasiveness of this has not been well studied. However, recent studies suggest that anxiety is an extremely common feature of the autism spectrum disorders.

Muris et al (1998) examined the presence of co-occurring anxiety symptoms in 44 children with autism spectrum disorder. The sample included 15 children with autism, and 29 with pervasive developmental disorder—not otherwise specified (PDD-NOS). They found that 84.1% of the children met criteria for at least one anxiety disorder. In descending order, the percentage of children meeting diagnostic criteria for an anxiety disorder were as follows: simple phobia (63.6%), agoraphobia (45.5%), separation anxiety (27.3%), overanxious (22.7%), social phobia (20.5%), avoidant disorder (18.2%), obsessive–compulsive disorder (11.4%), and panic disorder (9.1%). While the authors raised the caveat that anxiety symptoms were assessed via parental interview, they noted that parents often underreport internalizing symptoms, such as anxiety.

More recently, Gillott et al (2001) compared high-functioning children with autism to two control groups including children with specific language impairment and normally developing children on measures of anxiety and social worry. Children with autism were found to be more anxious on both indices. In fact, four of the six factors on the anxiety scale were elevated with obsessive–compulsive disorder and separation anxiety showing the highest elevations.

These studies do not provide much insight into the pervasiveness of anxiety in autism. Both clinical and parental reports indicate that not all children with autism demonstrate symptoms of anxiety. The DSM-IV summarizes that children with autism may exhibit 'a lack of fear in response to real dangers, and an excessive fearfulness in response to harmless objects' (American Psychiatric Association 1994, p 68). Wing & Gould (1979) highlighted the heterogeneity in the occurrence of anxiety in their classification system. Specifically, the active-but-odd subtype tend to exhibit extreme reactions to social situations, whereas the aloof subtype may be completely oblivious to environmental changes. Rescorla (1988) conducted a factor and cluster analysis using the Child Behavior Checklist (CBCL, Achenbach 1991), a general instrument of childhood behaviour, to distinguish boys with autism from other disorders. Among many differences, the analysis demonstrated that the more severe cases of autism were distinguished from the milder ones based on the presence or absence of anxiety.

The amygdala and anxiety

A number of recent studies have provided evidence that the amygdala may be dysregulated in emotional disorders such as anxiety and depression (Davidson et al 1999). Tillfors et al (2001), for example, demonstrated increased blood flow in the amygdala in social phobics anticipating a public presentation. Recently, Thomas et al (2001) used fearful faces as probes and demonstrated that anxious children showed heightened activity in the amygdala. De Bellis et al (2000) also showed that the right amygdala of children with generalized anxiety disorder was larger than in age-matched controls. These findings are consistent with the results of our studies in non-human primates in that removal of the amygdala produced animals that were less fearful of inanimate objects as well as other monkeys.

Conclusions

The amygdala has been proposed to play an essential role in the elucidation of normal social behaviour, and its dysfunction has been proposed to play a role in the social pathology of autism. Studies both in the rhesus monkey and data from human subjects with bilateral lesions of the amygdala indicate that the amygdala is not essential for many facets of normal social interaction. Rather, it appears that the amygdala may have a more selective role in detecting threats in the environment. If this proves to be correct, it would be unlikely that dysfunction of the amygdala alone could provide the substrate for the impairments of social interaction that are a hallmark feature of autism. If, however, the amygdala is indeed dysfunctional in autism, this could contribute to the abnormalities of fear and anxiety that appear to be a common feature of autism. If this were the case, one might expect the amygdala to be hyperfunctional in autism rather than hypofunctional as predicted by the current theories of the role of the amygdala in autism.

Acknowledgements

This original research described in this paper was supported, in part, by grants from the National Institute of Mental Health and by the base grant of the California National Primate Research Center. This work was also supported through the Early Experience and Brain Development Network of the MacArthur Foundation.

References

Abell F, Krams M, Ashburner J et al 1999 The neuroanatomy of autism: a voxel-based whole brain analysis of structural scans. Neuroreport 10:1647–1651

Achenbach TM 1991 Manual for the child behavior checklist/4-18 and 1991 profile. University of Vermont, Burlington

American Psychiatric Association 1994 Diagnostic and statistical manual of mental disorders: DSM-IV. American Psychiatric Association, Washington DC
Adolphs R, Tranel D, Damasio H, Damasio A 1994 Impaired recognition of emotion in facial expressions following bilateral damage to the human amygdala. Nature 372:669–672
Adolphs R, Tranel D, Damasio H, Damasio AR 1995 Fear and the human amygdala. J Neurosci 15:5879–5891
Adolphs R, Tranel D, Damasio AR 1998 The human amygdala in social judgment. Nature 393:470–474
Amaral DG, Price JL, Pitkanen A, Carmichael T 1992 Anatomical organization of the primate amygdaloid complex. In: Aggleton J (ed) The amygdala: neurobiological aspects of emotion, memory, and mental dysfunction. Wiley-Liss, New York, p 1–66
Ashwin C, Baron-Cohen S, Fletcher P, Bullmore E, Wheelwright S 2001 fMRI study of social cognition in people with and without autism. International Meeting for Autism Research, November, San Diego, CA (abstr B-32)
Aylward EH, Minshew NJ, Goldstein G et al 1999 MRI volumes of amygdala and hippocampus in non-mentally retarded autistic adolescents and adults. Neurology 53:2145–2150
Bachevalier J 1994 Medial temporal lobe structures and autism: a review of clinical and experimental findings. Neuropsychologia 32:627–648
Bachevalier J 1996 Brief report: medial temporal lobe and autism: a putative animal model in primates. J Autism Dev Disord 26:217–220
Baron-Cohen S, Ring HA, Wheelwright S et al 1999 Social intelligence in the normal and autistic brain: an fMRI study. Eur J Neurosci 11:1891–1898
Baron-Cohen S, Ring HA, Bullmore ET, Wheelwright S, Ashwin C, Williams SC 2000 The amygdala theory of autism. Neurosci Biobehav Rev 24:355–364
Bauman M, Kemper TL 1985 Histoanatomic observations of the brain in early infantile autism. Neurology 35:866–874
Brothers L 1990 The social brain: a project for integrating primate behaviour and neurophysiology in a new domain. Concepts Neurosci 1:27–51
Brown S, Schafer EA 1888 An investigation into the functions of the occipital and temporal lobes of the monkey's brain. Phil Trans R Soc Lond B 179:303–327
Corkin S, Amaral DG, Gonzalez RG, Johnson KA, Hyman BT 1997 H.M.'s medial temporal lobe lesion: Findings from magnetic resonance imaging. J Neurosci 17:3964–3979
Davidson RJ, Abercrombie H, Nitschke JB, Putnam K 1999 Regional brain function, emotion and disorders of emotion. Curr Opin Neurobiol 9:228–234
De Bellis MD, Casey BJ, Dahl RE et al 2000 A pilot study of amygdala volumes in pediatric generalized anxiety disorder. Biol Psychiatry 48:51–57
Dicks D, Myers RE, Kling A 1968 Uncus and amygdala lesions: effects on social behavior in the free-ranging rhesus monkey. Science 165:69–71
Emery NJ, Capitanio JP, Mason WA, Machado CJ, Mendoza SP, Amaral DG 2001 The effects of bilateral lesions of the amygdala on dyadic social interactions in rhesus monkeys (*Macaca mulatta*). Behav Neurosci 115:515–544
Gillott A, Furniss F, Walter A 2001 Anxiety in high-functioning children with autism. Autism 5:277–286
Howard MA, Cowell PE, Boucher J et al 2000 Convergent neuroanatomical and behavioural evidence of an amygdala hypothesis of autism. Neuroreport 11:2931–2935
Kanner L 1943 Autistic disturbances of affective contact. Nervous Child 2:217–250
Kling A, Cornell R 1971 Amygdalectomy and social behavior in the caged stumped-tailed macaque (*Macaca speciosa*). Folia Primatol (Basel) 14:190–208
Kling A, Steklis HD 1976 A neural substrate for affiliative behavior in nonhuman primates. Brain Behav Evol 13:216–238

Kling A, Lancaster J, Benitone J 1970 Amygdalectomy in the free-ranging vervet (*Cercopithecus aethiops*). J Psychiatr Res 7:191–199
Kluver H, Bucy PC 1938 An analysis of certain effects of bilateral temporal lobectomy in the rhesus monkey, with special reference to 'psychic blindness'. J Psychol 5:33–54
Kluver H, Bucy PC 1939 Preliminary analysis of functions of the temporal lobes in monkeys. Arch Neurol Psychiatry 42:979–997
Muris P, Steerneman P, Merckelbach H, Holdrinet I, Meesters C 1998 Comorbid anxiety symptoms in children with pervasive developmental disorders. J Anxiety Disord 12:387–393
Pierce K, Muller RA, Ambrose J, Allen G, Courchesne E 2001 Face processing occurs outside the fusiform 'face area' in autism: evidence from functional MRI. Brain 124: 2059–2073
Prather MD, Lavenex P, Mauldin-Jourdain ML et al 2001 Increased social fear and decreased fear of objects in monkeys with neonatal amygdala lesions. Neuroscience 106:653–658
Rescorla L 1988 Cluster analytic identification of autistic preschoolers. J Autism Dev Disord 18:475–492
Reynolds CR, Richmond BO 1994 Revised childrens manifest anxiety scale. Western Psychological Services, Los Angeles, p 1–45
Rosvold H, Mirsky A, Pribram K 1954 Influence of amygdalectomy on social behavior in monkeys. J Comp Phys Psychol 47:173–178
Thomas KM, Drevets WC, Dahl RE et al 2001 Amygdala response to fearful faces in anxious and depressed children. Arch Gen Psychiatry 58:1057–1063
Tillfors M, Furmark T, Marteinsdottir I et al 2001 Cerebral blood flow in subjects with social phobia during stressful speaking tasks: a PET study. Am J Psychiatry 158: 1220–1226
Wing L, Gould J 1979 Severe impairments of social interaction and associated abnormalities in children: epidemiology and classification. J Autism Dev Disord 9:11–29

DISCUSSION

Folstein: I'm trying to understand the difference in the response of babies to objects compared with other monkeys. In one case they are not anxious at all, and in the other case they are at least initially unduly anxious.

Amaral: I wish I had a good explanation for that! From the simple view that all fears originate in one region: the amygdala, this doesn't make any sense. However, child development experts such as Jerome Kagan, are not surprised that there are multiple fear systems. Separation anxiety comes in at a different time point developmentally, for example, than other fears. So our results do not surprise him at all. But we don't have a neural substrate for this response. The one concern is that by making a lesion early on we have altered the brain. A part of the brain that wouldn't normally subserve social fear may now be subserving it. We are planning PET experiments to evaluate this at the moment. We will be able to take these neonatal animals, put them into a novel social interaction and see whether brain regions are activated that wouldn't normally be activated in that encounter. We have looked at this in a variety of different ways, and it has something to do with novel social interactions because in these group interactions, while the operated monkeys are not as socially interactive—they are making more fear grimaces, for example—they are able to generate all the social gestures that you would expect of them.

Folstein: It is sort of similar to what is seen with autistic children. They have very little fear of objects: they immediately explore objects on the one hand, and on the other hand when they see a new person it takes them a long time to get used to them.

Amaral: I'd like to get more of a sense of that. How common is this?

Bailey: It's only true of one subset of individuals. With some children one can go into their home and they are sat on your lap within 5 seconds. There is a huge range of abnormal social behaviour in autistic children.

Amaral: Is anything else associated with that?

Bailey: They are disinhibited. They will come and put their arms around you.

Folstein: These children wouldn't be classically autistic.

Howlin: It is not just children when you go into their homes. There are children I have known that will follow complete strangers just because they are wearing a particular brand of trainers, for example, with no fear. There are also children who were withdrawn when they were very young, who then become very undiscriminating.

Rogers: To me it is the obliviousness to other people that is the most marked in autism, at least in our preschool children. It always amazes us to watch a new child with autism walk into a group of unfamiliar children as if no one was in the room except for the objects. There is no overt avoidance, but seeming complete obliviousness to the other children.

Lord: There is variability in the anxiety shown by children with autism. It is not clear that it corresponds to the behaviours of the monkeys.

Amaral: One gets a sense from the literature that it is a predominance of children that have one or more defined anxiety symptoms. I'm getting the sense here that this is not the case.

Hollis: The papers generally describe anxiety symptoms in high-functioning, older autistic children whereas people here are referring to pre-school children.

Rutter: Research with non-autistic populations has shown the necessity of differentiating between anxiety symptoms that are not accompanied by functional impairment, which are very common, and those leading to social impairment, which are much less frequent (Bird et al 1990, Simonoff et al 1997). Making this distinction in an autistic group is inevitably more complicated because of the uncertainty in deciding when impairment is due to anxiety, and when due to some other aspect of autism.

Charman: I can't remember what they are using in the Dutch paper, but the Spence anxiety self-report questionnaire would not be a good way of ascertaining in detail whether a child was anxious or not.

Lord: It was the DIS.

Charman: At least that is more intensive. But I'd be cautious about what proportion of disorders with ASDs have a clinical level of anxiety in their presentation. We don't know yet.

Amaral: What would you use?

Charman: We are trying to use the Child and Adolescent Psychiatric Assessment (CAPA) (Angold et al 1995) to do that. One of the problems in asking parents in detail about children's responses is that with a mood disorder it is parents who report the behaviour, but it is quite difficult to interpret the internal state of an autistic child because they are bad at communicating this. There is a subgroup of children who in adolescence are able to explicate internal anxious traits, but I have no idea what proportion this is.

Bolton: It is important to differentiate between social anxiety and mood disorder. What we are talking about here is a social-induced anxiety. If you look at the family studies in autism, there is a familial aggregation of affective disorders, but it doesn't seem to link to the broader phenotype of autism as we currently conceptualize it. However, there is also evidence that the relatives of people with autism are more 'withdrawn' and socially 'retiring' than expected and these indices of social anxiety are associated with other aspects of the broader phenotype. In future research therefore, we need to distinguish between mood disorders and social anxiety, as social anxiety seems much more akin to what you report in the animals.

Skuse: If we are going to measure anxiety in autistic children, we probably need to use neurophysiological measures rather than reports of behaviour or asking children how they feel. Using measurements such as skin conductance responses and heart rate would be a much more sensible way of doing this.

David Amaral, it was fascinating that you found, when selective lesions were made in the amygdalae of two-week-old macaques, that at 6 months or so of age the lesioned animals were less fearful of novel objects such as rubber snakes than controls, but they were substantially more fearful during dyadic social interactions.

I have been discussing the findings from your paper on these neonatal lesions with John Morris (Behavioural and Brain Sciences Unit, Institute of Child Health, London), and trying to come up with an explanation for this surprising dissociation. We did have an idea, which arises from work that suggests a very primitive response to social threat can be induced merely by eye contact. Even the eyes of a conspecific produce an initial pattern of brain activation as if they were the eyes of a predator. We've developed all sorts of cortical control mechanisms during evolution to modulate this activity.

John's idea is that the innate fear cue mediated by direct eye contact is initially processed by subcortical visual structures — superior colliculus and thalamic pulvinar nucleus. These subcortical fear responses are normally relayed to the amygdala, which also receives extensive neocortical inputs from sensory regions in temporal lobe and 'executive' regions in the prefrontal lobe.

Amygdala-mediated fear behaviour depends on an integration of all these influences. We suggest the 6–8 month old macaques showed less fear of novel objects, because this behaviour depends on neocortical inputs to amygdala—it was a fear that had at least in part to be learned. The reason why they showed more fear of social interactions was because the innate fear signals from collicular and pulvinar processing of eye contact could no longer be subject to prefrontal inhibitory modulation in amygdala, because it had been removed. The pulvinar projects directly to many subcortical and neocortical areas and the amygdala is bypassed.

Amaral: How do you know that they don't involve the amygdala?

Skuse: Other work that John has done has suggested this. It is a supposition. Nevertheless, it was interesting that you demonstrated that you could possibly get this fear response without the amygdala being there.

Amaral: The curious thing is that in the adult animals, when the amygdala is removed they do not have a physiological response to a social stressor. A normal monkey interacting for the first time with another animal has an increased cortisol response that isn't seen in the animals without an amygdala. The neonates, however, show a full social stress response only in these novel social interactions. If this were completely subcortical you would expect this in the adults as well, I would imagine.

Skuse: I suppose it depends on how the brain has rewired itself in the neonates who were subject to surgery. This wouldn't be a normal situation, but what we are suggesting is that if you were to look at the stages of brain activation in response to fear in those surgically treated animals, you see abnormal activity. It is conceivable that there are some direct connections between the pulvinar and other brain regions which wouldn't normally be active. Presumably, the amygdalectomized animals could not be fear conditioned. Animals who have been lesioned in the neonatal period could not, if our theory is correct, learn an association between the social behaviour of conspecifics and objects that normally evoke fear responses, such as snakes.

Bailey: David Amaral, you very nicely showed what you thought the problem isn't. Can you speculate on what you think the brain basis of social abnormalities in autism is?

Amaral: I suspect that other components of social cognition are located in the orbito-frontal cortex. Work that Jocelyn Bachevalier has done with destructive lesions suggest that you get more profound social deficits with orbitofrontal lesions. I suspect that we would be able to replicate Jocelyn's findings in this case. With respect to Jocelyn's studies, I should also emphasize that our animals that had neonatal lesions were all mother reared. This is the first time that animals have been subjected to amygdala lesions and put back on their mothers for normal rearing. In addition to the maternal rearing they have had normal group socialization. All the other monkey studies have been done with nursery-reared animals. We know that nursery-reared animals have their own social problems regardless of whether they have a lesion, and this can be a confounder.

Dawson: To address the question whether autism affects only processing of social information, it is important to control for several factors. For example, other animals tend to approach the animal being studied, whereas objects might not. Complexity is also important. There are lots of controls that need to be done.

Amaral: There are lots of studies that have been done with these animals. We have looked at their responses to a moving car, for example. We were worried about the fact that a conspecific is moving towards and away from the animals as opposed to a snake that just sits there. Even with a moving car their response seems to be more like that to an inanimate object as opposed to a conspecific. We have just completed a study where we have looked to see whether these animals with amygdala lesions will show normal mother preference. As these animals were being weaned we put the infant monkey in the middle and the mother and another female on two sides: we measured the amount of time the infant would spend with each. It turns out that while normal animals will spend the bulk of the time with their mothers, the hippocampal-lesioned animals will do also, but the amygdala animals have no preference. However, we believe that this is due to the fact that the amygdala-lesioned animals are not frightened by the novel enclosure and thus do not seek the comfort of their mothers.

Dawson: We also have to be careful not to say that this is evidence that the amygdala is not involved in social behaviour in autism. The analogy would be the old lesion studies in which a whole hemisphere is removed, and the result was that language could develop well elsewhere. It is possible for the brain to be substantially lesioned and rearranged and yet function as if relatively intact. Lesions are very different from a developmental brain abnormality. One of the best demonstrations of this was a study reported at a recent Cure Autism Now symposium. A young scientist was developing a mouse model to study the cerebellum and its effect on motor behaviour. We have always wondered: if the cerebellum is involved in autism, why don't you find frank cerebellar signs? People have interpreted these negative findings as meaning that the cerebellum is not critical in autism. This young scientist developed one knockout mouse in which there were no Purkinje cells, and this mouse had frank cerebellar signs. Of course, this is not what happens in autistic brains. But, he also was able to create a heterozygote mouse which had reduced not absent Purkinje cells. In this case he didn't get frank cerebellar signs but stereotyped motor behaviours, similar to that found in autism. This demonstrates I think that our animal models are going to have to be closer to what the actual abnormalities are, and we need to look at how the brain functions when those subtle abnormalities exist. This may give us a different kind of answer than what we get from lesion studies because the lesion itself promotes brain reorganization in some unusual way.

Amaral: I agree with what you say. Our program is moving to transient inactivation, which will get us around a lot of the problems with secondary effects. This is the next phase. There are interesting ways of doing transient inactivation. You can have animals socially interacting and then turn part of their brain off for a short period. It is certainly the case that when you damage the brain, particularly early on, it can reorganize so that other regions take on new functions. We can't control for that in these animals. However, if you took out the hippocampus either neonatally or in the mature animal, you would never have normal episodic memory processing. There are some brain regions that can't be compensated for. The sense is that as you get more subcortical, the uniqueness of the structure tends to make it more difficult to compensate. It may be that some other brain region has taken over the function of the amygdala in these animals, but I don't know what it would be.

Bauman: It struck me that the lesioned monkeys had a change in the modulation of fear responses. They were afraid of things that unlesioned monkeys weren't, and vice versa.

Amaral: This was only in infants. The infants were not fearful of inanimate objects, but they showed increased fear only in a novel social situation. They weren't confused.

Rutter: How long did that last for?

Amaral: We have tested up to a year periodically with novel social interactions. We started at six months. It looks like it persists, but we haven't gone beyond a year.

Rutter: Did you say that any novel social stimulus goes on evoking this response?

Amaral: Yes, even though they have a daily social interaction.

Rutter: That is interesting. How do you account for this?

Amaral: There must be some adaptation. They are gaining some familiarity with the animals that they see on a daily basis. There is some aspect of the novel situation that is probably highly ambiguous: they have to interpret this other animal and they don't know how to do this. They are in this ambiguous situation without an amygdala that allows them to determine whether this is potentially friendly or not. It may be this context that is creating the fear in these animals. Once they experience the same animals over time, they may be able to use learning and memory to become habituated to them. It doesn't seem as though they ever completely adapt. Even in the highly social situations we see more fear grimaces. We will follow these animals for 6–8 years, hopefully, and I don't know whether they will completely habituate. At this point the novel social encounter is always more evocative of a fear situation than their daily experience.

Rutter: The reason I find that so surprising and so interesting is that in normal humans, contact during that key phase with one set of individuals is associated with them not showing fear reactions to other unfamiliar individuals at a later point. This happens over periods of just a few months.

Bauman: Could one interpret this as a difficulty in reading social cues?

Amaral: When they are presented to a group or another monkey, they make an appropriate response. They seem to be interpreting and producing social gestures to the same extent as their non-lesioned peers.

Bauman: When you lesion the entire amygdala, in the autistic children that we have looked at the lateral nucleus is almost always preserved, and the medial nuclei are the ones that seem to be affected. There are a couple of cases where the entire amydgala is involved. In the future studies that you are planning, is it possible to be more precise in terms of which nuclei are lesioned? It might be interesting to see whether this makes a difference.

Amaral: That is a good point. We intend to make more selective inactivations of central versus lateral nuclei, for example. Presumably this could be modelled: say if you transiently inactivated the central nucleus, you could show that an animal had behavioural problems of fear but no autonomic fear response. If you did the lateral nucleus, though, you should turn off both the behavioural and autonomic responses.

Rutter: Do you think that your monkey findings are closer to the extreme social anxiety seen in some fragile X individuals who do not have autism, in whom social anxiety is very striking and very persistent?

Amaral: To the extent that we are beginning to think that the amygdala may be involved in generating social anxiety, that is a reasonable speculation. I don't know whether people have looked at the amygdala in fragile X.

Rutter: Just to complicate this still further, let me refer to a monozygotic twin pair from the sample that Susan and I first studied, who turned out to have fragile X (Le Couteur & Rutter 1988). One girl was mentally handicapped and clearly autistic, but she was also very socially disinhibited. The co-twin was not autistic, but showed marked social anxiety. Within a social interaction she showed fairly normal social behaviour. She was just incredibly anxious through all social encounters. The girls had the same genes but very different behavioural phenotypes.

Bishop: Could they have different patterns of X inactivation?

Skuse: Yes, monozygotic twins do.

Folstein: There were also pre- or perinatal differences in the pair.

Lipkin: I have seen David Amaral's impressive snake response in monkeys before. I have always been impressed that in primates this fear of snakes seems to be very primitive. Has anyone looked at responses to things that look snake-like in children with any of the disorders that we are discussing?

Bishop: You might have trouble with the ethics committee!

Sigman: We have looked at emotional responsiveness in autistic children. We measured heart rate changes in response to separation, social interaction with strangers and in response to an adult showing strong affect. In the last situation, there was no increase in the heart rate in the children with autism. In contrast, the matched group of developmentally delayed children showed an orienting response, a decline in heart rate, to someone showing very strong emotions. I did this study thinking that we would find more heart rate increases in the autistic group than in the control group in at least some of those situations, but we did not find that.

Amaral: The model that we are developing is that in the normal brain, responding to a provoker of anxiety, you should see increased activity in at least some portions of the amygdala. This has been found. If we extrapolate this to the autistic brain, in those people with autism who also suffer anxiety, then those same probes should show increased activation in the amygdala, but my friend Bob Schultz says that this doesn't occur.

Schultz: Yes, that is correct. We are doing studies involving discrimination of different types of facially expressed emotions with fMRI. Compared to healthy controls, our data so far suggest that persons with autism show less amygdala activation. However, this result is specific to this particular task involving evaluating emotions expressed on the face. Perhaps there would be different probes which provoke anxiety and a stronger than normal amygdala response in persons with autism.

Lipkin: I don't hear anyone talking about the nucleus accumbens. Could there be some reward circuit that is triggered by face contact or social interaction?

Schultz: Areas traditionally described as part of the reward circuit have not been shown to be activated by anything to do with faces, and we have often wondered why. That is, even in contrasts between faces and objects, such as furniture, there is not any extra activation in the reward circuits. Thus, faces alone, devoid of some task or context that inspires reward motivation, doesn't seem to trigger this circuitry.

Folstein: Does it in normals?

Schultz: No, I have never seen a report of this effect in normals either.

Dawson: This raises a question. We know that the amygdala has a lot to do with negative arousal and fear responses. We know so much less about what brain systems mediate social reward. The closest answer to this I have seen is from Kawashima et al's (1999) PET study in which they showed that during eye contact there is activation of the amygdala.

Amaral: I worry about some of those studies that use faces to stimulate rewards. Neutral faces in many respects are interpreted as either negative or potentially negative. I don't know this study you are referring to.

Dawson: It was a neutral-type face, but it was either looking at you or not. There was increased amygdala activation just when the eye contact was made. The authors interpreted it as a positive arousal mechanism.

Amaral: From my monkey studies, what has impressed me is that if the amygdala is important for generating social motivation, then you would expect some diminished social interaction in the lesioned monkeys with no amygdala. This isn't seen in either neonatals or adults, nor in human patients like SM. We have also worked with a patient HM who has no amygdala or hippocampus. He will engage other people socially and knows social norms. You just don't appreciate that there is a massive change in their ability to interact socially. I don't know what other brain regions might generate a social reward.

Lipkin: I'm amazed that he can interact without a hippocampus.

Amaral: At least two thirds of his hippocampus and his entire amygdala is missing. He is certainly not autistic. If you sat down and had a chat with him you wouldn't be able to detect anything problematic with him, unless you walked away and came back!

C. Frith: There are many people who have had temporal lobectomies (including the whole of the amygdala), but no one comments on this.

Amaral: There has been a population of people who had bilateral amygdalectomies early on. These were patients in India and Japan, during the psychosurgery years of the 1960s and 1970s. They were operated on for behaviour problems as early as four years of age. As many as 700 of these surgeries were done in India, and we are trying to follow-up some of these patients. We found three of them but they are so complicated it is difficult to make any sense of it. For example, the surgery was often done by space occupying lesions using paraffin wax. They may ultimately be of interest, but they are a complicated population. I'd like to ask the audience here a question, as a relative newcomer to autism and trying to deal with it from a systems neurobiological approach. Francesca Happé, your paper (p 198–207) is excellent in discussing these domains. But we tend to think of these domains as involving different parts of the domains. Is this an emergent process of some simpler systems going awry and hampering all these functions? Or is it more akin to the perspective that Simon Baron-Cohen presented to me a few weeks ago, that if you really want to study real autism then you study high-functioning autism that only has social impairment. Is there a real autism?

Charman: Clinically, there is a divergence of views from different people about whether or not there are people with a solely clear social impairment of an autistic nature, even if they are extremely bright and have structurally very good language. They typically also have communication problems, pragmatic problems and repetitive behaviours, to a certain extent. You don't get very severe social impairments in isolation. This is because we categorize these people by having a cluster of symptoms. It may be that out there in the population there are people who just have that degree of social impairment but don't have the other things.

Dawson: If you accept the viewpoint that people are discussing in this meeting, that there are these different domains in autism, and that people will have different combinations of strengths and weaknesses in these domains depending on genetic heterogeneity, then what is autism? With that viewpoint, if you were to study high-functioning people, you may be getting some aspect of social impairment but you are not necessarily going to be looking at the language component of autism. We have to abandon the concept of autism as a unitary thing, and also the idea of a core or primary deficit.

Happé: I am sympathetic to that view, but it is remarkable to me that the Romanian orphans show not only social and communicative problems, but also circumscribed interests. We still have to explain the coincidence. It is the same paradox that Dorothy Bishop finished with: they can be dissociated but they go together.

Dawson: There are a lot of surprising reasons why things go together, besides psychological reasons.

Rogers: There is a development perspective to this disorder that can be important. If you start with Asperger's as a pure case, then you can't diagnose that at age two. It comes on board differently from autism. If you start with Asperger's and a diagnosis sometime in childhood, then there is clearly a major problem in understanding mental states. If you go earlier in autism and ask where mental state knowledge comes from, developmentally we assume that it is constructed: it doesn't develop *de novo*, not as a module but rather is constructed out of earlier social knowledge. We can diagnose autism in two year olds, but at that age it is not the problem of understanding other peoples' mental states that stands out, but rather simpler social processes that are already not online, such as looking at faces, responding to names and imitating simple motor movements. These are simple early developmental processes that don't seem to imply much understanding of any other person as a mentalizing being. Yet we have excellent reliability in diagnosing young children whose mental ages are about 12 months and whose communicative levels are much earlier than understanding even joint attention behaviour.

Rutter: Most people would be fairly resistant to focusing on just one subgroup within the autism spectrum. What is really interesting about autism is the span of behaviour it encompasses.

Howlin: The other thing is that children change over time.

Rutter: We have all tended to focus on the three domains of symptomatology that are regarded as diagnostic of autism, but there are other features associated with autism. For example, many individuals with autism have an unusual gait.

Charman: Did you find features such as unusual gaits or sensory interests in any of the Romanian orphans?

Rutter: I don't think that there were any that showed unusual gaits of the kind that you would associate with autism.

Folstein: Gait loaded on one of the factors. I think it was sensory aversions. Another story like that is that one of the perfect pitch musicians with Asperger's syndrome was going up the stairs behind one of my research assistants to do some tests. He said that he wished he could walk quietly too!

C. Frith: We always used to say that one of the best markers for schizophrenia was the length of the trousers!

References

Angold A, Prendergast M, Cox A, Harrington R, Simonoff E, Rutter M 1995 The Child and Adolescent Psychiatric Assessment (CAPA). Psychol Med 25:739–753

Bird HR, Yager TJ, Staghezza B, Gould MS, Canino G, Rubio-Stipec M 1990 Impairment in the epidemiological measurement of childhood psychopathology in the community. J Am Acad Child Adolesc Psychiatry 29:796–803

Kawashima R, Sugiura M, Kato T et al 1999 The human amygdala plays an important role in gaze monitoring: a PET study. Brain 122:779–783

Le Couteur A, Rutter M 1988 Fragile X in female autistic twins. J Autism Dev Disord 18:458–460

Simonoff E, Pickles A, Meyer JM et al 1997 The Virginia Twin Study of Adolescent Behavioral Development: influences of age, gender and impairment on rates of disorder. Arch Gen Psychiatry 54:801–808

Cognition in autism: one deficit or many?

Francesca Happé

Social, Genetic and Developmental Psychiatry Research Centre, Institute of Psychiatry, De Crespigny Park, Denmark Hill, London SE5 8AF, UK

Abstract. The aim of this paper is to provoke discussion concerning the nature of the cognitive impairments that characterize autism. Autism spectrum disorders appear to be heterogeneous at the biological and behavioural levels, but it is currently unclear whether one or more cognitive abnormalities may be universal to people with autism. In addition, it is unknown whether one cognitive deficit is primary and causal, or whether several complimentary accounts are needed to explain the full range of behavioural features. From research to date, it seems that the psychological abnormalities that characterize autism may be dissociable, and it is uncertain whether the degree of social and non-social impairments is related. Possible reasons for the co-occurrence of social and non-social cognitive abnormalities in autism are discussed. One implication is that searching for the biological bases of specific social and non-social deficits may be more profitable than searching for the aetiology of autism *per se*.

2003 Autism: neural basis and treatment possibilities. Wiley, Chichester (Novartis Foundation Symposium 251) p 198–212

The diversity encompassed by the term 'autism spectrum disorders' is truly striking. One person with autism may be silent, socially aloof, apparently lost in simple stereotypies and unaware of the world around. Another person with the same diagnosis may be highly verbal, indeed pedantically so, keen to make friends, albeit through inappropriate approaches, and fascinated by some abstruse academic topic. In the face of this diversity, the notion of aetiological diversity seems natural. Current consensus appears to be that autism spectrum disorders are the result of a range of different interacting factors, and that different biological causes may apply in different individuals with autism.

If there is diversity at the biological and behavioural levels, is there also diversity at the cognitive level? One possibility is that the different interacting causal factors, and the associated neural abnormalities likely to be found in multiple brain regions, map on to distinct abnormalities at the cognitive level. If this is the case, then we should, perhaps, be searching not for genes that predispose for autism, but for genes that predispose for these different and distinct cognitive dysfunctions. It is therefore important to know whether the uneven profile of abilities and disabilities in autism spectrum disorders is the result of one or many cognitive deficits, and the nature of the relationship between these cognitive deficits. Ultimately, the answer to the question of whether autism is the result of one or many different cognitive deficits should be of practical importance both for discovery of the genetic and neural bases of autism, and for targeted educational interventions.

Current cognitive theories

There are many accounts of autism at the psychological level. This paper cannot attempt to do justice to all of these nor to review the field. It may be helpful, however, to mention some current theories, to illustrate the range of symptoms they attempt to explain. The 'theory of mind' (ToM) deficit account suggests that people with autism are impaired in everyday 'mindreading'—that is, the ability to attribute thoughts (e.g. beliefs and desires) in order explain and predict behaviour. This theory is able to explain why aspects of social and communicative behaviour that require social insight are impaired in autism, while other elements of social interaction and affection (e.g. attachment) are not. The theory has also been influential in suggesting directions for intervention and early screening, as well as brain imaging investigations of the neural basis of social cognition (see chapters in Baron-Cohen et al 2000). This theory provides a convincing account of two of the defining autistic impairments (social and communicative functioning, including imagination), but does not explain the non-social aspects of autism such as repetitive behaviour.

The executive dysfunction account, by contrast, addresses precisely the non-social aspects of autism, having much less to say about the social and communicative aspects (see chapters in Russell 1997). Autism shows parallels with frontal lobe damage, in terms of deficits in planning, controlling and monitoring goal-directed behaviour, especially in novel circumstances. People with autism have been shown to fail tests of at least some executive functions (EFs). However, the specificity of these deficits is unclear; many other developmental disorders also show executive deficits. Executive dysfunction can explain non-social deficits, but fails to account for either the social deficits or the areas of spared and superior ability in autism.

Responding in part to the challenge of explaining assets in autism (in, for example, rote memory and jigsaw-type tasks), the notion of weak 'central coherence' (CC) describes a cognitive style rather than deficit (see Frith 1989, Frith & Happé 1994). People with autism appear to show an information processing bias favouring featural over configural processing (for review see Happé 1999). This leads to superior performance on tasks where resisting gestalt is useful (e.g. Block Design, Embedded Figures), but deficits where meaning must be integrated in context (e.g. using sentence context to disambiguate homographs). This account attempts to explain non-social assets in autism, and certain task difficulties, but does not attempt to replace the ToM deficit account of social and communicative impairments.

These, and other current psychological accounts, remain in debate, and in particular it is uncertain which might be primary, specific or universal in autism.

Reducing down to one primary deficit?

It is clear that the three accounts described above are to some extent complimentary: a ToM deficit accounts well for the social and communication problems in autism, while executive dysfunction may explain some of the non-social difficulties, and weak coherence some of the non-social strengths. Thus the three theories do different jobs — but should they really be considered distinct and co-occurring deficits? Parsimony is naturally appealing; it would be satisfying if we could reduce down these three to one primary, causal problem. Considering autism as a *developmental* disorder, and taking seriously the importance of secondary 'knock-on' effects in development (Frith & Happé 1998), might social deficits in autism be derived from primary non-social deficits, or vice versa?

Primary EF deficits have been postulated by Russell (1996), who argues that basic executive processes underlie the infant's discovery of his/her own agency, from which ToM later develops. Weak coherence might be expected to impact social development through a number of effects. Featural (versus configural) processing might hamper emotion recognition and other aspects of face processing, and could conceivably disrupt normal joint attention (since what the child and parent attend to and find of interest are unlikely to be the same). Impaired face processing and joint attention might result in poor ToM, and the failure to integrate information in context would make it hard to process complex social situations.

In the opposite causal direction, poor ToM (and precursors such as joint attention) will significantly compromise socially mediated learning (e.g. through observation and imitation). This, in combination with impairments in attributing intentions, would seriously hamper word learning (Bloom 1997) and might be sufficient in some cases to explain low measured intelligence and language delay (Frith & Happé 1998). Inability to reflect on one's own mental states might hamper development of executive skills such as planning and inhibitory control (see Perner et al 1999). In order to derive weak coherence from primary social deficits, it is necessary to posit a default to local processing in the absence of socially directed attention and interests. These possible causal accounts are discussed at greater length in Happé (2001).

Whatever the face plausibility of these types of stories (and this writer, at least, remains unconvinced), evidence is needed concerning causality. Relevant evidence would include information about: developmental priority (what develops first?), longitudinal effects (which measures predict which?), and dissociations between deficits. This last point is especially relevant to the question of single or multiple cognitive deficits in autism, and their possible neural bases.

Can deficits in ToM, EF and CC be dissociated?

Does degree of ToM impairment relate to degree of executive dysfunction or weak coherence? Correlations between measures of these three constructs might argue for a common cause (although threshold effects might mask this), while dissociations between the three characteristics would suggest separable bases.

Do all children with ToM problems show executive deficits or weak coherence? On the face of it the answer would appear to be, no. Here the question is clouded by the fact that any test is at best an indirect probe of an underlying cognitive ability. The evidence is that some people who do not have autism (e.g. children with general developmental delay, Yirmiya et al 1998; deaf children not exposed to early signing, Peterson & Siegal 1995) may fail ToM tests. These groups have not been reported to show specific executive dysfunction or anomalies of coherence, although these have not been systematically investigated. However, whether such individuals really lack the ability to attribute mental states may be questioned; in autism ToM task failure is validated by everyday life 'mind-blindness', which does not appear to be the case in these other groups (e.g. Frith et al 1994).

A dozen papers testing social ability and EFs in young typically developing children show, in most cases, significant correlations between performance on tests of these different abilities, which in many cases remain significant when age and even verbal ability are covaried (e.g. Hughes 1998, 2001). Perner & Lang's (1999) useful meta analysis of recent studies with normal children aged between 2 and 7 years found a strong positive correlation between performance on ToM tasks and EF tasks. Fewer studies have been conducted with people with autism, but to date these also report significant correlations between EF and ToM. For example, Dawson et al (2002) recently reported an association between performance on EF tests linked to ventromedial prefrontal cortex and assessments of joint attention in three- and four-year-old children with autism.

On the other hand, the existence of EF deficits in a wide array of developmental disorders not accompanied by qualitative social impairment (e.g. Attention Deficit/Hyperactivity Disorder [ADHD], Tourette's) appears to argue against a causal link from executive to social skills. However, the term EF covers a wide range of somewhat fractionable abilities (including planning, monitoring, inhibition, generativity), and autism may be unique in its particular profile of dysfunctions (Pennington & Ozonoff 1996, Sergeant et al 2002). Specifically, autism may be characterized by deficits in shifting set and planning, in contrast to, say, ADHD in which deficits on tests of inhibitory control are notable. The precise timing of executive function impairments may also determine their knock-on effects and long-term impact.

Weak coherence seems to characterize people with autism regardless of their ToM ability (Happé 2000), at least as measured by the standard dichotomous measures. People with autism who pass false belief tests still show weak CC. For example, ToM task performance is related to performance on the Comprehension subtest of the Wechsler scales (commonly thought to require pragmatic and social skill), but not to performance on the Block Design subtest (Happé 1994). Jarrold et al (2000), however, have evidence of an inverse relation between ability to ascribe mental states to faces (interpreted as tapping ToM) and segmentation ability (interpreted as evidence of weak coherence, Shah & Frith 1983) in normally developing and autism groups. Individuals with Williams Syndrome, on the other hand, have been suggested to show detail-focused visuospatial processing, without deficits in ToM.

Rhonda Booth, Rebecca Charlton, Claire Hughes and I have recently completed a study comparing boys with autism and boys with ADHD on a range of EF tests and tests of CC. Results so far suggest that deficits in EF and CC are indeed dissociable; the boys with ADHD, despite their EF impairments, did not show a featural processing style. Examination of IQ subtest profiles, for example, showed distinct patterns for autism and ADHD; the former characterized by peak performance on Block Design, and the latter characterized by poor performance on the Mazes subtest. Block Design is a marker for weak coherence, because of the need to segment the whole design into its constituent blocks; Mazes performance is compromised by impairment in EFs such as planning, inhibitory control and monitoring. In addition to IQ profile, the dissociation between EF and CC in these groups was seen on specially devised tasks, such as a planning drawing task, where EF deficits were seen in failure to plan for inclusion of a new element, and weak CC was seen in piecemeal and detail-focused drawings; these two variables were independent and characteristic of the ADHD and ASD groups respectively (Booth et al 2003). In this study, parents were also asked to rate their sons on social (ToM) and non-social (both EF and CC) behaviours. Interestingly, ratings on these three dimensions were correlated in both the autism group and the typically developing controls, and correlations between EF, ToM and CC remained significant in the autism group even when full-scale IQ was partialled out.

Studies of the broader autism phenotype may be especially relevant for answering the question whether deficits in ToM, EF and CC can be dissociated.

In autism, diagnostic criteria ensure that children so diagnosed have social and non-social abnormalities. This limits the possibilities for finding dissociations among traits and among underlying cognitive capacities. In relatives, however, no such pre-selection applies, and dissociations can be fairly sought. Uta Frith, Jackie Briskman and I examined CC in the parents and siblings of boys with autism, dyslexia or no disorder (Briskman et al 2001). Our experimental tests of coherence showed detail-focused processing leading to superior performance in approximately half of the fathers and a third of the mothers in the autism families, and only two individuals among the Control and Dyslexia families (4%). In addition, we asked parents to rate themselves on questions we put together to tap both social and non-social preferences and abilities (with an emphasis on ToM and CC, respectively). Parents' self-ratings showed a significant correlation in the autism group only. This correlation was still significant after partialling out IQ. The correlation reflects the fact that parents who rated themselves as relatively lacking in social interest and skills, also rated themselves as more detail-focused in their preferences and abilities. However, individual parents could be found who rated themselves as sociable and socially able while having very detail-focused interests and abilities, and vice versa. This suggests that CC and ToM are separable. Support for this, and reassurance that these findings do not apply to self-report alone (where degree of insight might be questioned), comes from examination of test performance; self-ratings of detail-focus (coherence) were significantly predictive of performance on tests of coherence such as EFT and Block Design, but self-report of social abilities and preferences were not (Briskman et al 2001).

Why might social and non-social deficits co-occur?

If deficits in ToM, EF and CC are dissociable, as seems likely from the review of existing (albeit sparse) literature above, why do they appear to co-occur in autism spectrum disorders? My own attempts at drawing causal links from social to non-social deficits have left me unconvinced. Here I would like to sketch two possibilities, for the purpose of encouraging discussion. The first possibility would place the locus of connection not at the psychological level, but at the biological level. It may be, for example, that key brain regions or pathways are shared by these different functions, hence brain abnormalities that disrupt one process are also likely to disrupt the other functions.

What is known about the neural substrates of ToM, EF and CC? Might common regions be found? The two major sources of information at present appear to be functional neuroimaging, and neuropsychological studies of acquired brain damage. Both methods are, however, only just beginning to be used in this area. To date, functional imaging has been used to explore ToM in ordinary volunteers in half a dozen studies, using diverse methods and materials (see Frith 2003, this volume). All have found, not surprisingly, some involvement of frontal regions, and many have found ToM-related activation in medial frontal cortex (Frith & Frith 2000). It remains to be seen whether these regions are also active in the types of EF tasks people with autism fail. Work by Petrides, for example, suggests conditional reasoning tasks may activate certain medial frontal regions pinpointed in ToM studies (Petrides 1982, 1995). Imaging studies of coherence have scarcely begun; Ring et al (1999) used the Embedded Figures Test in an imaging study with normal and autism spectrum volunteers. They found relatively reduced frontal activity in the autism group, but the control task used was not closely matched, leaving a number of questions unanswered. Links have been reported between the amount of stereotyped behaviour and reduction of volume in the cerebellar vermis (lobules VI–VII) in one study of volunteers with autism (Pierce & Courchesne 2001). There is clearly scope for many more studies linking structural and functional imaging to individual differences in social and non-social impairments in people with autism spectrum disorders.

Studies of acquired deficits in previously healthy individuals suffering brain damage are also in their infancy as regards autistic deficits. A number of recent papers have reported ToM deficits following damage to either frontal (e.g. Stuss et al 2001) or right hemisphere regions (e.g. Happé et al 1999). The former is clearly interesting for the link with EF, although more specific localization information is needed. Some cases of acquired damage underscore the independence of ToM and EF (e.g. Rowe et al 2001). Right hemisphere damage has long been known to be associated both with social and communicative deficits, and with problems of global processing in both visuospatial and linguistic domains (see Brownell et al, in Baron-Cohen et al 2000). This suggests, perhaps, that the non-dominant hemisphere may be important in both ToM and CC.

A third line of evidence, which we will be pursuing, involves behaviour genetic studies of normal individual differences in social and non-social traits possibly related to autism. We are currently examining the range of normal variation in, for example, social insight and insistence on sameness in a population sample of 7-year-old twins. These data should allow examination of the phenotypic correlation between such traits in the normally developing population, and the possible genetic or environmental bases of relationships between individual differences in these characteristics.

The second possibility that I would like to raise for discussion explains the apparently systematic co-occurrence of anomalies in ToM, EF and CC in terms of compensation. Is it possible that deficits in, say, ToM are not sufficient to cause autism, and indeed may escape clinical notice, unless accompanied by a cognitive style (weak coherence) or deficit (executive dysfunction) that severely limits the child's ability to compensate for their social insight problems? If this were the case, autism spectrum disorders would only arise when several or all of these cognitive abnormalities happened to occur together. Whether this co-occurrence arises at above chance rates would be hard to establish if single deficits in, say, ToM, were difficult to identify and qualitatively different in their manifestations when compensatory abilities are uncompromised. If this idea is worth testing, it may be time for another epidemiological survey in the tradition of Wing and Gould's ground-breaking work — to establish the co-occurrence not only of socio-communicative and rigid/repetitive behaviours, but also of the cognitive deficits and characteristics that may underlie these different features of autism.

Conclusions

This paper aims to raise questions for discussion, and to prompt further questions for empirical exploration. Definitive evidence is lacking to answer important questions such as whether degree of social and non-social abnormalities in autism are strongly related, whether all elements of the 'cognitive phenotype' of autism can be dissociated, and whether and which of the associated deficits might be primary and causal. One implication for future research, however, would appear to be that we might make more progress by searching for the neural and genetic causes of impaired ToM, EF and weak CC, than by searching for the neural substrates of autism *per se*.

References

Baron-Cohen S, Tager-Flusberg H, Cohen DJ (eds) 2000 Understanding other minds: perspectives from autism and developmental cognitive neuroscience. 2nd edn, Oxford University Press, Oxford

Bloom P 1997 Intentionality and word learning. Trends Cogn Sci 1:9–12

Booth R, Charlton R, Hughes C, Happé F 2003 Disentangling weak coherence and executive dysfunction: planning drawing in autism and ADHD. Philos Lond Trans R Soc B Biol Sci 358:387–392

Briskman J, Happé F, Frith U 2001 Exploring the cognitive phenotype of autism: weak 'central coherence' in parents and siblings of children with autism. II Real-life skills and preferences. J Child Psychol Psychiatry 42:309–316

Brownell H, Griffin R, Winner E, Friedman O, Happé F 2000 Cerebral lateralization and theory of mind. In: Baron-Cohen S, Tager-Flusberg H, Cohen D (eds) Understanding other minds: perspectives from autism and developmental cognitive neuroscience. 2nd edn, Oxford University Press, Oxford, p 306–333

Dawson G, Munson J, Estes A et al 2002 Neurocognitive function and joint attention ability in young children with autism spectrum disorder versus developmental delay. Child Dev 73:345–358

Frith C 2003 What do imaging studies tell us about the neural basis of autism? In: Autism: neural basis and treatment possibilities Wiley, Chichester (Novartis Found Symp 251) p 149–176

Frith C, Frith U 2000 The physiological basis of theory of mind: functional neuroimaging studies. In: Baron-Cohen S, Tager-Flusberg H, Cohen D (eds) Understanding other minds:

perspectives from autism and developmental cognitive neuroscience. 2nd edn, Oxford University Press, Oxford, p 334–356

Frith U 1989 Autism: explaining the enigma. Blackwell Publishers, Oxford

Frith U, Happé F 1994 Autism: beyond 'theory of mind'. Cognition 50:115–132

Frith U, Happé F 1998 Why specific developmental disorders are not specific: on-line and developmental effects in autism and dyslexia. Dev Sci 1:267–272

Frith U, Happé F, Siddons F 1994 Autism and theory of mind in everyday life. Social Dev 3: 108–124

Happé FGE 1994 Wechsler IQ profile and theory of mind in autism: a research note. J Child Psychol Psychiatry 35:1461–1471

Happé F 1999 Autism: cognitive deficit or cognitive style? Trends Cogn Sci 3:216–222

Happé F 2000 Parts and wholes, meaning and minds: central coherence and its relation to theory of mind. In: Baron-Cohen S, Tager-Flusberg H, Cohen D (eds) Understanding other minds: perspectives from autism and developmental cognitive neuroscience. 2nd edn, Oxford University Press, Oxford, p 203–221

Happé F 2001 Social and non-social development in autism: where are the links? In: Burack JA, Charman T, Yurmiya N, Zelazo PR (eds) The development of autism: perspectives from theory and research. Lawrence Erlbaum Associates Inc, Mahwah, New Jersey, p 237–253

Happé FGE, Brownell H, Winner E 1999 Acquired 'theory of mind' impairments following stroke. Cognition 70:211–240

Hughes C 1998 Executive function in preschoolers: links with theory of mind and verbal ability Br J Dev Psychol 16:233–253

Hughes C 2001 Executive dysfunction in autism: its nature and implications for the everyday problems experienced by individuals with autism. Burack JA, Charman T, Yurmiya N, Zelazo PR (eds) The development of autism: perspectives from theory and research. Lawrence Erlbaum Associates Inc, Mahwah, New Jersey, p 255–275

Jarrold C, Butler DW, Cottington EM, Jimenez F 2000 Linking theory of mind and central coherence bias in autism and in the general population. Dev Psychol 36:126–138

Pennington BF, Ozonoff S 1996 Executive functions and developmental psychopathology. J Child Psychol Psychiatry 37:51–87

Perner J, Lang B 1999 Theory of mind and executive function: is there a developmental relationship? Baron-Cohen S, Tager-Flusberg H, Cohen D (eds) Understanding other minds: perspectives from autism and developmental cognitive neuroscience. 2nd edn, Oxford University Press, Oxford, p 150–181

Perner J, Stummer S, Lang B 1999 Executive functions and theory of mind: cognitive complexity or functional dependence? In: Zelazo PD, Astington JW, Olson DR (eds) Developing theories of intention: social understanding and self-control. Lawrence Erlbaum Associates, Hillsdale, New Jersey, p 133–152

Petrides M 1982 Motor conditional associative-learning after selective prefrontal lesions in the monkey. Behav Brain Res 5:407–413

Petrides M 1995 Impairments on nonspatial self-ordered and externally ordered working memory tasks after lesions of the mid-dorsal part of the lateral frontal cortex in the monkey. J Neurosci 15:359–375

Peterson CC, Siegal M 1995 Deafness, conversation and theory of mind. J Child Psychol Psychiatry 36:459–474

Pierce K, Courchesne E 2001 Evidence for a cerebellar role in reduced exploration and stereotyped behavior in autism. Biol Psychiatry 49:655–664

Ring H, Baron-Cohen S, Wheelwright S et al 1999 Cerebral correlates of preserved cognitive skills in autism — a functional MRI study of embedded figures task performance. Brain 122:1305–1315

Rowe AD, Bullock PR, Polkey CE, Morris RG 2001 'Theory of mind' impairments and their relationship to executive functioning following frontal lobe excisions. Brain 124:600–616
Russell J 1996 Agency: its role in mental development. Erlbaum, Hove, UK
Russell J (ed) 1997 Autism as an executive disorder. Oxford University Press, New York
Sergeant JA, Geurts H, Oosterlaan J 2002 How specific is a deficit in executive functioning for Attention Deficit/Hyperactivity Disorder? Behav Brain Res 130:3–28
Shah A, Frith U 1983 An islet of ability in autistic children: a research note. J Child Psychol Psychiatry 24:613–620
Stuss DT, Gallup GG, Alexander MP 2001 The frontal lobes are necessary for 'theory of mind'. Brain 124:279–286
Yirmiya N, Erel O, Shaked M, Solomonica-Levi D 1998 Meta-analyses comparing theory of mind abilities of individuals with autism, individuals with mental retardation, and normally developing individuals. Psychol Bull 124:283–307

DISCUSSION

Skuse: In our Social Communication Disorders clinic, where we see high-functioning children with communication problems, we have established a database on over 200 school-age children over the last couple of years. Bear in mind that this is a tertiary referral centre and there could be a certain amount of referral bias — these are children whose diagnosis has foxed other people — but what we find is that they almost invariably have social-interaction deficits typical of autism. We don't necessarily find these are associated with significant pragmatic language disorders, although they very often are. We sometimes find pragmatic disorders without significant social communication problems, but this is rare. The dimension of the autistic spectrum that we don't see so often is the stereotyped interests and repetitive behaviours. I suspect that this is a function of the fact that we are seeing children with good ability. We see children who are generally at mainstream schools and who have good language skills. Our findings fit with what Tony Bailey and others have described as being the broader autistic phenotype, If we were to do an epidemiological study of the sort you are describing, we would probably find that social interaction and pragmatic language deficits are much more common than stereotyped and repetitive behaviours, and that they are probably associated with a rather heterogeneous group of psychiatric problems.

Bishop: If you are going to say that they really are completely independent, in order for autism to occur at a frequency of 1 in 1000, each of these would have to occur in 1 in 10.

Skuse: The stereotyped interest/repetitive behaviours are relatively unusual, and they rarely occur on their own, independent of the other two components of the autistic triad of impairment.

Bishop: You have to multiply the probabilities to get the probability of core autism. If you are multiplying by one behaviour that is particularly rare you would need an even higher rate of these other ones.

Skuse: I agree with your point. I believe from our own preliminary research that social interaction and pragmatic language impairments are relatively common.

Happé: I was pointing this out as an alternative to the idea that there was a common biological substrate that explained the coincidence. I mentioned it mainly because I found myself perplexed by the fact that, if one thought that these things occurring in isolation wouldn't look like a problem, then how would we ever know the rate of occurrence of each of the deficits alone. But I take your point that the rate of each would have to be pretty high, and this may render it unlikely.

C. Frith: I was wondering about the 'executive problems', since this is a broad term that covers many things. Work by James Russell suggests that only a small subset of executive processes are impaired in autism (Russell et al 1999). Has anyone else found that? These findings would restrict the number of problems you are looking for.

Happé: The recent review by Sergeant et al (2002) suggests that there may be problems in set shifting and planning, and possibly not so much in working memory. It is certainly not a blanket deficit.

Dawson: I really liked Tony Bailey's point about emergent qualities. If you think about these as underlying styles or proclivities, maybe they are not so profound as to be maladaptive in the clinical sense. But if they are combined so that there are two of them, they also act in a synergistic way functionally. If you have a child with a social and language impairment, the social interaction dysfunction will affect the language, and vice versa. Deficits will be more severe when they are in combination.

Folstein: In genetics this is called epistasis.

Baird: Where would you fit into these models an aspect of function that is quite significantly impaired, particularly in young children, which is the different sensory sensitivities? This can be one of the earliest signs that one sees in the first year. How does one integrate this kind of information in these models?

Happé: We still lack information about how widespread those difficulties are, in autism and also among the non-clinical population. In the very early years I would have to see them as additional problems. Later on, one can see some of it resulting from a failure to process information in context, and therefore have your sensory perception modulated. There is also some degree of social shaping: we decide what sounds, tastes and so on are disgusting partly by what everyone else thinks. Children with autism don't do this.

Lord: It feels to me as if this is starting in the middle rather than at the beginning. When I think about a two year old with autism, except for central coherence, which you could operationalize at that age, it seems like your beginning point is something that has already gone wrong with the two year old: the joint attention and the impaired relatedness. This still has to be explained.

Happé: I agree: we have to step back from that to ask why it broke down. I don't see people out there trying to explain this. In Geraldine Dawson's recent papers (Dawson et al 2002a, 2002b) she talks about more basic processes still, in terms of reassigning values to associations and so on. This would be a step back. I don't know whether the ultimate step has to be back to the brain.

Rutter: In considering how particular cognitive deficits could underlie autism we need to ask how they might operate. Let us take 'theory of mind' as an example. Initially, it was postulated that a lack of an ability to understand what another person was likely to be thinking might directly cause the social impairment evident in autism. The strong association with autism tended to justify the inference. However, the finding from Geraldine Dawson's home movie studies show that many children exhibit manifestations of autism in the first year of life. If this is so, it raises queries, because that is well before theory of mind (as we ordinarily understand it) is evident. It could well be that a precursor of theory of mind was operative but it is less plausible that the mentalizing limitation itself caused the social impairment.

Happé: The term 'theory of mind' is used in a number of different ways. It certainly shouldn't be equated with performance on false belief tests. Alan Leslie's first analysis of theory of mind was focusing much more on pretend play at 18 months, and the ability to represent thoughts and intentions. It would be this ability that I would see as crucial. You can ask whether joint attention difficulties that would precede that 18 month point are really the crucial thing. If they are, are they the first signs of this system going wrong, or are they actually the important thing and the theory of mind difficulties are secondary? From the literature I would speculate that the theory of mind difficulty—in the sense of the ability to represent others' mental states—is the crucial thing. This is because other children may have joint attention problems that they grow out of. Joint attention problems *per se* don't lead to this cascade of difficulties that we see in autism, but joint attention problems that lead to or are bound up with an inability to represent others' mental states do seem to lead to bad knock-on effects. There is still the question as to whether preceding 18 months of age there are things that are markers of autism, and we have to explain those. Behaviours such as orienting to name-calling are very much in the flavour of recognizing speakers' intention. We are not talking about a conscious process of recognition of intention, but a very basic mechanism—although not one you would expect to see in animals.

Dawson: In the home videotape studies of infants at age 1, when we did a discriminant functional analysis, 'looking at others' categorized 71% of the normal babies versus babies with autism correctly. The other behaviours that discriminated were pointing and showing, and orienting to name. Looking at others was the strongest discriminator, however.

Charman: You need to be cautious about getting tripped up by the fact that a lot of the early theory of mind work in the 1980s concentrated on the 3–4 year age period. In the normative developmental literature, one area of immense interest is to try to track how notions about social orienting, reading intentions and distinguishing between inanimate and animate objects before the end of the first year of life may actually relate to what we understand as social understanding in the second year of life. This may well relate to the onset and individual differences in theory of mind ability by the age of three or four. This is where one would want to look for key deficits. I would shy away from the notion of using the term 'primary' deficits. If you have early developmental impairments that mean that you are not going to orient to voices, or not going to jointly attend when someone shifts their gaze, this will have all sorts of sequelae in terms of how you will develop social understanding of the world and other people as agents in the world. It isn't the thing that is causing autism. Theory of mind is a post-cursor of joint attention abilities. Joint attention impairments in autism are a post-cursor of something else that has gone wrong earlier on. They are staging posts rather than causal factors. We need to be careful not to think about what is primary and what is causal. These are manifestations that we can measure as psychologists.

Rutter: I accept all the points you have made, but at the end of the day surely we do want to get to a causal model. What is clear is that infants are intensely social, long before they are one year old. If one is putting forward a cognitive account, it has to be a feature operating in infancy, unless it is argued that the home movies are misleading. The issue of regression in language in the second year of life seen in about a quarter of children with autism must also be accounted for. If Peter Hobson was here, he might argue that these findings all suggest that the cognitive deficits are secondary to a lack of empathy.

Sigman: One of the questions I have concerns how much you can separate the cognitive from the social in the infant. Although cognitive and social functions are certainly related at older ages, in infants these functions are very highly intertwined. I don't think these things can be taken apart so clearly as they can in adults. After Peter Mundy and I were convinced that we had found a deficit in joint attention, we came to England and met with Uta Frith, Simon Baron-Cohen and Alan Leslie. We told them that we thought we had found the developmental precursor to their theory of mind deficit. But it has been hard to document the link between joint attention and theory of mind in our longitudinal samples because not enough children have acquired sufficient language skills so that their theory of mind can be assessed.

Rutter: They are closely intermingled, so that early social behaviour involves anticipation of another person's actions. It may not involve reading other peoples' minds in quite the sense that is shown later, but certainly it involves real responsivity to the social context. Do we have to accept that these different skills cannot be separated?

Dawson: Let me give an example from our recent *Child Development* paper (Dawson et al 2002a), in which we showed a strong correlation between performance on a ventromedial prefrontal neuropyschological task and joint attention. The example I gave in the discussion section of this paper is the social referencing situation. This is an experimental paradigm used to assess a child's ability to pick up on social cues and modulate their own behaviour in reaction to the other person's behaviour. Typically, the child enters and sees a novel object. The child will look up to monitor the other person's emotional expression, and depending on that expression, they will approach or avoid the novel object. In other words, the child sees a stimulus and forms a representation about what that stimulus is like (positive or negative reward value). Let's say that the child's initial assessment of the object is different from that of the other person. Then the child has to hold his initial representation online, get some more information—someone's facial cue—and use this to inhibit his first representation, using the new representation to guide his behaviour. This is a good example of cognition and social interaction in concert. How could you separate them?

Monaco: I would be interested in the correlations between the different aspects. In your studies, have you seen so far that they are more highly correlated in males than females? If you are talking about having deficits in two domains leading across a threshold for becoming more clinically susceptible, you would expect that males would be more highly correlated than females.

Dawson: We have never looked at that.

Rutter: In terms of Francesca Happé's study, that is something else to be looked at. The idea of a synergistic effect among deficits makes a lot of sense. How can you use epidemiological data to refute this idea?

Charman: One way of determining this would be to take a measure of the cognitive function, such as executive function, a measure of theory of mind, and also a measure of impairment or severity. Then you multiply across a population your score on theory of mind and executive function, and use some sort of threshold.

Rutter: That's a good suggestion. I take it that you are saying that if the features are independent of one another, there should be no multiplicative effect on impairment, whereas if there were synergism, a multiplicative effect is exactly what would be predicted.

Happé: Within the tasks themselves, presumably, you would predict that impairment on one or other task does occur separately, but when you get them together you will be worse on both tasks. This requires going off and not just asking about behavioural level features, but also giving tests to try to tap in to the core cognitive processes underlying these—there are too many routes to the same behaviours.

Charman: One of the things that is interesting in infancy research, in the normal development literature, is that people are very keen on intentionality in the first year. There are infancy researchers who are doing looking preference paradigms and talking about 'false belief' in 14-month-old children. They are also looking at animate–inanimate distinctions and reading intentionality at 6 and 8 months. There is a huge variation between people who are making claims about what 4- and 6-month-olds can do, with other people having data that completely refute this. In some ways one of the problems is that our understanding of how to measure these early things is less well established than our understanding of how to measure it at 3 or 4 years of age. One of things about the theory of mind literature is that we do know quite a lot about what performance on a false belief task tells us, but we know much less about how to measure aspects of intentionality between 6 and 12 months of life.

References

Dawson G, Munson J, Estes A et al 2002a Neurocognitive function and joint attention ability in young children with autism spectrum disorder versus developmental delay. Child Dev 73:345–358

Dawson G, Carver L, Meltzoff AN, Panagiotides H, McPartland J, Webb SJ 2002b Neural correlates of face recognition in young children with autism spectrum disorder, developmental delay, and typical development. Child Dev 73:700–717

Russell J, Jarrold C, Hood B 1999 Two intact executive capacities in children with autism: implications for the core executive dysfunctions in the disorder J Autism Dev Disord 29:103–112

Sergeant JA, Geurts H, Oosterlaan J 2002 How specific is a deficit in executive functioning for attention-deficit/hyperactivity disorder? Behav Brain Res 130:3–28

Autism and specific language impairment: categorical distinction or continuum?

Dorothy V. M. Bishop

Department of Experimental Psychology, University of Oxford, South Parks Road, Oxford OX1 3UD, UK

Abstract. Traditionally, autism and specific language impairment (SLI) are regarded as distinct disorders, with differential diagnosis hinging on two features. First, in SLI one sees isolated language impairments in the context of otherwise normal development, whereas in autism a triad of impairments is seen, affecting communication, social interaction and behavioural repertoire. Second, there are different communication problems in these two conditions. Children with SLI have particular difficulty with structural aspects of language (phonology and syntax). In contrast, abnormal use of language (pragmatics) is the most striking feature of autism. However, recently, this conventional view has been challenged on three counts. First, children with autism have structural language impairments similar to those in SLI. Second, some children have symptoms intermediate between autism and SLI. Third, there is a high rate of language impairments in relatives of people with autism, suggesting aetiological continuities between SLI and autism. One interpretation of these findings is to regard autism as 'SLI plus', i.e. to assume that the only factor differentiating the disorders is the presence of additional impairments in autism. It is suggested that a more plausible interpretation is to regard structural and pragmatic language impairments as correlated but separable consequences of common underlying risk factors.

2003 Autism: neural basis and treatment possibilities. Wiley, Chichester (Novartis Foundation Symposium 251) p 213–234

Language impairment is a central feature of autistic disorder. The 1970s saw researchers focusing on whether language was *the* central feature of autism, i.e. asking whether the other symptoms of this disorder were secondary consequences of limited language skills. To this end, comparisons were made between children with autism and those with specific language impairment (SLI). The answer seemed clear-cut: the syndrome of autism could not be attributed to language difficulties: symptoms were more severe, more extensive, and different in kind from those seen in SLI. Consequently, contemporary diagnostic frameworks draw a sharp dividing line between autism and SLI and emphasize the differential diagnosis of these conditions. Nevertheless, in recent years, this neat division has been questioned. Cases have been described who show an intermediate clinical picture. Furthermore, family studies have suggested possible aetiological overlap between SLI and autism. In this chapter, I review this recent evidence, and consider the implications for studies of the aetiology of developmental disorders.

Language and communication in SLI and autism: the conventional view

SLI is defined when a child fails to acquire language at the normal rate for no apparent reason. Non-verbal ability is within normal limits, and there is no indication of physical or sensory handicaps that could account for the language difficulties. Although it is widely accepted that SLI is heterogeneous (Bishop 1997), for most children the principal difficulties are with structural aspects of language, i.e. mastery of phonology (speech sounds) and syntax. It is usually assumed that children with SLI have normal non-verbal communication and social use of language—or if there are problems, these are simply secondary consequences of the structural language difficulties.

The language abilities of children with autism vary tremendously. Around 50% do not learn to talk and have severe comprehension problems. Others acquire language late and do not progress beyond simplified speech. Distinctive features of autistic language are most readily observed in children of normal nonverbal ability—cases of so-called high-functioning autism (HFA). Many of these do acquire speech and may talk in long and complex sentences. However, their use of language is abnormal (see Table 1). Lord & Paul (1997) noted that whereas in SLI, children who talk most tend to be the most competent communicators, in autism, it is often the most talkative children in whom communicative abnormalities are especially apparent. Problems in the appropriate use of language in context come under the domain of pragmatics, and are the most striking feature in autism. On the basis of these contrasting phenotypes, SLI and autism are usually thought of as distinct disorders with different aetiologies, as illustrated in model A (see Fig. 1).

Comparisons of communication in SLI and HFA

Although the textbook accounts of these disorders suggest a clear divide, some studies suggest that the boundaries between these disorders are not so sharp. On the one hand, on standardized tests of structural language skills, children with HFA often have deficits similar to those seen in SLI. On the other, there is evidence of pragmatic difficulties in some non-autistic children with language impairments.

TABLE 1 Typical characteristics of language and communication in verbal children with autism (based on Lord & Paul 1997)

First words acquired late
Marked impairment in language comprehension
Articulation normal or even precocious
Abnormal use of words and phrases with idiosyncratic meanings
Use of made-up words (neologisms)
Pedantic and over-precise speech
Dissociation between mastery of grammar and functional use of language
Echolalia
Confusion and interchanging of personal pronouns, such as I/you
Abnormal vocal quality
Abnormal intonation and stress
Failure to use contextual information in comprehension
Over-literal interpretation without appreciation of speaker's intention
Low rate of spontaneous initiation of communication
Little reference to mental states
Persistent questioning
Poor at judging what a listener needs to be told
Difficulty in making causal statements
Lack of cohesion
One-sided talk rather than to-and-fro conversation

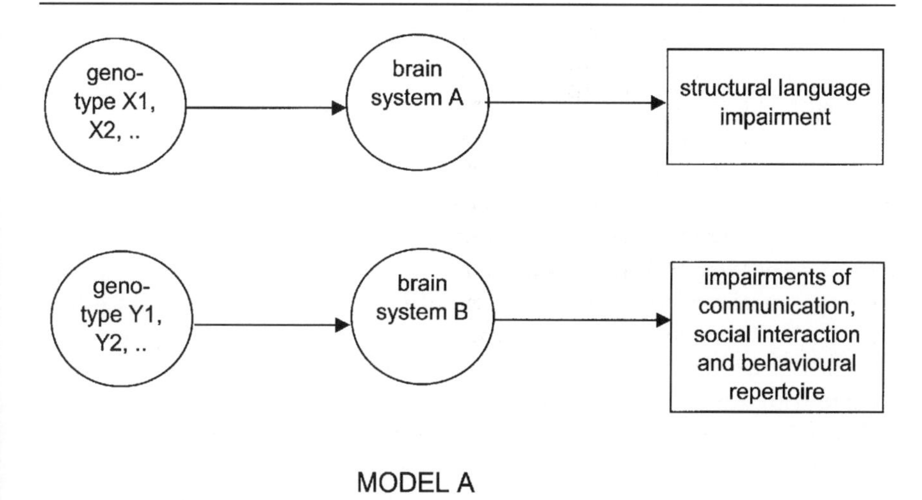

FIG. 1. Model A: distinct causal routes for SLI and autism. Both disorders are highly heritable, but it is assumed that different sets of genetic risk factors are implicated in the two disorders.

Sigman & Capps (1997) summarized the communicative features of autism by concluding that phonological, semantic, and grammatical development tend to follow a normal course (albeit at a slow rate in children with low IQ), but language use does not, and is aberrant. Nevertheless, in most children with autism, structural language skills are at least as poor, if not worse, than those of children with SLI. Figure 2 shows illustrative test data from Lincoln et al (1993). In a much larger study, Fein et al (1996) compared language scores of preschool children of normal non-verbal ability who had autism or SLI. Although profiles of language scores were more uneven for the children with autism, on no test did they significantly outperform children with SLI. Kjelgaard & Tager-Flusberg (2001) did not directly compare autism and SLI, but they used a broad range of language measures with a large group of children with autism. They found that in general these children had impaired expressive and receptive language, and there was a clear relationship between IQ level and language skills. Articulation skills were almost always unimpaired, but on a test of non-word repetition, in which the child repeats back meaningless strings of sounds such as 'blonterstaping', many children with autism did very poorly. They noted that poor non-word repetition in children is frequently seen in SLI (e.g. Bishop et al 1996). Overall, then, studies using standardized language measures suggest children with autism have many of the same impairments as are seen in SLI: these, however, tend to be overlooked because the pragmatic difficulties are more severe and unusual.

Evidence for cases intermediate between SLI and autism

A landmark study in this field was initiated by Bartak et al (1975). They recruited boys aged 4.5–9 years who had broadly normal non-verbal IQ but severe comprehension problems, and found that most of them could be categorized as cases of autism or receptive SLI ('developmental dysphasia'). A detailed psychometric assessment was carried out, together with a parental interview. The main conclusion was that children with autism have distinctive pragmatic difficulties not seen in SLI (though, as in the studies reviewed above, structural language impairments similar to those in SLI were also present). Nevertheless, five of the 47 children recruited to the study could not be unambiguously classified in either category: their symptoms were intermediate and tended to change with age. Furthermore, when the sample was followed up in middle childhood (Cantwell et al 1989) and adulthood (Howlin et al 2000), the distinction between groups became blurred. Many cases from the language-impaired group developed autistic-like symptoms in non-language domains. This study suggested that the boundaries between autism and SLI might be less clear-cut than originally thought.

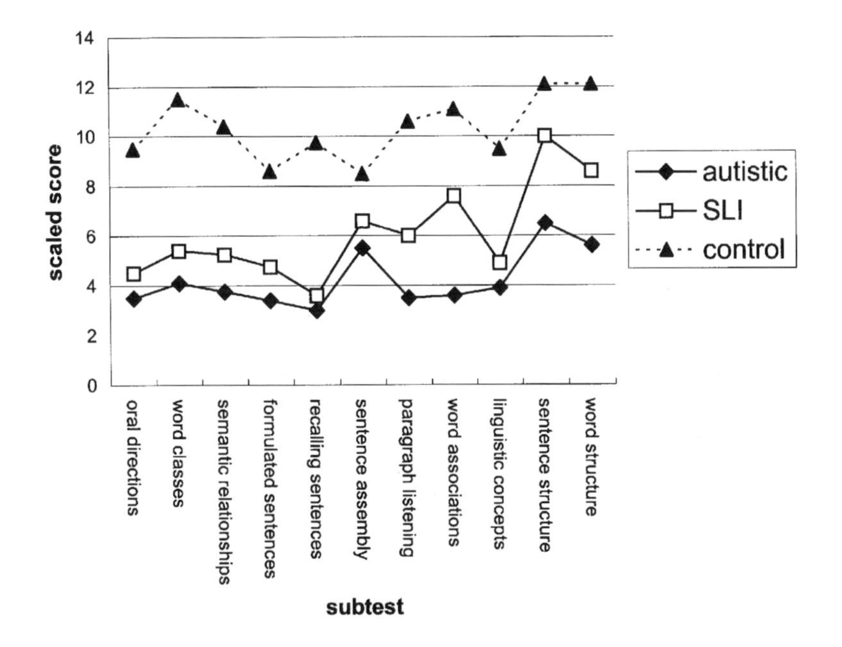

FIG. 2. Scores obtained on the Clinical Evaluation of Language Functions — Revised by 10 high-functioning children with autism and 10 children with receptive language impairment in a study by Lincoln et al (1993). Children were aged from 8–14 years.

The same conclusion was suggested by other research on subtypes of language impairment. Rapin & Allen (1983) coined the term 'semantic pragmatic deficit syndrome' to refer to children who used fluent and complex language, but had abnormalities of language use, producing tangential or irrelevant utterances. Bishop (2000), who described similar cases, suggested the term 'pragmatic language impairment' (PLI) is preferable. The diagnostic status of these children has been the matter of some debate, because their language difficulties are reminiscent of those in HFA, yet, according to both Rapin & Allen (1983) and Bishop (1998) this language profile can be seen in children who are sociable and do not show major autistic symptomatology. Bishop & Norbury (2002) used standardized autism diagnostic instruments that assessed both current status and past history with a group of children recruited from special schools for those with communication impairments. None had a definite diagnosis of autism, though some had been described as 'on the autistic spectrum'. Twenty-eight children had evidence of pragmatic difficulties on the Children's Communication Checklist

(Bishop 1998), and were designated as the PLI group. The remaining 17 children did not have evidence of pragmatic difficulties and formed the typical SLI group. Table 2 shows how these children scored in relation to cut-offs for autism and its milder variant, Pervasive Developmental Disorder Not Otherwise Specified (PDDNOS) on both a parent report measure and direct observation. Although five of the children with PLI met diagnostic criteria for autism on both measures, the majority had autistic features that were not pervasive or severe enough to merit a diagnosis of autism. Six cases had no significant autistic symptomatology on either parental report or direct observation. There was also evidence that children's symptoms changed with age, with autistic symptoms declining as they matured. An unexpected finding from this study was the relatively high rate of autistic symptomatology seen in the SLI-T group on the parental report measures.

Family studies and the 'broader phenotype'

Superficial similarities between autism and SLI do not on their own provide particularly convincing evidence for common origins. After all, the same pattern of behaviour can arise for different reasons. However, evidence that the symptom overlap may reflect deeper commonalities comes from studies of relatives of people with autism. Although relatives meeting diagnostic criteria for autism are rare,

TABLE 2 Numbers of children with typical SLI and pragmatic language impairment (PLI) categorized according to parental report and direct observation[a,b]

		Diagnosis from parental report		
		Unaffected	PDDNOS	Autism
Diagnosis from direct observation	Unaffected	7 SLI 6 PLI	5 SLI 4 PLI	2 SLI 4 PLI
	PDDNOS	3 PLI	1 SLI	1 SLI 2 PLI
	Autism	1 SLI 2 PLI	2 PLI	5 PLI

[a]The Autism Diagnostic Interview — Revised (Lord et al 1994) or the Social Communication Questionnaire (Berument et al 1999) were used to obtain parental report of autistic symptoms, focusing largely on the period when the child was aged 4–5 years. These two measures gave good diagnostic agreement. The Autism Diagnostic Observation Schedule — Generic (Lord et al 2000) was used for diagnosis based on observation of current behaviour.
[b]From Bishop & Norbury (2002) studies 1 and 2 combined.

cases of subthreshold symptomatology are common, including people who have linguistic and communicative difficulties resembling SLI and/or PLI (Bolton et al 1994). In the only large-scale study that directly compared family histories of children with autism and those with SLI, Rapin (1996) reported that rates of siblings affected with SLI were as high for children with autism as they were for those with SLI (see Table 3).

Autism as 'SLI plus'

The overlap between autistic and SLI symptomatology, both within individuals and within families, raises questions about model A as an accurate depiction of the relationship between these disorders. An alternative would be to treat these disorders as points on a continuum of severity: mildly impaired cases have only structural language problems, and more severely impaired people have structural and pragmatic impairments, often accompanied by non-linguistic symptoms of autism. This simple view can readily be rejected, because it predicts that the most severe structural language problems should be seen in those with pragmatic

TABLE 3 Rates of SLI and autism in first degree relatives of children with SLI or autism

	Child diagnosis[a]			
	SLI (n=192)	high-functioning autism (n=51)	low-functioning autism (n=120)	non-autistic controls with low IQ (n=105)
% with an affected immediate family member	33.3% SLI 2.1% autism	19.6% SLI 3.9% autism	24.2% SLI 5.0% autism	21.0% SLI 0 autism
% with affected parent	19.4% SLI 0.5% autism	7.8% SLI 0 autism	10.4% SLI 0 autism	15.5% SLI 0 autism
% with affected sibling[b]	22.8% SLI 1.7% autism	19.4% SLI 6.5% autism	22.0% SLI 6.6% autism	11.0% SLI 0 autism

Based on Rapin (1996).
[a]Rapin (1996) used conventional diagnostic criteria for SLI, but preferred the term 'developmental language disorder' to refer to this group.
[b]Excludes 10% of SLI, 39% of high-functioning autistic, 27% of low-functioning autistic and 26% of non-autistic control group who had no siblings.

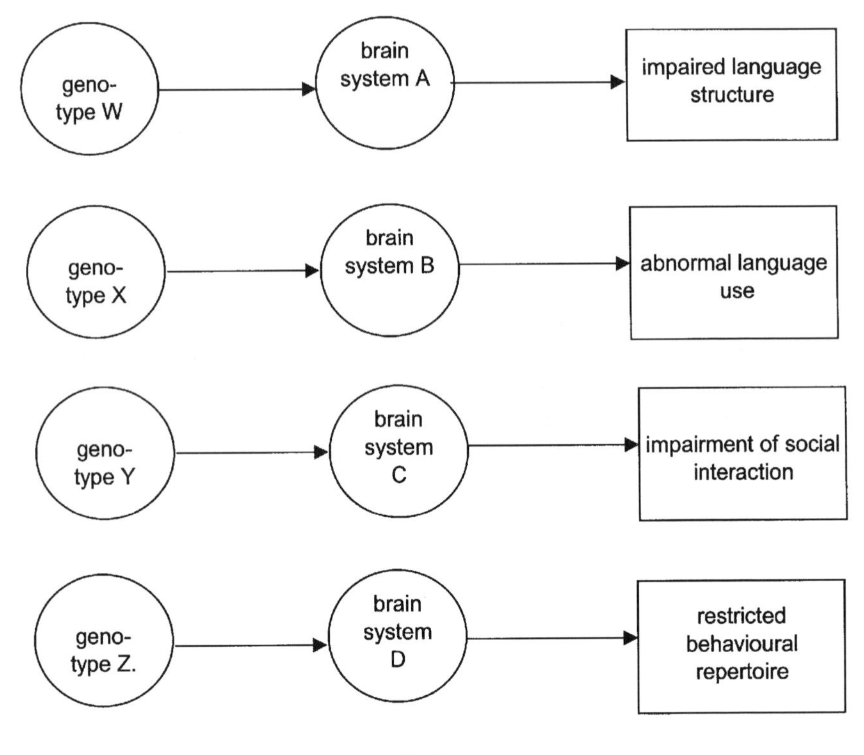

FIG. 3. Model B: separate causal paths for different components of autism. In SLI, only brain system A is affected; in classic autism, systems A–D are affected; other phenotypes with a partial autistic profile are also possible.

difficulties, and this is not the case, either for those with autism or with PLI (Bishop 2000). Indeed, some children with HFA score within normal limits on structural language tests (Kjelgaard & Tager-Flusberg 2000). Thus though structural and pragmatic language difficulties tend to co-occur, there can be double dissociation — some have pragmatics unaffected and poor structure, and others have structure unaffected and poor pragmatics. This indicates that one deficit is not logically dependent upon the other, and implies distinct neurological bases for these aspects of communication. Model B (Fig. 3) depicts one way of accounting for this pattern, in which autistic disorder involves multiple underlying impairments, each with its own cause. According to this model, there might be a range of genetic risk factors, each of which affects a different brain system and leads to a specific set of symptoms. Autism would result when a child was unfortunate enough to inherit a particularly disadvantageous constellation of alleles (W, X, Y and Z), leading to the full clinical picture. Model B is compatible with the heterogeneous symptomatology seen in developmental disorders affecting communication: for instance, it is possible to have cases of Asperger syndrome with good structural language skills, in whom only X, Y and Z are implicated, or relatively pure cases of PLI, where only risk allele X is present. On this model, the familial association between SLI and autism is explained by assuming that the pathway from genotype W to structural language impairment is usually implicated in autism.

Reconceptualizing the relationship between autism and SLI in this way would have major consequences for how we conduct genetic studies. Rather than working with a phenotype defined in categorical, clinical terms, we would move to a more dimensional approach, in which we look for quantitative trait loci linked to different aspects of autism, including language structure and use. An advantage of this approach is that we could assess these traits in parents and other relatives, rather than having to rely on multiplex families with more than one child who meets full diagnostic criteria for autism. Furthermore, model B implies that, in studies of autism, rather than relying on conventional standardized language tests, we should use language measures that have been shown to be sensitive phenotypic markers of heritable SLI, such as non-word repetition and tests of tense marking (Rice 2000).

Model B is attractive, in that it implies we might see clearer genotype–phenotype relationships if we move to a quantitative, dimensional view of autism. However, before we become too enthused with such a model, it is important to note several arguments against it. First, there are phenotypic differences between SLI and autism that raise difficulties for this account. As noted above, children with autism usually do well on articulation tests, whereas those with SLI often make errors in producing speech sounds, despite normal motor control of the articulators — i.e. they have a phonological impairment. This, then, is one aspect of communication where one cannot make the generalization that children with autism have the same deficits as SLI.

Second, the model entails that the different domains of impairment in autism have independent origins, and their co-occurrence in autism is a matter of chance. For this to be the case, the risk alleles implicated in genotypes W, X, Y and Z would need to be very frequent in the population, and cases of isolated symptoms should be extremely common. We do not have epidemiological figures on rates of autistic symptomatology in the general population, or on co-occurrence of symptoms, but it seems unlikely that symptoms are independent of one another.

Third, different symptoms tend to co-occur within families. If the combination of risk alleles determined the pattern of symptoms, then monozygotic (MZ) twins should be phenotypically identical for autistic symptoms. LeCouteur et al (1996) found this was not so: variation between two members of a MZ twin pair was as great as that seen between twin pairs. Clearly some factor other than genetic makeup influences the symptom profile. Furthermore, phonological problems, which are not usually part of the autism phenotype, are seen in *relatives* of people with autism.

The final argument is based solely on precedent — in general, single gene disorders that affect brain development do not influence a discrete brain system and cause a distinctive modular impairment. Although it is often possible to identify a prototypical behavioural phenotype associated with a genotype, there is often substantial individual variation within a given genetic syndrome. Neurofibromatosis type I is an instructive example (Reiss & Denckla 1996): the same genetic mutation can affect different brain regions, and lead to very different clinical pictures, e.g. a parent with mild symptoms may have a severely affected child. To complicate matters even further, there are also ample instances of *different* genes leading to the *same* phenotypic outcome, e.g. phenotypically identical manifestations of tuberous sclerosis are caused by genetic mutations on chromosomes 9 and 16 (Udwin & Dennis 1995).

Language impairment and autistic disorder associated through pleiotropy

Such considerations suggest we need to consider an alternative account, model C (Fig. 4), in which autism and SLI have at least partially distinct neurological bases, but common aetiological factors affect both of them. For instance, suppose that there are genes that disrupt processes of neuronal migration, leading to abnormal brain structure. The precise outcome of such a process will depend on which brain systems are implicated, and this might be affected by the genetic background (i.e. other genes interacting with the risk genes), systematic environmental influences, or chance events.

This model differs from model B in that it predicts that some cases of SLI will have symptoms that go beyond what is seen in autism. It can accommodate the apparently paradoxical finding that symptoms that frequently go together need not necessarily do so, by assuming that different symptoms have different brain bases, but common aetiological factors can disrupt their neurodevelopment. However, although this model may be plausible, it suggests that the enterprise of discovering the genetic basis of autism is going to be considerably more difficult than we might have imagined. It also raises questions about where we draw the boundaries of the phenotype. On the basis of both family and behavioural data, I have argued for continuities between language impairment and autism, but one could ask why stop there? There is considerable comorbidity between SLI and attention deficit disorder (Beitchman et al 1996), literacy impairments (McArthur

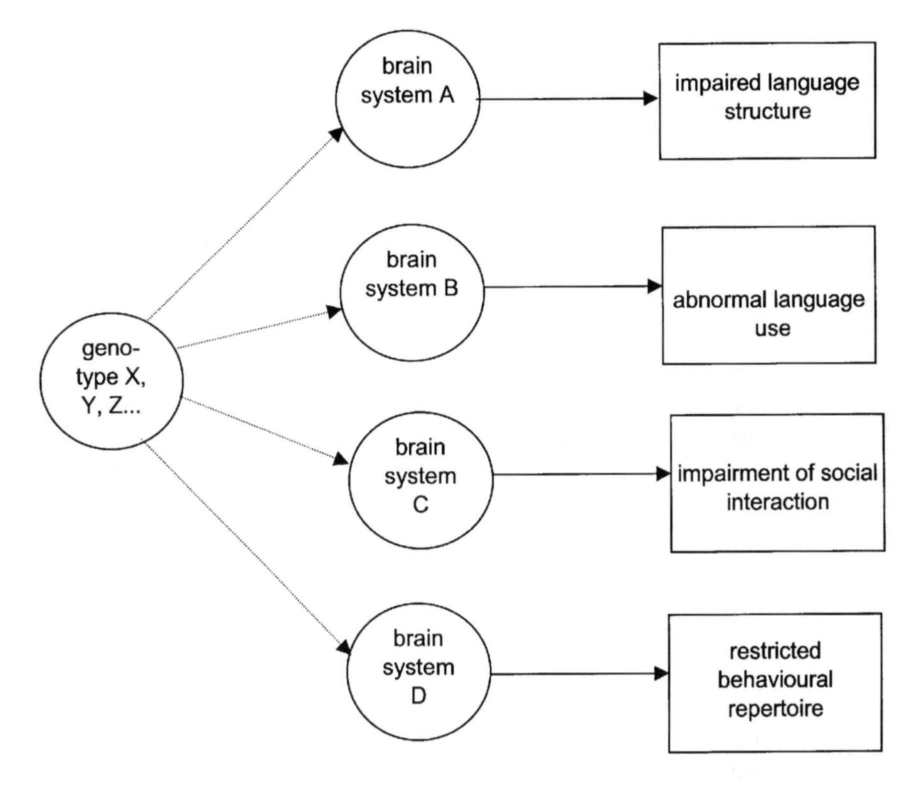

FIG. 4. Model C: a range of genetic risk factors (X, Y, Z, etc) is implicated in the aetiology of developmental disorders, each of which can affect separate brain systems: the dotted path from genotype to brain systems indicates a probabilistic influence. The phenotype will depend on which systems are affected, and this could be a function of genetic background, systematic environmental influences, or random factors. Correlations between symptoms may reflect involvement of adjacent brain systems.

et al 2000) and motor immaturity (Bishop 2002). Are these also influenced by the same genes that pose a risk for autism?

Model C implies that two kinds of study need to be at the top of our research agenda. First, we need behaviour genetic studies that can clarify which developmental disorders are heritable, and how far comorbid traits are influenced by the same genes (Rutter 2000). Such studies depend on the availability of reliable, quantitative methods of assessment that can act as sensitive markers of underlying genotype. Second, studies of MZ twins teach us that, even though autism is a strongly genetic disorder, non-genetic factors are important in determining the phenotype. Consider the MZ twin pair described by Rapin (1996), one of whom had classic autism and the other SLI. The difference may be due to chance influences early in development that determine which brain regions are subject to genetic effects, but before accepting that, we need to look for more systematic environmental influences that may play a part. Within-family comparisons of affected relatives, and contrasts between children who follow different developmental courses have the potential to throw light on environmental factors that may influence severity and pattern of autistic symptomatology in genetically at risk individuals.

Conclusions

Autistic disorder and SLI have traditionally been regarded as distinct disorders, but recent work suggests some overlap both at the phenotypic and the aetiological level. One way forward would be to view autism as a form of SLI in which a broader range of impairments is present, and to look for genetic correlates of specific components of the autistic triad. However, we need to be cautious about assuming a simple one-to-one relation between genotype and phenotype: it is more likely that there are genetic risk factors that have the potential to compromise brain development, but their precise impact depends on the genetic background, environmental influences and chance factors, sometimes leading to SLI, sometimes to autism, and sometimes to an intermediate clinical picture. The answer to the question, are autism and SLI on a continuum, depends on the level of description. Phenotypically the pragmatic communication deficits seen in autism are not continuous with structural language impairment: they can show a pattern of double dissociation, indicating they are logically separable. However, aetiologically, they appear to share common risk factors.

Acknowledgements

The author is supported by a Wellcome Trust Principal Research Fellowship.

References

American Psychiatric Association 1994 Diagnostic and statistical manual of mental disorders, 4th edn. American Psychiatric Association, Washington, DC

Bartak L, Rutter M, Cox A 1975 A comparative study of infantile autism and specific developmental language disorder: I. The children. Br J Psychiatry 126:127–145

Beitchman JH, Brownlie EB, Inglis A et al 1996 Seven-year follow-up of speech/language impaired and control children: psychiatric outcome. J Child Psychol Psychiatry 37:961–970

Berument SK, Rutter M, Lord C, Pickles A, Bailey A 1999 Autism screening questionnaire: diagnostic validity. Br J Psychiatry 175:444–451

Bishop DVM 1997 Uncommon understanding: development and disorders of language comprehension in children. Psychology Press, Hove

Bishop DVM 1998 Development of the Children's Communication Checklist (CCC): a method for assessing qualitative aspects of communicative impairment in children. J Child Psychol Psychiatry 39:879–891

Bishop DVM 2000 Pragmatic language impairment: a correlate of SLI, a distinct subgroup, or part of the autistic continuum? In: Bishop DVM, Leonard LB (eds) Speech and language impairments in children: causes, characteristics, intervention and outcome. Psychology Press, Hove, UK, p 99–113

Bishop DVM 2002 Motor immaturity and specific speech and language impairment: evidence for a common genetic basis. Am J Med Genet 114:56–63

Bishop DVM, Norbury CF 2002 Exploring the borderlands of autistic disorder and specific language impairment: a study using standardized diagnostic instruments. J Child Psychol Psychiatry 43:917–929

Bishop DVM, North T, Donlan C 1996 Nonword repetition as a behavioural marker for inherited language impairment: evidence from a twin study. J Child Psychol Psychiatry 37:391–403

Bolton P, MacDonald H, Pickles A et al 1994 A case-control family history study of autism. J Child Psychol Psychiatry 35:877–900

Cantwell D, Baker L, Rutter M, Mawhood L 1989 Infantile autism and developmental receptive dysphasia: a comparative follow-up into middle childhood. J Autism Dev Disord 19:19–31

Fein D, Dunn M, Allen DA et al 1996 Language and neuropsychological findings. In: Rapin I (ed) Preschool children with inadequate communication: developmental language disorder, autism, low IQ. Clinics in developmental medicine No. 139. Mac Keith Press, London, p 123–154

Howlin P, Mawhood L, Rutter M 2000 Autism and developmental receptive language disorder—a follow-up comparison in early adult life. II. Social, behavioural, and psychiatric outcomes. J Child Psychol Psychiatry 41:561–578

Kjelgaard MM, Tager-Flusberg H 2001 An investigation of language impairment in autism: implications for genetic subgroups. Lang Cog Processes 16:287–308

Le Couteur A, Bailey A, Goode S et al 1996 A broader phenotype of autism: the clinical spectrum in twins. J Child Psychol Psychiatry 37:785–802

Lincoln AJ, Courchesne E, Harms L, Allen MH 1993 Contextual probability evaluation in autistic, receptive developmental language disorder, and control children: event-related brain potential evidence. J Autism Dev Disord 23:37–58

Lord C, Paul R 1997 Language and communication in autism. In: Cohen DJ, Volkmar FR (eds) Handbook of autism and pervasive developmental disorders, 2nd edn. Wiley, New York, p 195–225

Lord C, Rutter M, LeCouteur A 1994 Autism Diagnostic Interview—Revised: a revised version of a diagnostic interview for caregivers of individuals with possible pervasive developmental disorders. J Autism Dev Disord 24:659–685

Lord C, Risi S, Lambrecht L et al 2000 The Autism Diagnostic Observation Schedule-Generic: a standard measure of social and communication deficits associated with the spectrum of autism. J Autism Dev Disord 30:205–223

McArthur GM, Hogben JH, Edwards VT, Heath SM, Mengler ED 2000 On the 'specifics' of specific reading disability and specific language impairment. J Child Psychol Psychiatry 41:869–874

Rapin IC 1996 Historical data. In: Rapin I (ed) Preschool children with inadequate communication: developmental language disorder, autism, low IQ. Clinics in developmental medicine No. 139. Mac Keith Press, London, p 58–97

Rapin I, Allen D 1983 Developmental language disorders: nosologic considerations. In: Kirk U (ed) Neuropsychology of language, reading, and spelling. Academic Press, New York, p 155–184
Reiss AL, Denckla MB 1996 The contribution of neuroimaging: Fragile X syndrome, Turner syndrome, and neurofibromatosis I. In: Lyon GR, Rumsey JM (eds) Neuroimaging. Brooks, Baltimore, p 147–168
Rice ML 2000 Grammatical symptoms of specific language impairment. In: Bishop DVM, Leonard LB (eds) Speech and language impairments in children: causes, characteristics, intervention and outcome. Psychology Press, Hove, UK, p 17–34
Rutter M 2000 Genetic studies of autism: from the 1970s into the millennium. J Abnorm Child Psychol 28:3–14
Sigman M, Capps L 1997 Children with autism. Harvard University Press, Cambridge, MA
Udwin O, Dennis J 1995 Psychological and behavioural phenotypes in genetically determined syndromes: a review of research findings. In: O'Brien G, Yule W (eds) Behavioural phenotypes. Mac Keith Press, London, p 90–208

DISCUSSION

Fombonne: I was interested in your comments that perhaps you are seeing fewer diagnoses of SLI. Do you have evidence for this?

Bishop: We need to do a study on this. I don't think this is just a case of diagnostic labels: it is more my impression of the children I find when I go into schools. When I did my PhD I saw 80 children, and about 10 of these looked as if they had obvious pragmatic problems. Now when I go to schools it is very hard to find children with classic SLI who don't have some sort of syndrome. It could just be a case of who is getting into which schools. If it is the case that rates of autistic spectrum disorder are increasing, it could be that those children are pushing out the classic SLI cases, who therefore end up in mainstream schools instead. However, I'm intrigued by another fascinating possibility, namely that autism and SLI may involve the same core disorder and it is just the way it is manifesting that has changed.

Fombonne: In your ADI scores, in the autistic upper right corner, with the children with SLI diagnosed with autism by the ADI, were the scores very high like in autism samples, or were they just meeting criteria?

Bishop: In general most of these children were just meeting criteria. We are talking about children who are coping in a special school which didn't have particular facilities for dealing with massive behavioural problems. What was low in this sample was repetitive behaviour of any severity.

Fombonne: If you start with SLI problems or pragmatic syndromes, have you looked at the family data in terms of rates of autism?

Bishop: We do have some of these data but I haven't analysed them yet. There are a couple of children in this sample with pragmatic problems who have older siblings with classic autism.

U. Frith: To my mind there is a pressing need to study variable or discordant cases, either in twins or in multiplex families. It is extremely important to verify how different they really are. One possible outcome is that these cases look very different on the surface, but are not essentially different in terms any neurocognitive deficits. The alternative outcome, namely that there are differences in the basic neurocognitive deficits even in genetically related cases, is extremely interesting. Susan Folstein gave a hint of this possibility. She analysed different items on the ADI and showed cross correlations on different items between two affected siblings. Mostly these were high. However, there was one dimension where there was no cross-correlation on ADI measures between affected siblings. Strangely enough, this was in the social communication scale. This needs to be followed up, because after all social communication impairment is core to autism. Could it be that as currently assessed on the ADI it is less subject to genetic influence than other signs of autism? If we measured social impairment in different ways (for example by laboratory tests) would the same lack of cross-correlation be obtained?

Bishop: I take your point in general. Moving away from whether or not subjects meet criteria, to exploring to what extent these symptoms are similar or different within children, is absolutely key. We have only just begun to do this. The other issue that concerns me is whether it is appropriate to be looking at the symptom level, and whether we shouldn't perhaps be using some of these measures of underlying processes such as Francesca Happé was talking about. My worry is that I am not convinced that we have good measures with adequate psychometric properties.

Folstein: I also mentioned that both Peter Szatmari and Jeremy Silverman, using two separate autism data sets found the same thing: most of the ADI components do have sib–sib correlations, but not social intent.

Rutter: What do you conclude from that?

Folstein: First I thought it was related to the birth order severity effect. There are, however, differences in the age at which the ADI was done. Older children were given more severe ratings by their parents. I don't know whether that has something to do with optimism or different interventions in different cohorts. Another thing that I thought is that social interaction is the sum of a lot of different parts. It is not a separate entity. I also feel that when I see the parents, sometimes. They are unsociable for several different reasons, or unsuccessful in their sociability, rather than intending to be unsociable.

Fombonne: There might be some contrast effects, as well. If the same informants report on two different children of their own, they might artificially increase the contrast between the two, as has been shown in twin studies.

U. Frith: In this case you should also find low correlations in the other dimensions.

Lord: Not necessarily. In the ADI, the codes for the social questions are more clearly pegged to the questions. Frequently, in the last sections of the ADI, the examiner asks the parent whether the child ever behaved in a certain way, and the parent responds by describing a behaviour that is coded under other headings. If you ask, 'Does your child have any usual interests?' The parents might say, 'Yes, he spins everything that he sees'. I wonder if there you might get more similarity, not in terms of what the parents say but in terms of the examiner who codes it, than in social where the code is more clearly prescribed.

Bishop: I was very surprised that people were not reporting or making use of the data from the ADI on current functioning as opposed to earlier functioning. We re-scored our ADI data in terms of what subjects are like now rather than what they were like in the preschool period. It was clear that a lot of the children improved quite markedly over time. This might be a useful thing to do more generally. Because autism is regarded as a lifelong disorder, once it is diagnosed people seem to lose interest in how symptoms may change. But certainly for these marginal children they can change a lot.

Folstein: When we just put in the 'ever' codes there was not enough variation in the coding to get any sensible factors. Most autistic subjects have most of the symptoms at one time or another, which would be coded as 'ever'. We didn't want just to put the current ones in, because somehow this didn't seem to give a good view, so we put both in.

Bishop: I'm thinking merely in terms of documenting the natural history of this disorder. I would have liked to know how our children compared with others in terms of changes over time. I could not find anything in the literature on this.

Sigman: We have looked at change. We see stability in the low-functioning children, and parents report decreased severity of symptoms in the high-functioning children.

Buitelaar: Might Susan Folstein's finding of an absent sibling correlation for social deficit scores be an artefact due to the fact that both siblings have high deficit scores in the social domain, and that there is reduced variability?

Folstein: There is a broad distribution of the scores on the social factor.

Rutter: You would have to postulate that you don't have that methodological problem with the other symptoms. I would be surprised if that was the case.

Bailey: It is worth adding that in the IMG SAC sample we have exactly the same finding: there is no familial clustering in the ADI social domain, but there is clustering in the non-verbal communication and repetitive domains. This is correcting for IQ and age when the ADI was carried out.

Dawson: It is a question of variability. We would anticipate more variability in general in language and repetitive behaviours.

Bailey: No. The possible ADI score is much higher in the social domain.

Folstein: There are more items there, but what I did was to make them all come out to a maximum of 1, to account for the fact that different numbers of items were loaded on different factors. This wasn't an issue in our analysis.

Rutter: Chris Hollis, could you say something about Judy Clegg's further follow-up under your supervision?

Hollis: This adds to Francesca Happé's discussion about the similarities and differences between cognitive processes involved in autism and SLI. We continued Mike Rutter and Lyn Mawhood's follow-up of the receptive SLI group into their mid-30s. Rutter and Mawhood had previously found that the SLI group had quite marked and unexpected social impairments in their early 20s. We had two contrasting groups: an IQ matched control group and siblings without a history of language disorder. We assessed them in terms of their social function, language and literacy, phonological processing ability (non-word repetition) and three different measures of theory of mind. The SLI group had significant impairments on both phonological processing and theory of mind measures—but these two domains of impairment were not correlated. This finding suggests some independence between on the one hand, phonological processing, language and literacy and on the other, theory of mind, social cognition and social functioning.

Bishop: I'd be happy with that on the basis of my viewpoint.

Hollis: For various reasons we didn't re-assess the autism group. So, I am interested in whether cognitive measures of language processing, such as non-word repetition are abnormal in autism.

Bishop: Kjelgaard & Tager-Flusberg (2001) did that and found that many of the children with autism were very poor at non-word repetition. If you looked at the group as a whole on average they performed very poorly, but there were still some with autism who were doing fine. Nicola Botting and Gina Conti-Ramsden have found the same with pragmatically impaired children (Botting & Conti-Ramsden 2002): some are severely impaired and some score within normal limits on non-word repetition. It is a messy picture. There are variable symptoms that occur probabilistically without seeming to have very strong causal links to one another.

Rutter: You have a fascinating set of data and a persuasive model. What I find really puzzling, though, is why this language group was so relatively normal in their social behaviour early on, and yet the individuals developed quite marked problems later. Whether measured psychometrically, socially or behaviourally, this was the case. Why so late?

Bishop: One answer could be that this is not like autism at all and that this is the consequence of being stuck in the big wide world not understanding much of what is going on around you, in a rather unsympathetic environment. It may be that this is a symptom whose manifestation depends on the intervention that the children receive. The reason I have tended not to favour this explanation is that children with profound deafness don't seem to end up looking like children with autism, despite their poor understanding of oral language. Of course, many deaf children have exposure to a rich language community through sign language, but that is not always true in the UK. Some deaf children are stranded in mainstream schools without other deaf people around them, yet they do not become autistic We tend to regard social impairment as a consequence of poor understanding, but if you interact with deaf children you find that they are socially so normal.

Hollis: I don't think the social difficulties seen in some children with SLI can just be explained as a consequence of their language problems. The finding of increasing social difficulties while language function improves over time argues against this. We matched the SLI group with controls with equivalent IQ and found the controls were functioning socially far better. We then compared the SLI group with a performance-IQ matched sample from the National Child Development Cohort (NCDS). Again, the SLI group were functioning much worse than IQ matched controls. While at one level this may look like evidence for a possible causal link between language and social impairments, the underlying cognitive mechanisms involving phonological processing and social cognition appear to be independent. This suggests that both types of cognitive deficit may be required to produce social impairment and SLI, whereas a specific phonological processing deficit may result in SLI or dyslexia without significant social impairment.

Bishop: You said you matched on IQ. Is that the same as language functioning? Was this verbal IQ?

Hollis: They were matched on performance IQ.

Bishop: When you were conversing with them, were they able to understand at speed and formulate language at speed in a social interaction as fluently as other people? I do think sometimes that children can look all right in the test situation but still not be able to perform so well in everyday life when they are under time pressure.

Rutter: Very few of these adults with a developmental language disorder would be regarded as showing autism. Their social behaviour was closer to autism than was the case when they were young but the groups with autism and specific language impairment continued to be different in important ways (Howlin et al 2000).

Charman: What are the most important differences between an adult high-functioning autistic sample and SLI?

Rutter: There is less in the way of repetitive and stereotyped behaviour with SLI. In terms of language, as in the earlier follow-up, there was less language abnormality as distinct from poor communication.

Bishop: What is their non-verbal communication like?

Hollis: It is difficult. They weren't initially selected as being a pragmatic-impairment group, so probably not all of them would also fit your pragmatic-impaired group.

Bishop: Certainly, within our pragmatically impaired group we see some who have dreadful non-verbal communication as well. We have some who have good eye contact and ability to use facial expression: they look very normal non-verbally, but they come out with odd things and use rather stereotyped and odd intonation that are classic for high-functioning autism.

Monaco: With regard to the late-onset of the systems in the pragmatically impaired group, could you not look at the families and use variance component modelling to get at whether this is 'environmentally' induced or is primarily genetic? If you are going to break the measures into some kind of distribution, you can then compare this with the variance of the siblings and attribute this to environmental or genetic causes.

Bishop: The trouble is, it could be genetic and late-onset.

Monaco: But if they are purely environmental it will come out.

Bishop: I don't see how this relates to early or late onset. It would just tell us whether things are heritable.

Folstein: Helen Tager-Flusberg's sample has been alluded to. I have been involved in this study, and one of my roles was to do a psychiatric interview with the parents. I was struck by how often the children with SLI had social phobias that were particularly related to speaking in public, such as talking on the telephone or asking a stranger for directions, even after they had acquired quite adequate language capabilities. Their mothers would say that for so long they were unable to speak well that they began to avoid it and became fearful of it. On the other hand, if they had to be in a play they weren't as bad because then they memorised and practised their lines. Their problem is with spontaneous speech. Helen and I are now doing a study comparing the language phenotypes—including pragmatic language—of the parents and one randomly chosen sibling in autism and SLI families. The probands are matched on verbal IQ. We still have children in the old 'mixed' language/autism group.

Rutter: While we are discussing age manifestation differences, can anyone help explain the findings in our study of Romanian adoptees. In the published paper (Rutter et al 1999) we compared them with one of Cathy Lord's longitudinal studies. The two groups were indistinguishable on the ADI at age 4 but they were already different at 6 years. We are now about three-quarters of the way through a further follow-up at age 11. The findings so far suggest appreciable further change. The circumscribed interests that were so striking early on have gone completely in some cases, and have faded in the majority. The language abnormalities are also much less evident. With some important exceptions, their social behaviour would no longer be regarded as autistic-like. The children have plenty of social problems, but they seem to have more in common with disinhibited attachment. I am as puzzled over why the autistic-like behaviour in this group diminished with age, as I am with the increase with age in such behaviour in the SLI group.

Skuse: We have to come back to the notion of a sensitive period in early postnatal life, during which an experience that would normally have occurred didn't occur. There is no reason to think that this particular group was genetically predisposed in any way to be autistic, and they weren't abandoned there because they were autistic-like.

Rutter: The average age of admission was just a few weeks, so that would have been extremely unlikely.

Skuse: I think what we are therefore looking at is some environmental deficit that occurred at a period when such experience would be ubiquitous in the general population. It was this deficit which perturbed the development of a neural circuit that has eventually righted itself, insofar as the autistic features are ameliorating in later childhood. What was that experience? Following on from what I was suggesting yesterday about the importance of eye contact in social development, I wonder whether any of these infants had any significant face–face interaction with their 'care givers' during that early period. Infants have, as a matter of course, intense interest in eye-to-eye contact during the postnatal period.

Rutter: We know that they didn't have much interaction, as shown by many reports from people visiting the institutions. The children were fed by bottles being stuck in their mouths and left there, and there were no toys. There was just one staff member for 30–40 children. One can be fairly sure that there was little opportunity for face–face interaction.

Skuse: We propose that there is a very primitive neural circuit that is anticipating such eye contact occurring, and this arouses huge interest in the infant within hours after birth. If this eye contact doesn't happen for some sustained period of time, then there will be perturbations in this circuit. Some autistic features could well be consequential to that. The importance of this early face to face contact for neurodevelopment is pointed up by the LeGrand et al (2001) study of infants with congenital cataracts, who had impaired face processing abilities in later childhood.

Lord: It wouldn't necessitate as much eye contact as middle-class children get in western societies. There are such cultural differences in how much time babies spend, for example, being carried on backs.

Skuse: I don't know of any cultures in which mothers habitually carry their infants on their backs without making eye contact with their infants during feeding, for example.

Sigman: I have looked at this in African children because of some anthropological literature that suggested there is very little face–face eye contact.

In fact, if you count the amount of eye contact that African babies have with their siblings, then their experience of eye contact is equivalent to that of American babies. A paper describing these results is in press in the *Journal of Cross-Cultural Psychology*.

Bailey: The Romanian adoptees are a very interesting sample. One assumes that the vast majority of babies went into the institution with normal brains, and that what followed then dramatically altered the course of development. What has struck me in hearing about them is the almost total environmental deprivation they were subjected to. They were just left alone for the vast majority of the day. I have no trouble in accepting that various experience-dependent mechanisms cause deficits in social interaction and communication. What I find particularly interesting is the presence of circumscribed interests. This comes to the point that Frankie Happé was raising earlier: there might be multiple routes to get to circumscribed interests. In autism it may be because particular brain systems are affected. It looks as if environmental deprivation—and its knock-on effects on social skills and communication—is another means for reaching circumscribed interests. It is interesting that the behaviour was there and that it has disappeared as the environment has become enriched.

Rutter: There are two caveats. First, we don't know about prenatal alcohol exposure. This is likely to have been a problem in some. The other is the possible effects of severe stress on the mothers during pregnancy. I agree, though: in general we can assume that most of them will have had normal brains.

Howlin: Did the Romanian babies have eye contact with each other?

Rutter: They were mostly in separate cribs.

Rogers: As the children become ambulatory, leave their cribs and start to interact, can they begin to become social partners for each other in the way that siblings are in large families? Can the peers themselves begin to take the roles of other important people?

Rutter: That may well happen post-adoption, but I doubt whether that will have occurred pre-adoption because they were so delayed in development. Very few of them were even walking. The degree of deprivation in this group was profound.

References

Botting N, Conti-Ramsden G 2002 Clinical markers in relation to pragmatic language impairment and other disorders of communication. Symposium presentation at Symposium for Research on Child Language Disorders, July 2002, Madison

Howlin P, Mawhood L, Rutter M 2000 Autism and developmental receptive language disorder—a follow-up comparison in early adult life. II: Social, behavioural, and psychiatric outcomes. J Child Psychol Psychiatry 41:561–578

Kjelgaard MM, Tager-Flusberg H 2001 An investigation of language impairment in autism: implications for genetic subgroups. Lang Cogn Processes 16:287–308
Le Grand R, Mondloch CJ, Maurer D, Brent HP 2001 Neuroperception. Early visual experience and face processing. Nature 410:890
Rutter M, Andersen-Wood L, Beckett C et al 1999 Quasi-autistic patterns following severe early global privation. J Child Psychol Psychiatry 40:537–549

Why have drug treatments been so disappointing?

Jan K. Buitelaar[1]

University Medical Center Utrecht, Department of Child and Adolescent Psychiatry, B.01.324, PO Box 85500, 3508 GA Utrecht, The Netherlands

Abstract. The title of this contribution involves two consecutive questions: have the effects of medication in autism indeed been disappointing? And if so, why? The answer to the first question depends on whether one focuses on the core social and communicative deficits of autism, or on various complicating behaviour problems. Attempts over the past decades to develop drugs that specifically improve social and communicative functioning have failed. Among the most ambitious attempts were medical interventions in the endogenous opioid system that were motivated from animal models on the involvement of this system in various aspects of social behaviour. By contrast, medications such as the newer antipsychotics, psychostimulants, presynaptic noradrenergic blocking agents (clonidine and guanfacine) and selective serotonin reuptake inhibitors were shown to reduce impairing complicating symptoms of affective instability, irritability, hyperactivity and inattentiveness, aggression, self-injury and stereotypies. The explanation for the medication-refractory status of social and communicative deficits should be sought in at least two related factors: (1) the as yet unidentified neurochemical basis of autism, and (2) the obvious lack of involvement of the main neurotransmitter systems (dopamine, noradrenaline and serotonin) in the pathophysiology of social and communicative behaviour.

2003 Autism: neural basis and treatment possibilities. Wiley, Chichester (Novartis Foundation Symposium 251) p 235–249

Why have drug treatments been so disappointing?

The aim of this chapter is to address the two consecutive questions that are implied in the provocative title. First, have the effects of medication in autism indeed been disappointing? And if so, why? Whether the effects of medication treatment in autism can considered to be disappointing depends to a large extent on the choice

[1]Present address: Department of Psychiatry, University Medical Center St Radbound, PO Box 9101, 6500 HB Nijmegen, The Netherlands.

of the outcome variables. The crucial distinction here is between the effects of medication on the core social and communicative deficits versus the effects on various complicating behaviour problems.

Medication and social and communicative behaviour

Among the most direct and specific attempts to affect the core social and communicative deficits of subjects with autism have been interventions in the endogenous opioid system. These interventions were guided by preclinical data indicating that opioid systems in the brain appeared to be involved in the neuroregulation of social behaviour (Benton et al 1988). Morphine and the endogenous opioid β-endorphin tend to stimulate the frequency of social interactions of animals in low doses, while higher doses cause a reduction of social approach behaviour. Opioids are involved in maternal–infant attachment in animal studies by influencing feelings of social comfort and blocking separation distress reactions (Panksepp et al 1978). These effects of opioids appear to be codependent on prior experiences as social conflict and social isolation. This led to the hypothesis that excessive activity of opioid systems in the CNS would prevent the formation of social bonding in humans and contributes to the pathogenesis and maintenance of autistic symptoms (Deutsch 1986, Panksepp 1979). Direct support for this hypothesis however was lacking since data on plasma and cerebrospinal fluid (CSF) levels of β-endorphin in subjects with autism were inconsistent (Buitelaar 2002). Therapeutic interventions in the endogenous opioid system have included the neuropeptide Org 2766 and naltrexone.

Org 2766 is a synthetic ACTH 4–9 analogue that exclusively affects the functioning of the brain and has lost its peripheral activity on the adrenal cortex. Org 2766 was shown to normalize experimentally induced changes in social behaviour of rats by influencing the integration of sensory information in the amygdala (Wolterink et al 1989). The effects of Org 2766 have been examined in three placebo-controlled clinical trials in 74 children with autistic spectrum disorders in total (Buitelaar et al 1990, 1992, 1996). Detailed observations of the social interactions of these children revealed that treatment with 20–40 mg of Org 2766 per day during 4–6 weeks was associated with significant improvements in mutual eye-contact and in social reciprocity. These changes, however, did translate in only modest improvements in overall functioning as measured by rating scales that were completed by parents and teachers. Furthermore, in a number of children the social activating effects of Org 2766 seemed to be neutralized by concomitantly increased irritability and agitation.

Naltrexone is a potent opiate antagonist which can be administered orally. Initial enthusiasm about naltrexone treatment of autism based on open-label data dampened when the results of placebo-controlled studies became available. The social and communicative core deficits of autism were not ameliorated in three larger treatment studies with dosages of about 1.0 mg naltrexone/kg per day (Buitelaar 2002). Controlled studies that used lower dosages were able to report modest improvements of autistic symptoms (Buitelaar 2002). Higher baseline levels of β-endorphin in plasma were associated with a better treatment response but the reduction of β-endorphin plasma levels *per se* was unrelated to the effect of treatment. Open-label continuation treatment for a period of 6 months of five autistic children who showed a clear individual response in a placebo-controlled trial of 4 weeks, did not reveal later-appearing therapeutic effects on social and communicative functioning (Willemsen-Swinkels et al 1999).

To end this section, a short note on secretin treatments. Secretin is a polypeptide that is present in the so-called 'S cells' in the mucosa of the upper small intestine in an inactive form, prosecretin. An anecdotal report about a child with autism whose condition markedly improved after an open-label treatment with a single dose of secretin led to inflated claims by the media and on the internet. Subsequently, thousands of children with autism may have been treated by secretin injections. There may be found some clues in the neuroscience literature that provide a theoretical rationale for using secretin in autism. For example, secretin receptors are present in the hippocampus, and secretin has been further found to bind to receptors of another peptide (vasoactive intestinal peptide) in the hypothalamus, cortex and hippocampus. As well as a direct neuropeptidergic action of secretin, indirect influences may also be involved, such as on the activity of brain neurotransmitter systems, brain circulation or gastrointestinal permeability (Horvath et al 1998). Carefully conducted and controlled studies in larger samples of children with autistic spectrum disorders failed to observe any benefit of a single dose of secretin (Sandler et al 1999).

Medication and complicating and comorbid behaviour problems

Medications of all sorts of psychotropic classes have been examined in autism, including conventional and newer antipsychotics, old and new antidepressants, psychostimulants, presynaptic noradrenergic blocking agents, mood stabilizers, antiepileptics, β-adrenergic blocking agents and anxiolytics. By and large, these medications have been shown to reduce various complicating behaviour problems such as aggression, hyperactivity, impulsivity, irritability, stereotypies and rigidity, self-injury, negativism and anxieties. In many cases, these effects on complicating problems did lead to improvement of overall functioning, as reflected in better scores on the Clinical Global Impression severity and improvement scales. This does not necessarily mean, however, that the severity of autism was changed. The effects on the core autistic symptoms seem to be indirect and limited in scope and effect size. Moreover, few studies using these medications have attempted to measure directly details of social and communicative functioning.

Haloperidol has been the best-studied conventional antipsychotic in treating autism. Dosages of 0.25–4.0 mg per day for 4 weeks were effective in decreasing motor stereotypies, hyperactivity, withdrawal and negativism in 2–8 year old autistic children. Side-effects such as dystonic reactions, dyskinesias, parkinsonism, akathisia, and autonomous and cardiovascular signs and symptoms limit the use of this drug (Campbell et al 1996). The long-term efficacy of haloperidol is not well documented, however, and the risk for serious long-term side effects such as tardive dyskinesias, induction of anxiety and depression, and weight gain is of great concern. Similar considerations apply to other potent conventional antipsychotics.

The newer antipsychotics have received much interest over recent years, given their lower propensity to induce extrapyramidal side effects at therapeutic doses. In addition, the positive effects of the atypical neuroleptics on the negative symptoms of schizophrenic patients seem promising as a potential strategy to improve the core social deficits of subjects with autism. Pharmacologically, these newer antipsychotics block both dopamine as well as serotonin (5HT) receptor systems. A series of open-label studies in children, adolescents and adults have documented promising clinical improvements following treatment with risperidone. This was replicated in a large multisite National Institutes of Health-sponsored placebo-controlled study regarding 101 children with autism between 5 and 17 years (McCracken et al 2002). All participants were selected for high pretreatment scores on the irritability scale of the Aberrant Behavior Checklist. Risperidone in dosages between 0.5 and 3.5 mg/day was associated with a significant decrease in agitation, self-injury and aggression scores and with significant overall improvement. Objective changes in social and communicative behaviour were not observed, however. Risperidone was well tolerated, the most common side-effects were mild transient sedation and weight gain. There were no serious adverse events as measured by electrocardiogram (ECG) and laboratory tests. The risk for extrapyramidal side effects and tardive dyskinesias when administering risperidone, however, is not totally absent and necessitates low dose treatment, preferentially below 4.0 mg/day.

The psychostimulants are the first-line treatment of symptoms of hyperactivity, inattentiveness and impulsivity. Since subjects with autism are often hyperactive and highly distractible, treatment with stimulants would appear an obvious strategy. A double-blind cross-over study in 10 autistic children aged 7–11, using placebo and two dosages of methylphenidate (10 mg and 20 mg b.i.d.), showed that both dosages of methylphenidate resulted in a significant decrease in hyperactivity. Troublesome side-effects were found to be absent, particularly the worsening of stereotypic movements (Quintana et al 1995). Negative effects of stimulants are thought to occur mostly with mentally retarded children with IQs below 45 or mental ages below 4.5 years. Presynaptic adrenergic blocking agents like clonidine have also been examined in autism and were able to reduce hyperactivity, impulsivity and irritability on a short-term basis. Clinical experience, however, indicates that many patients will develop tolerance to the therapeutic effects of clonidine, which seem to limit its applicabitity in clinical practice.

Specific serotonin-reuptake inhibitors (SSRIs) have been tried in autism, given that these medications are effective in obsessive–compulsive disorder (OCD) and because of the phenotypical similarities between the rigid behaviour patterns in autism and OCD. Fluvoxamine, one of the SSRIs, proved significantly more effective than placebo in a recent 12-week double-blind placebo-controlled trial with thirty autistic adults (McDougle et al 1996a). From the 15 patients who received fluvoxamine eight were categorized as responders. Roughly equivalent results have been obtained with other SSRIs in autism. A similar study with fluvoxamine in subjects with autism younger than age 18, however, could not establish a significant treatment response but documented a high rate of side effects and adverse behavioural activation. Since fluvoxamine given in similar dosages to children with OCD was found to be devoid of distressing behavioural activation, children with autism may be particularly sensitive to serotonin-reuptake blockers. This suggests that the serotonergic system is differentially involved in autism compared to OCD. The results with the SSRIs are encouraging in adults with autism who are characterized by strong behavioural rigidity and OCD-like symptoms. Changes in social behaviour following treatment with SSRIs, though, seem to be secondary rather than primary.

Why have drug treatments been so disappointing?

The relative success of medication in psychiatric disorders other than autism, such as attention-deficit/hyperactivity disorder, mood disorder, OCD, anxiety disorder and schizophrenia seems to be related to two connected facts. First, the medication does affect one or more of the monoamineregic neurotransmitter systems (dopamine, noradrenaline, serotonin) in a powerful way. Second, there is at least some evidence, not only *ex iuvantibus* but derived from for example brain imaging, neurochemical or genetic studies, that one or more of the transmitter systems is critically involved in the pathophysiology of these disorders. Unfortunately, this second consideration does not apply to autism.

Interventions in the dopamine, noradrenaline or serotonin system at best exert modulating effects on social and communicative functioning by influencing basic processes of attention, impulse control, activity level, reward dependence and

TABLE 1 Summary of involvement of the monoamine neurotransmitter systems in autism

Noradrenaline	*Serotonin*	*Dopamine*
CSF: MHPG= (DeLong 1977) Plasma: NE ↑ (Lake et al 1977; Launay et al 1987, Leventhal et al 1990) MHPG= (Young et al 1981, Minderaa et al 1994) Urine: noradrenaline and adrenaline inconsistent: = (Minderaa et al 1994), ↑ (Martineau et al 1994)	CSF: 5HIAA= (Narayan et al 1993, Ross et al 1985, Gillberg et al 1987, 1983, Komori et al 1995) Whole blood: 5HT ↑ in 30% of cases (Naffah-Mazzacoratti et al 1993, Leventhal et al 1990, Cook et al 1990, Minderaa et al 1987, Anderson et al 1987, Hanley et al 1977) ↑[a] (McBride et al 1998) Urine: 5HIAA= (Minderaa et al 1987)	CSF: HVA inconsistent (both=and ↑) (Narayan et al 1993, Ross et al 1985, Gillberg et al 1987, 1983) (Komori et al 1995) Plasma: dopamine ↑(Martineau et al 1994) Urine: HVA inconsistent (both=and ↑) (Martineau et al 1994, Minderaa et al 1989, Garreau et al 1988, Launay et al 1987, Martineau et al 1992) DA ↓ (Martineau et al 1992, 1994)

[a]age effect (only in younger children).
↓ significantly decreased levels compared to control group.
↑significantly increased levels compared to control group.
= not significantly different from control group.

emotionality. Vice versa, studies into the functioning of these neurotransmitter systems in autism have produced few consistent findings. Table 1 summarizes findings of the monoamine transmitters and their metabolites in plasma, CSF and urine. The potential relevance of the dopaminergic system for understanding the pathophysiology of autism comes from observations in animal studies in which the dopaminergic system was found to be involved in hyperactivity and stereotyped behaviours. The levels of homovanillic acid (HVA), the major metabolite of dopamine, in CSF and in urine of subjects with autism have been found to be equal as well as increased, when compared to those of control and contrast groups. The excretion of dopamine in urine has been reported to be lowered, whereas higher levels of dopamine were measured in whole blood of subjects with autism. A recent positron emission tomography (PET) scan study suggested a low activity of the frontal dopamine system in autism.

Current interest in the neurochemistry of autism is focused foremost on the 5HT system. An elevation of the concentration of 5HT in whole blood of individuals with autism compared to normal controls is one of the most robust and well-replicated findings in the neurobiology of autism. The elevation is commonly observed in over 30% of all subjects with autism and the magnitude of the difference in mean level is about 25%. The importance of hyperserotoninaemia in autism, however, had remained unclear for at least two reasons. First, the CSF levels of 5-hydroxy-indoleacetic acid (5HIAA), the breakdown product of serotonin, were not found to differ between subjects of autism and controls. Second, hyperserotoninaemia has also been reported in non-autistic subjects with mental handicaps. A recent attempt to resolve these inconsistencies regarding hyperserotoninaemia in autism pointed to the importance of pubertal and racial factors, when interpreting serotonin levels.

Hyperserotoninaemia appears to be a function of pubertal status (measured in prepubertal but not in postpubertal autistic subjects) and was not found to be present in mentally retarded or cognitively impaired control subjects without autism (McBride et al 1998). Though the mechanisms underlying hyperserotoninaemia have not been fully clarified, increased activity of the 5HT transporter of platelets and decreased binding to the 5HT receptor have been observed (Cook et al 1996). Preliminary findings from candidate-gene studies indicate that the short variant of the promoter of the serotonin transporter gene in one report, the long variant in another report but not a polymorphism of the 5HT receptor gene have been significantly associated with autism (Cook 2001). The clinical relevance of hyperserotoninaemia for autism is further strengthened by reports of positive correlations of whole-blood 5HT with clinical severity and negative correlations with verbal-expressive abilities in autistic probands and their first-degree relatives. Findings concerning the central activity of 5HT metabolism are mixed. Measurements of 5HT metabolites in CSF of autistic subjects have failed to demonstrate consistent abnormalities but neuroendocrine responses to pharmacological probes of the 5HT system were found to be blunted, suggesting a low central tonus of the 5HT system (McBride et al 1989). Using radioactive L-tryptophan as a tracer for serotonin synthesis with positron emission tomography, Chugani et al (1997) observed unilateral alterations of serotonin synthesis in the dentatothalamocortical pathway in autistic boys. Further, acute dietary depletion of tryptophan, a precursor of 5HT, was associated with an exacerbation of stereotyped behaviours rather than with changes of social unrelatedness in drug-free adults with autism (McDougle et al 1996b).

Conclusion and perspectives

It seems fair to conclude from this short review of the neurochemistry of autism that the noradrenaline and dopamine system only play a rather secondary role, if any at all, and that the role of the serotonin system is unclear. This leaves us with an as yet unidentified neurochemical basis of autism. An important caveat should be made, however. Autism basically differs from other psychiatric disorders in being foremost a disorder of brain development that has onset very early in life, during fetal phases of development or shortly after birth. For example, any implication of

5HT in the pathogenesis of autism would be of great interest, given the critical role of 5HT during embryogenesis and maturation of the brain and the modulatory effects of 5HT on a variety of important processes, such as sensory perception, motor function, learning, memory and sleep, which are all often perturbed in autism (Whitaker-Azmitia 2001). That is, the neuromaturational role of 5HT rather than its role as a circumscript neurotransmitter may be relevant to autism. One could imagine further that neurochemical systems that are involved in the early programming of brain development are sensitive to pharmacological manipulations only during those early phases of brain development. Medication treatment at preschool age and thereafter may be outside the critical time window. By the same logic could it be that early neurochemical alterations leave no biochemical fingerprints at later age. The implication is that ongoing efforts to lower the diagnostic threshold age of autism below age 24 months should be combined with research into the early development of the brain and into the effects of medication.

To take the issue a step further, autism may be a disorder in the basic regulatory processes of brain development in general rather than a disorder of a specific neurochemical system or specific brain circuit. By consequence then, neurochemical alterations should be sought at the early regulatory level, and put into terms of the altered expression of certain neurotrophic factors and neuropeptides. Some preliminary evidence for this possibility has been found (Nelson et al 2001).

Autism, finally, stands out from other psychiatric disorders by its high heritability (Cook 2001). Currently, internationally well-concerted efforts are underway to unravel the genetic mechanisms of autism and to localize and ultimately identify genes involved. This has the prospect of potentially documenting new biochemical pathways to normal and abnormal social and communicative behaviour. In turn, this may give new clues to developing effective and safe medications for subjects with autism. Over the past decades, unfortunately, little progress has been made in developing new and effective pharmacotherapies for autism. Medication research in autism has evolved in the slipstream of the psychopharmacology of other psychiatric disorders rather than as a result of targeted investigational activities.

References

Anderson GM, Freedman DX, Cohen DJ 1987 Whole blood serotonin in autistic and normal subjects. J Child Psychol Psychiatry 28:885–900

Benton D, Brain PF 1988 The role of opioid mechanisms in social interaction and attachment In: Rodgers RJ, Cooper SJ (eds) Endorphins, opiates and behavioural processes. Wiley, New York, p 215–235

Buitelaar JK 2002 Miscellaneous compounds: beta-blockers, opiate-antagonists, and others. In: Martin A, Scahill L, Charney DS, Leckman JF (eds) Pediatric psychopharmacology: principles and practice. Oxford University Press, New York, p 353–362

Buitelaar JK, Van Engeland H, van Ree JM, De Wied D 1990 Behavioral effects of ORG 2766, a synthetic analog of the adrenocorticotrophic hormone (4-9), in 14 outpatient autistic children. J Autism Dev Disord 20:467–478

Buitelaar JK, Van Engeland H, De Kogel CH, De Vries H, Van Hooff JARAM, van Ree JM 1992 The adrenocorticotrophic hormone (4-9) analog ORG 2766 benefits autistic children: report on a second controlled clinical trial. J Am Acad Child Adolesc Psychiatry 31:1149–1156

Buitelaar JK, Dekker M, van Ree JM, Van Engeland H 1996 A controlled trial with ORG 2766, an ACTH-(4-9) analog, in 50 relatively able children with autism. Eur Neuropsychopharmacol 6:13–19

Campbell M, Schopler E, Cueva JE, Hallin A 1996 Treatment of autistic disorder. J Am Acad Child Adolesc Psychiatry 35:134–143

Chugani DC, Muzik O, Rothermel R et al 1997 Altered serotonin synthesis in the dentatothalamocortical pathway in autistic boys. Ann Neurol 42:666–669

Cook EH Jr 2001 Genetics of autism. Child Adoles Psychiatr Clin N Am 102:333–350

Cook EH, Leventhal BL 1996 The serotonin system in autism. Curr Opin Pediatr 8:348–354

Cook EH Jr, Leventhal BL, Heller W, Metz J, Wainwright M, Freedman DX 1990 Autistic children and their first-degree relatives: relationships between serotonin and norepinephrine levels and intelligence. J Neuropsychiatry Clin Neurosci 2:268–274

DeLong GR 1977 Lithium carbonate treatment of select behavior disorders in children suggesting manic-depressive illness. J Pediatr 93:689–694

Deutsch SI 1986 Rationale for the administration of opiate antagonists in treating infantile autism. Am J Ment Defic 90:631–635

Garreau B, Barthélémy C, Jouve J, Bruneau N, Muh JP, Lelord G 1988 Urinary homovanillic acid levels of autistic children. Dev Med Child Neurol 30:93–98

Gillberg C, Svennerholm L 1987 CSF monoamines in autistic syndromes and other pervasive developmental disorders of early childhood. Br J Psychiat 151:89–94

Gillberg C, Svennerholm L, Hamilton-Hellberg C 1983 Childhood psychosis and monoamine metabolites in spinal fluid. J Autism Dev Disord 13:383–396

Hanley HG, Stahl SM, Freedman DX 1977 Hyperserotonemia and amine metabolites in autistic and retarded children. Arch Gen Psychiatry 34:521–531

Horvath K, Stefanatos G, Sokolski KN, Wachtel R, Nabors L, Tildon JT 1998 Improved social and language skills after secretin administration in patients with autistic spectrum disorders. J Assoc Acad Minor Phys 9:9–15

Komori H, Matsuishi T, Yamada S, Yamashita Y, Ohtaki E, Kato H 1995 Cerebrospinal fluid biopterin and biogenic amine metabolites during oral R-THBP therapy for infantile autism. J Autism Dev Disord 25:183–193

Lake CR, Ziegler MG, Murphy DL 1977 Increased norepinephrine levels and decreased dopamine-beta-hydroxylase activity in primary autism. Arch Gen Psychiatry 34:553–556

Launay JM, Bursztejn C, Ferrari P et al 1987 Catecholamine metabolism in infantile autism: a controlled study of 22 autistic children. J Autism Dev Disord 17:333–347

Leventhal BL, Cook EH, Morford M, Ravitz A, Freedman DX 1990 Relationships of whole blood serotonin and plasma norepinephrine within families. J Autism Dev Disord 20:499–508

Martineau J, Barthélémy C, Jouve J, Muh JP, Lelord G 1992 Monoamines (serotonin and catecholamines) and their derivatives in infantile autism: age-related changes and drug effects. Dev Med Child Neurol 34:593–603

Martineau J, Herault J, Petit E et al 1994 Catecholaminergic metabolism and autism. Dev Med Child Neurol 36:688–697

McBride PA, Anderson GM, Hertzig ME et al 1989 Serotonergic responsivity in male young adults with autistic disorder. Results of a pilot study. Arch Gen Psychiatry 46: 213–221

McBride PA, Anderson GM, Hertzig ME et al 1998 Effects of diagnosis, race, and puberty on platelet serotonin levels in autism and mental retardation. J Am Acad Child Adolesc Psychiatry 37:767–776

McCracken JT, McGough J, Shah B et al 2002 Risperidone in children with autism and serious behavioural problems. New Engl J Med 347:314–321

McDougle CJ, Naylor ST, Cohen DJ, Volkmar FR, Heninger GR, Price LH 1996a A double-blind, placebo-controlled study of fluvoxamine in adults with autistic disorder. Arch Gen Psychiatry 53:1001–1008

McDougle CJ, Naylor ST, Cohen DJ, Aghajanian GK, Heninger GR, Price LH 1996b Effects of tryptophan depletion in drug-free adults with autistic disorder [see comments]. Arch Gen Psychiatry 53:993–1000

Minderaa RB, Anderson GM, Volkmar FR, Akkerhuis GW, Cohen DJ 1987 Urinary 5-hydroxyindoleacetic acid and whole blood serotonin and tryptophan in autistic and normal subjects. Biol Psychiatry 22:933–940

Minderaa RB, Anderson GM, Volkmar FR, Akkerhuis GW, Cohen DJ 1989 Neurochemical study of dopamine functioning in autistic and normal subjects. J Am Acad Child Adolesc Psychiatry 28:190–194

Minderaa RB, Anderson GM, Volkmar FR, Akkerhuis GW, Cohen DJ 1994 Noradrenergic and adrenergic functioning in autism. Biol Psychiatry 364:237–241

Naffah-Mazzacoratti MG, Rosenberg R, Fernandes MJ et al 1993 Serum serotonin levels of normal and autistic children. Braz J Med Biol Res 26:309–317

Narayan M, Srinath S, Anderson GM, Meundi DB 1993 Cerebrospinal fluid levels of homovanillic acid and 5-hydroxyindoleacetic acid in autism. Biol Psychiatry 33: 630–635

Nelson KB, Grether JK, Croen LA et al 2001 Neuropeptides and neurotrophins in neonatal blood of children with autism or mental retardation. Ann Neurol 49:597–606

Panksepp J 1979 A neurochemical theory of autism. Trends Neurosci 2:174–177

Panksepp J, Herman B, Connor R, Bishop P, Scott JP 1978 The biology of social attachments: opiates alleviate separation distress. Biol Psychiatry 13:607–618

Quintana H, Birmaher B, Stedge D et al 1995 Use of methylphenidate in the treatment of children with autistic disorder. J Autism Dev Disord 25:283–294

Ross DL, Klykylo WM, Anderson GM 1985 Cerebrospinal fluid indolamines and monoamine effects in fenfluramine treatment of autism. Ann Neurol 18:394–396

Sandler AD, Sutton KA, DeWeese J, Girardi MA, Sheppard V, Bodfish JW 1999 Lack of benefit of a single dose of synthetic human secretin in the treatment of autism and pervasive developmental disorder. New Eng J Med 341:1801–1806

Whitaker-Azmitia PM 2001 Serotonin and brain development: role in human developmental diseases. Brain Res Bull 56:479–485

Willemsen-Swinkels SHN, Buitelaar JK, Van Berckelaer-Onnes IA, Van Engeland H 1999 Brief report. Six-months continuation treatment of naltrexone-responsive children with autism: an open-label case-control design. J Autism Dev Disord 29:167–169

Wolterink G, van Ree JM 1989 Opioid systems in the amygdala can serve as substrate for the behavioral effects of the ACTH-(4-9) analog ORG 2766. Neuropeptides 14:129–136

Young JG, Cohen DJ, Kavanagh ME, Landis HD, Shaywitz BA, Maas JW 1981 Cerebrospinal fluid, plasma, and urinary MHPG in children. Life Sci 28:2837–2845

DISCUSSION

Lipkin: The studies that you were describing were 8 weeks in duration. Presumably these were looking at the effects of the drugs in isolation. I was wondering: if you reduce some of these comorbid symptoms, might you get an improvement in response to other types of interventions, such as behavioural modification? If you continue with these strategies to reduce comorbidity over a long period, then perhaps in a year you might be able to see these sorts of changes. Is there funding available to do those types of studies?

Buitelaar: To my knowledge, not yet. This was one of the reasons why we took the open-label extension approach. We have six significant clinical responders, and the parents were very ambitious and enthusiastic, and so we continued for six months. This might have been too short a time, but at this time there were no additional changes in social and communicative behaviours. One of the problems with this study is that if you dose children with naltrexone every child was improved in some outcome measure. Our hypothesis was that there would be some common change in those kinds of social interaction and contact measures, which, however, did not turn out to be the case.

Howlin: Rates of prescribing in the UK and The Netherlands are hugely conservative compared with rates of prescribing in the USA. For example the study by Martin et al (1999) of high functioning individuals found that over 60% had been on at least one form of medication during their lives. Which approach is the right one?

Buitelaar: I think the correct approach lies somewhere in the middle. In Europe we are sometimes too soft. An autistic child may be fairly hyperactive and irritable, and because of this uncontrolled behaviour they might not be able to attend school. Perhaps in the Netherlands they wait for months, trying out behavioural treatments, before they use drugs. The other side is that when you prescribe medication you have to monitor actively, so this takes lots of time. On the other hand we need to recognize the enormous pressure from parents to try anything. They will use vitamins and all kinds of alternative medication. The parental pressure is much stronger in the USA.

Folstein: I was thinking about the fact that the social intentionality scale is not a genetically viable scale. Perhaps it is everything else added up together. I was also thinking about the way that I approach treatment: that is, I try to make specific psychiatric diagnoses for each case and treat accordingly. I would never consider treating autism as an entity. The most common thing is a child who won't stay seated in school because they are so active, but a surprising number of them do respond to low doses of Ritalin. Occasionally the children have some appetite loss, but not often. They don't change in their autism, but it makes them better able to concentrate. The other common thing is the combination of mood disorder, anxiety and OCD. Again, these things take up a lot of time and the child is usually miserable. Mood abnormalities in autistic children are easy to spot. If they have periods of elevated mood, you need to give them a mood stabilizer before you give an SSRI. Barring that, they will respond to SSRIs. I don't think that SSRIs improve their social cognition, but they are more willing to use the speech that they have for social purposes. While we can treat specific symptoms, we don't have a clue about how to approach psychopharmacology for autism as such.

Buitelaar: I agree, but we should not be fatalistic and passive when faced with this cohort of symptoms. We should consider the possibilities without promising too much for the core symptoms and giving false hope.

Folstein: You can almost never get rid of the compulsive symptoms altogether, but medication does reduce its 'driven' nature, so that the behaviour can be more easily interrupted.

Baird: There are a number of things that are very interesting about this whole area. The striking response to placebo in this group is very interesting, and we should bear this in mind whenever we are embarking on treatment. I agree with Susan that it is helpful to target symptoms and then look at the drug that will make a difference to that symptom. We have had huge problems with side-effects, and this is a matter of concern, particularly with the newer antipsychotics. The timing at which one might use these or not use these is important. It is a matter of concern in the UK about putting young children on some of the newer antipsychotics.

Buitelaar: The weight gain issue is one that should be handled by proactive information to the parents. It is not automatic but depends on lifestyle factors. There have been extension trials of conduct disorder with risperidone for over two years. They have shown that sedation is mainly limited to the first two or three weeks of treatment. Prolactin changes are no longer seen after 10 months. There are no changes in ECG. So safety has been researched pretty well, and better than for any other psychotropic medication for children on the market.

Hollis: One of the things that struck me is the large variability in drug response, and the difficulty of predicting in advance which drugs might be most effective in each child. The question is to what extent this variability in clinical response reflects underlying genetic variability. You talk about waiting for advances in genetics of autism, but might genetics not also benefit from utilising this variability in drug response to create possibly more homogeneous sub-groups for analysis?

Buitelaar: Our peptide and naltrexone studies were prior to the era of genomics. There is now a possibility of using pharmacogenomics in terms of predicting side effects and efficacy of medication.

Monaco: Can you comment on the epilepsy in autism? Has anyone considered doing preventive treatments for this epilepsy, and do they have any effect on the other symptoms? The emphasis is placed on the neurotransmitter systems that regulate the connectivity and we think it is a connectivity problem.

Buitelaar: The clinical experience is that carbamazepine has a better effect on the behavioural disturbances on autism than other drugs when there is an association with epilepsy and autism. There has never been a trial though.

Bailey: Clearly, it is possible that there are specific genetic risk factors for epilepsy. When we compared the sibs in the multiplex sample there was familial clustering for epilepsy. It would be a significant advance, to identify a subgroup of individuals who are at increased risk of developing seizures.

Fombonne: I wanted to focus on the results with Fluvoxamine. There is a study showing clear beneficial effects. The same author also published a study on tryptophan depletion that led, in his sample of high-functioning autistic adults, to a marked worsening of behaviour, including of some social behaviours. In the Fluvoxamine trial, there was a clear improvement in social relatedness. It seems therefore that the SSRIs target some symptoms that are not too far from the core social deficits characteristic of autism. In this context I am surprised by the negative study you reported. It could well be that this study was underpowered, or that there were too many side effects due to the choice of Fluvoxamine. It would be incorrect to conclude that SSRIs don't work at the present stage. There is a lot of work ongoing at this point in time looking at the efficacy of other SSRIs. One strategy I would like you to comment on is looking at the response to SSRIs in relation to serotonin levels. The hyperserotoninaemia in autism applies to just about 30% of subjects; could that marker be used to select more efficiently SSRI responders?

Buitelaar: In the data from McDougle et al in children and adolescents, there was no relation between baseline serotonin levels, change in serotonin levels and change in symptomatology.

Bailey: What do you conclude from this?

Buitelaar: You can conclude that the core social and communicative deficits of autism are not driven by the serotonin system.

Bishop: I don't know enough about the neurotransmitters to know whether this is possible, but is there any increase in autism in populations such as children whose mothers were addicted to cocaine or other drugs?

Buitelaar: Not that I know of.

Skuse: Jan Buitelaar made a point that was a little provocative. He suggested that because certain classes of drug don't seem to be effective in modulating certain symptoms in autism, we could thereby gain some idea what neurotransmitter systems those drugs are targeting, and he seemed to be implying that we could therefore rule out those neurotransmitter systems as having anything to do with the symptomatology of autism. Is that true? And is the converse true: if you did show that certain symptoms responded to certain classes of drug, and you had some idea what neurotransmitter systems those drugs were targeting, would that tell you those symptoms were related to this neurotransmitter system?

Buitelaar: To some extent pharmacological manipulations are not the only way to refute the biological underpinnings. However, at least you have a problem where you have a very powerful intervention, for example in the dopamine system, and this intervention is not effective. Then you have a very long way to go before you prove the involvement of dopamine in social behaviour. Of course, you should look for converging evidence for that statement, from the likes of biochemical and physiological studies. Outside the serotonin system there is no clear evidence of biochemical alterations in autism. This leaves us struggling to explain the involvement of other neurotransmitter systems. On the other hand, you might say that in ADHD there has been some success with stimulants which affect dopamine and noradrenaline. This doesn't mean *per se* that ADHD is a dopaminergic disorder. We should look for converging evidence.

Bolton: I wanted to pick up on some of Susan Folstein's earlier comments and highlight the fact that it can be extremely difficult to evaluate the core symptoms of the syndrome as well as the associated psychiatric disturbances, particularly when there are communication problems of a marked degree. How do you, for example, evaluate the presence of anxiety or mood disorder in individuals that cannot speak? Moreover, we haven't yet developed good measures of change in either the core symptoms or associated features. We need to do more work on developing measures that will better help us evaluate the benefits and disadvantages of drug treatments.

Rutter: Jan, let me press you a bit about looking ahead. Amongst other things, you were arguing for more multicentre studies. Why would one want to do these? I agree that from a practical clinical point of view, the fact that someone is autistic should not prevent one using pharmacotherapy to deal with whatever other problems are associated. But it is not in the least bit obvious to me why, given that the existing studies show so little effect on core symptoms, given that there are no good leads, and given that clinical experience leads to the same conclusions, one would want a large multicentre study?

Buitelaar: Even if you look for drug effects on comorbid symptoms, such as anxiety or depression, you need a sample of at least 100 for proper assessment and analysis. It is very difficult to get a population of this size in just one centre. This is because there is a huge intrasample variation in the response. Too many studies have been underpowered and this has prevented them giving a clear answer.

Rutter: So this is in relation to non-core symptoms.

Buitelaar: Yes, even for that purpose. If you have your hands on some interesting compound for the core symptom, this could be worked out in a small study and then this could be extended.

Rutter: My other query is that you are arguing for earlier and earlier drug treatment, and you are bringing in notions of programming. What would be the justification for this? Is there any reason to suppose that very early drug treatment would work?

Buitelaar: If you take the idea that autism involves altered development of brain systems seriously, this is the critical period. Changing brain biochemistry in the first year of life may have the advantage in that you are working with maximum neuroplasticity.

Charman: This would make sense if we did have drug treatments that could target and alleviate the core symptoms that we thought might be related to brain pathology. In terms of what we know drugs can do with the associated symptoms, one of the difficulties about treating very young children when they are first diagnosed is that it is quite difficult in young children to make a good assessment of whether they are hyperactive or anxious, if they are autistic.

Lipkin: Nonetheless, if we develop the insights and tools required to identify associated symptoms at earlier timepoints, the use of pharmaceuticals to modify those symptoms could be profoundly important. Examples would include enhanced ability to participate in and benefit from behavioural interventions in individual and group settings.

Rogers: I think we can assess many of those secondary symptoms in young children.

Reference

Martin A, Scahill L, Klin A, Volkmar FR 1999 Higher-functioning pervasive developmental disorders: rates and patterns of psychotropic drug use. J Am Acad Child Adolesc Psychiatry 38:923–931

Can early interventions alter the course of autism?

Patricia Howlin

Department of Psychology, St. George's Hospital Medical School, Cranmer Terrace, London SW17 0RE, UK

Abstract. Interventions for autism have come a long way since the condition was described by Kanner in the 1940s. At that time, autism was considered to be closely linked to schizophrenia, and inadequate parenting was viewed as the principal cause. Psychoanalysis was often the therapy of choice, but there was also widespread use of the drugs and even electroconvulsive treatments that had been developed for use in schizophrenia. Over the years, as autism has come to be recognized as a developmental disorder, interventions have focused instead on enhancing developmental skills and on ways of ameliorating behavioural difficulties. Recognition of the role that language deficits in particular play in causing behaviour problems has led to a focus on the teaching of more effective communication skills. The need for early support for families and appropriate education is also widely acknowledged. Nevertheless, follow-up studies indicate that the prognosis for the majority of individuals with autism remains poor. And despite claims to the contrary, there is little evidence that very early, intensive interventions can significantly alter the long-term course of the disorder. The paper discusses findings from follow-up studies over the years and assess the impact of different intervention procedures on outcome.

2003 Autism: neural basis and treatment possibilities. Wiley, Chichester (Novartis Foundation Symposium 251) p 250–265

Changing views of autism

Recognition and understanding of autistic spectrum disorders have come a long way since Kanner first described the condition in the early 1940s. At that time, autism was believed to be an early form of schizophrenia, which in turn was generally viewed as psychogenetic in origin (Kanner 1943) and indeed, the view that poor parenting was a principal cause persisted over several decades. Autism was also considered to be a very rare condition occurring only in approximately 2 to 3 children per 10 000 (Lotter 1966). Now it is recognized as a genetic disorder (International Molecular Genetics, Study of Autism Consortium 2001) with a prevalence much greater than indicated in early epidemiological studies. In a recent comprehensive review, Fombonne (2002) concludes that a conservative estimate for the combined rates of autism, Asperger syndrome and other pervasive developmental disorders is around 27 to 28 per 10 000, far in excess of the original estimates. Indeed, more detailed diagnostic studies have suggested that prevalence rates for autism and related pervasive developmental disorders may exceed 62 per 10 000 children (Chakrabarti & Fombonne 2001). Whilst such data cannot be used to support claims of an 'epidemic' of autism (California Department of Developmental Services 1999) it is clear that the frequency of autistic spectrum disorders is many times greater than originally thought. The practical implications of this for paediatric, child psychology and psychiatry, and educational services are considerable.

Changing approaches to intervention

Recognition of the pervasiveness, persistence and complexity of the problems associated with autism has led to a continuing search for ways of overcoming, or at least ameliorating these difficulties. Here, too, there have been many changes in approaches over the years. In the 1940s and 50s, psychoanalytic treatments, with both the child and the family, were frequently the treatment of choice, but medical interventions including electroconvulsive therapy (ECT) as well as drugs were also commonly used. In the 1960s autism was viewed more as a behavioural disorder, and operant based procedures began to increase in popularity. These focused primarily on the elimination of 'undesirable' behaviours, rather than the establishment of spontaneous social and communication skills; there was heavy reliance on food rewards, and aversive procedures (including the use of electric shock) were widely employed. In the 1970s, much influenced by the work of Michael Rutter (Rutter 1972), autism became to be recognized as a cognitive and developmental disorder. Therapy moved towards the use of individually based treatments, with the emphasis on a structured approach to teaching, the involvement of parents in therapy, and the importance of naturalistic intervention programmes (Hemsley et al 1978, Schopler 1978, see also Koegel & Koegel 1995). Over the last two decades there has been an increasing emphasis on home and school based programmes; a strong movement against the use of aversive procedures, and a focus on inclusive approaches to education. Recognition of the impact of communication deficits on children's functioning has also led to a very different approach to dealing with so-called 'challenging' behaviours (Durand & Merges 2001).

There are many other claims for the effectiveness of specific approaches to treatment, some of which are even claimed to result in 'recovery' from autism (e.g. Perry et al 1995). These include holding therapy, scotopic sensitivity training, facilitated communication, conductive education, 'Daily Life Therapy',

Gentle teaching, the 'Son Rise Programme' and the Walden Approach (see Howlin 1998, Lord & McGee 2001). Various dietary and medical interventions have also been proposed. Amongst the most recent of these are the reduction of mercury levels in the child's body plus a 36 ingredient vitamin/mineral/antioxidant supplement (Rimland 2001) or secretin infusions (Horvath 1998). Other therapies lauded by the popular media have involved swimming with dolphins, swinging around in nets or cranial osteopathy. Few of these treatments have been subject to any form of experimental investigation. Moreover, of those that have, facilitated communication is now generally discredited (Mostert 2001); auditory integration therapy has been found to have no positive effects (Mudford et al 2000, Dawson & Watling 2000); Fenfluramine, a drug previously widely used in the USA, has been virtually withdrawn because of adverse side effects (Campbell & Cueva 1995); and control trials of secretin (Owley et al 2001) indicate no advantages over placebo.

The outcome of early intervention programmes

In contrast, early intervention studies, using a variety of developmental, educational or behavioural approaches have been found to have positive effects, with significant improvements being reported in language and social behaviours, and self care, motor and academic skills. (Dawson & Osterling 1997, Rogers 1998). When control groups have been involved (which is by no means always the case) the gains made by the experimental children have generally been greater, and comparatively more have subsequently entered mainstream school. Nevertheless, although children's overall level of functioning appears to be enhanced by programmes of this kind, there is less evidence of a marked reduction in autistic symptomatology. This conclusion holds even for the very early, intensive behavioural interventions of Lovaas and his colleagues (Lovaas 1996). These have been reported as bringing about major changes in children's cognitive ability (sometimes by as much as 30 IQ points or more), and it is claimed that around 40% of the children involved become 'indistinguishable' from their normally developing peers. However, such claims have been disputed and it is evident that the way in which IQ is measured, both prior to and following intervention, can have a significant impact on results (Magiati & Howlin 2001). The limitations of the outcome measures used have also been criticised (Gresham & MacMillan 1998).

Generally, the number of cases involved in evaluative studies of early behavioural/educational interventions remains very small, and blind, randomized control trials are virtually non-existent (Lord 2000). Many other questions remain to be answered concerning the *specific* effects of these early programmes, since the content is often very eclectic and the relative importance of the different components of treatment is unknown. There are also questions about the relative merits of one-to-one vs. group teaching, or home-based vs. school-based programmes. Similarly, surprisingly little is known about the characteristics of the children who appear to respond best to programmes of this kind. Thus, although several studies indicate that IQ and language levels are important predictive variables (Harris & Handleman 2000), with the most able children making greater progress, this is not invariably the case (Koegel 2000). The optimal length of time in therapy is another issue. Lovaas (1996) has proposed that 40 hours a week of therapy, over two years or more is required, although positive results have been reported for less intensive behavioural programmes (Gabriels et al 2001). On the whole, the more successful early intervention programmes appear to involve a minimum of around 15 to 20 hours a week, last at least 6 months, and require a relatively high adult:child ratio (Rogers 1996). Finally, but perhaps most importantly, longer-term evaluations, covering many different aspects of functioning are still required in order to evaluate the true effectiveness of early intervention programmes.

The recent New York State Department of Health Review (1999) of interventions for pre-school children with autism also highlights the poor quality of much of the research in this area. Of the several hundred published papers assessed only a minority met basic criteria for experimental research. It was concluded that there was no evidence for the effectiveness of many therapies, including sensory integration, touch therapy or auditory integration. The use of facilitated communication was 'strongly discouraged'. Of the various medical or dietary interventions reviewed (secretin; immunoglobulin injections; anti-yeast treatments; vitamin or dietary manipulations) again there was little, if any evidence of effectiveness, and on the whole, because of serious concerns about side-effects, or long-term sequelae, these could not be recommended. There was more evidence in favour of pharmacological treatments, but little information on the likely benefits and disadvantages for specific sub-groups of children, and serious concerns were raised about the long-term effects, particularly when used with very young children. The most positive findings related to behavioural and educational programmes, particularly those with a focus on reducing behavioural problems, improving communication, or enhancing social interaction. The involvement of parents in therapy was also viewed as important. Interventions of moderate intensity (around 20 hours a week) appeared to produce better results than shorter programmes, but there was no evidence that programmes of 40 hours or more a week provided significantly greater benefits. Overall, although applied behavioural techniques were deemed to be useful, as found in similar reviews no single, specific programme could be recommended. Thus, Sheinkopf & Siegel (1998) concluded that early behavioural/educational interventions are a good option for children with autism, and certainly far better than no intervention, or non-specialist school placements. However they found no evidence in favour of any one approach, any one level of intensity, or any particular degree of structure. Similarly, Prizant & Rubin (1999) note that, given the current state of research in the field, no one approach has been demonstrated to be superior to all others, or to be equally effective for all children.

Effective components of intervention

Although it is apparent that no specific treatment has the advantage over all others, research into psychological or educational interventions more generally has highlighted what seem to be important components of any programme for children with autism (Carr et al 1999, Koegel 2000, Lord 2000, Matson et al 1996, Rogers 2000, Schreibman 2000). Thus, the approaches that have tended to offer most are those that:

- Take account of the characteristic behavioural patterns of children with autism (i.e. their specific social and communication deficits, and stereotyped patterns of behaviour/interest) in developing intervention approaches
- Emphasise the development of skills rather than deficits
- Employ a structured, behaviourally based approach to intervention (i.e. utilize prompting, shaping and reinforcement strategies to enhance skill levels)
- Utilise a 'functional analysis' approach to understanding behaviour problems
- Focus on the development of *effective* communication skills, both verbal and non-verbal, in order to minimise behavioural problems and increase social interaction
- Modify the environmental setting (and the behaviour of carers/teachers) in order to enhance communication and understanding, reduce stress, and facilitate learning
- Use naturally occurring opportunities for teaching and reinforcement
- Recognize the importance of predictability, routine and consistency as important elements in the teaching of new skills and in the reduction of problem behaviours
- Foster integration with typically developing peers

Practical advice for parents, as and when they need it, is also essential. Many parents admit to seeking out alternative treatments because they are simply unable to get the information they need about locally based facilities. Indeed, it can often be far easier to find out, from newspapers or television, about interventions offering the opportunity for swimming with dolphins, than it is for parents to find out about local nursery provision.

The importance of early intervention

Whilst the relative effectiveness of different treatments for children with autism has generated considerable debate and argument, there has been little research on the importance of early intervention *per se*. It is widely accepted that early intervention is vital in helping the child to develop essential skills in the earliest years and in preventing the escalation of later behavioural difficulties. The claims, particularly by Lovaas (1996), that intervention is most effective if it can begin between the ages of 2–4 years has led to a push towards earlier and earlier educational and behavioural programmes. Moreover, in order to ensure access to pre-school intervention, much research over the last decade has concentrated on improving early identification and diagnosis. However, what, in fact is the evidence that early intervention, particularly in the pre-school years, does confer advantages compared with later therapy? Certainly, it makes sense to assume this is the case but in reality there are few data to support this view. Thus, although several studies have reported positive outcomes for children enrolled in intervention programmes prior to age 4 (Anderson et al 1987, Birnbrauer & Leach 1993, Sheinkopf & Siegel 1998) they did not include systematic comparisons between children of different ages. Lovaas (1993) noted that the younger children in his intervention studies did much better than those who were older, but again there was no direct comparison between children who began therapy at the recommended age (i.e. around 2) and those who started later, at around 4 or 5 years. Fenske et al (1985) conducted a small scale comparative study of 18 children, half aged under 5 years, and half aged over 5 when therapy begun. Six children in the early treatment group went onto mainstream school, compared with only one child in the later intervention group. This was despite the fact that the younger children were in therapy for a shorter period of time. However, school placement was the only outcome measure utilized, and the children were not matched prior to the onset of treatment. Recent findings by Stone & Yoder (2001), in showing that the amount of time in language therapy from the age of 2 tends to predict outcome at 4, might also be cited in support of the argument that 'earlier=better'. Similarly, Harrison & Handleman (2000) found that children admitted to a specialist pre-school programme before the age of 3.5 were more likely subsequently to be placed in a regular educational classroom than those who were aged on average 4.5 when pre-school intervention began. However, outcome, in terms of later educational placement was also significantly related to pre-school IQ measures, and the relative importance of IQ vs. age was not explored.

Rogers (1998) notes that 'the hypothesis that age at start of treatment is an important variable in determining outcome has tremendous implications for the field and needs to be tested with methodologically rigorous designs'. Unfortunately, no such designs have yet been employed!

Nevertheless, in the absence of any evidence to the contrary, it would seem to make common sense (and in reality that is all we have to go on) to ensure that families of young children with autism are offered appropriate help as soon as they require this. Given the rigid behaviour patterns of children with autism, it is clear that once certain behaviour patterns are established it can be very difficult to change these. Thus, the earlier effective management strategies and appropriate patterns of behaviour can be put in place, the less are the chances of inappropriate behaviours developing in the future. It is also evident that certain behaviours, whilst entirely acceptable in very young children, become increasingly less so as individuals grow older. Informed advice to parents, on what types of behaviour can lead to potential problems can also be important in preventing difficulties at a later age (Howlin 1998).

There is a danger, however, that focus on the importance of early pre-school intervention could have a negative impact for older children. It is evident that, despite improvements in the age of diagnosis over recent years, many children, particularly those who are more able, do not receive a definitive diagnosis until they reach junior school, or even later. For them, or for the thousands of children who for a variety of other reasons have no access to early intervention, an assumption of 'better late than never' is more appropriate than 'early intervention or nothing'. Harris & Handleman (2000) for example, noted that even the older children in their study made important progress and they are explicit that their data should 'not be taken to suggest that children 4 years of age and older should be denied intensive treatment'. Whatever the age at which the child's problems are recognized, then appropriate strategies can and should be put into place to help deal with these. After all, there is nothing inherent in behavioural principals that suggests they only work up to a certain age. Indeed the clinical psychology literature is full of concrete examples to the contrary. In addition, there is significant research indicating that many individuals with autism show considerable improvements with age (Gilchrist et al 2001, Howlin 2003, Mawhood et al 2000, Piven et al 1996). Moreover, for some, particularly those who are more able, adolescence can often be a period of remarkable improvement and change (Kanner 1973). This is an age at which some children, at least, become more aware of their difficulties and of how they can moderate their own behaviours in order to change the responses of people around them. There are very few intervention programmes geared specifically for the needs of this age group, and this may mean that both they, their families, and the professionals involved in their care are missing out on a crucial opportunity for change.

Studies of young adults with autism also indicate that the input available at school leaving age and beyond can have a major impact on outcome. Venter et al (1992), in their follow-up of higher functioning young adults, suggest that it may be the level of support services available, as much as inherent intellectual ability, that is predictive of good outcome. Another long-term follow-up of adults into their late twenties and thirties (Howlin et al 2003) indicates that although early IQ has some predictive value, above a certain threshold (IQ 70+) outcome is influenced by many other factors, including the educational and employment opportunities that are on offer. Certainly, specialized job schemes, such as those described by Keel et al (1997) or Mawhood & Howlin (1999), can significantly change the course of adult life. It is also apparent, from accounts by individuals with autism themselves, that help at various different stages in life can be crucial in affecting outcome. Wendy Lawson, for example, was a 42 year old mother with four children before being diagnosed with Asperger syndrome. Finally able to 'put away' the earlier misdiagnoses of intellectual disability and schizophrenia, she developed the self-awareness and self-confidence that has allowed her to make use of her considerable skills, and her writings are now an important source of support and advice for families with an autistic child (Lawson 2001).

In conclusion, returning to the original question posed by this paper: can early childhood intervention alter the course of autism? The answer must be 'probably, but we still lack the data to prove this'. Moreover, failure to gain access to specialist intervention programmes at this age should not be taken to mean that all is lost in the future. Autism is a life-long condition, and what evidence there is suggests that at whatever stage in an individual's life *appropriate* support is offered, then positive changes are likely to result.

References

Anderson SR, Avery DL, DiPietro EK, Edwards GL, Christian WP 1987 Intensive home-based early intervention with autistic children. Educ Treat Children 10:352–366

Birnbrauer JS, Leach DJ 1993 The Murdoch early intervention program after 2 years. Behav Change 10:63–74

California Department of Developmental Services 1999 Changes in the population of persons with autism and pervasive developmental disorders in California's Developmental Services System: 1987 through 1998. A report to the legislature. Sacremento, March 1999. *http://www.dds.ca.gov/autism/pdf/autism_report_1999.pdf*

Campbell M, Cueva JE 1995 Psychopharmacology in child and adolescent psychiatry: a review of the past seven years Part II. J Am Acad Child Adolesc Psychiatry 34:1262–1272

Carr EG, Horner RH, Turnbull AP et al 1999 Autistic behaviour support for people with developmental disabilities. American Association on Mental Retardation Monograph Series, Washington DC

Chakrabarti S, Fombonne E 2001 Pervasive developmental disorders in preschool children. J Am Med Assoc 285:3093–3099

Dawson G, Osterling J 1997 Early intervention in autism. In: MJ Guralnick (ed) The effectiveness of early intervention. Brookes Publishing Co, Baltimore, p 307–326

Dawson G, Watling R 2000 Interventions to facilitate auditory visual and motor integration in autism: a review of the evidence. J Autism Dev Disord 30:415–422

Durand VM, Merges E 2001 Functional Communication Training: a contemporary behavior analytic intervention for problem behavior. Focus Autism Other Dev Disord 16:110–119

Fenske EC, Zalenski S, Krantz PJ, McClannahan LE 1985 Age at intervention and treatment outcome for autistic children in a comprehensive intervention program. Analysis Interv Dev Disabil 5:5–31

Fombonne E 2002 Epidemiological trends in rates of autism. Mol Psychiatry 2:S4–S6

Gabriels RL, Hill DE, Pierce DE, Rogers SJ, Wehner B 2001 Predictors of treatment outcome in young children with autism: a retrospective study. Autism 5:407–429

Gilchrist A, Green J, Cox A, Rutter M, Le Couteur A 2001 Development and current functioning in adolescents with Asperger syndrome: a comparative study. J Child Psychol Psychiatry 42:227–240

Gresham FM, Macmillan DL 1998 Early Intervention Project: can its claims be substantiated and replicated? J Autism Dev Disord 28:5–13

Harris SL, Handleman JS 2000 Age and IQ at intake as predictors of placement for young children with autism: a four- to six-year follow-up. J Autism Dev Disord 30: 137–142

Hemsley R, Howlin P, Berger M et al 1978 Treating autistic children in a family context. In: Rutter M, Schopler E (eds) Autism: a reappraisal of concepts and treatment. Plenum Press, New York, p 379–412

Horvath K, Sefanatos G, Sokolski KW, Wachtel R, Nabors L, Tildon JT 1998 Improved social and language skills after secretin administration in patients with autistic spectrum disorders. J Assoc Acad Minor Phys 9:1–15

Howlin P 1998 Treating children with autism and Asperger syndrome: a guide for parents and professionals. Wiley, Chichester

Howlin P 2003 Outcome in high-functioning adults with autism with and without early language delays: implications for the differentiation between autism and Asperger syndrome. J Autism Dev Disord 33:3–13

Howlin P, Goode S, Hutton J, Rutter M 2003 A cognitive and behavioural study of outcome in young adults with autism. J Child Psychol Psychiatry, in press

International Molecular Genetic Study of Autism Consortium, IMGSAC 2001 A genomewide screen for autism: strong evidence for linkage to chromosome 2q 7q and 16p. Am J Hum Genet 69:570–581

Kanner L 1943 Autistic disturbances of affective contact. Nervous Child 2:217–250

Kanner L 1973 Childhood psychosis: initial studies and new insights. Wiley, New York

Keel JH, Mesibov G, Woods AV 1997 TEACCH—Supported employment programme. J Autism Dev Disord 27:3–10

Koegel LK 2000 Interventions to facilitate communication in autism. J Autism Dev Disord 30:383–391

Koegel RL, Koegel LK 1995 Teaching children with autism: strategies for initiating positive interactions and improving learning opportunities. Jessica Kingsley, London

Lawson W 2001 Understanding and working with the spectrum of autism—an insider's view. Jessica Kingsley, London

Lord C 2000 Commentary: achievements and future directions for intervention research in communication and autism spectrum disorders. J Autism Dev Disord 30:393–398

Lord C, McGee JP (eds) 2001 Educating children with autism. Committee on educational interventions for children with autism. Division of behavioral and social sciences and education. National Research Council. National Academy Press, Washington DC

Lotter V 1966 Epidemiology of autistic conditions in young children I: Prevalence. Soc Psychiatry 1:124–137

Lovaas OI 1993 The development of a treatment-research project for developmentally disabled and autistic children. J Appl Behav Anal 26:617–630

Lovaas OI 1996 The UCLA young autism model of service delivery. In: Maurice C (ed) Behavioral intervention for young children with autism. Pro-ed, Australia, p 241–250

Magiati I, Howlin P 2001 Monitoring the progress of young children with autism enrolled in early intervention programmes: problems in cognitive assessment. Autism 5: 399–406

Matson JL, Benavidez DA, Compton LS, Paclawskyj T, Baglio C 1996 Behavioral treatment of autistic persons: a review of research from 1980 to the present. Res Dev Disabil 17: 433–465

Mawhood LM, Howlin P 1999 The outcome of a supported employment scheme for high functioning adults with autism or Asperger syndrome. Autism 3:229–254

Mawhood LM, Howlin P, Rutter M 2000 Autism and developmental receptive language disorder — a follow-up comparison in early adult life I: Cognitive and language outcomes. J Child Psychol Psychiatry 41:547–559

Mostert MP 2001 Facilitated communication since 1995: a review of published studies. J Autism Dev Disord 31:287–313

Mudford OC, Cross BA, Breen S et al 2000 Auditory integration training for children with autism: no behavioral benefits detected. Am J Ment Retard 105:118–129

New York State Department of Health 1999 Clinical practice guideline: report of recommendations. Autism/pervasive developmental disorders. Assessment and intervention for young children (0–3 Years). New York State Department of Health Early Intervention Program. Albany, New York. Available at: *http://www.health.state.ny.us/nysdoh/eip/menu.htm*

Owley T, McMahon W, Cook EH et al 2001 Multisite double-blind placebo-controlled trial of porcine secretin in autism. J Am Acad Child Adolesc Psychiatry 40:1293–1299

Perry R, Cohen I, DeCarlo R 1995 Case study: deterioration autism and recovery in two siblings. J Am Acad Child Adolesc Psychiatry 34:232–237

Piven J, Harper J, Palmer P, Arndt S 1996 Course of behavioural change in autism: a retrospective study of high-IQ adolescents and adults. J Am Acad Child Adolesc Psychiatry 35:523–529

Prizant BM, Rubin E 1999 Contemporary issues in interventions for autism spectrum disorders: a commentary. J Assoc Pers Sev Handic 24:199–208

Rimland B 2001 Mercury detoxification report nearing completion. Autism Res Rev Intl 15: 1–2

Rogers SJ 1996 Brief report: early intervention in autism. J Autism Dev Disord 26: 243–246

Rogers SJ 1998 Empirically supported comprehensive treatments for young children with autism. J Clin Child Psychol 27:168–179

Rogers SJ 2000 Interventions that facilitate socialization in children with autism. J Autism Dev Disord 30:399–409

Rutter M 1972 Childhood schizophrenia reconsidered. J Autism Child Schizophr 2: 315–337

Schopler E 1978 Changing parental involvement in behavioral treatment. In: Rutter M, Schopler E (eds) Autism: a reappraisal of concepts and treatment. Plenum Press, New York, p 413–422

Schreibman L 2000 Intensive behavioral/psychoeducational treatments for autism: research needs and future directions. J Autism Dev Disord 30:373–378

Sheinkopf SJ, Siegel B 1998 Home-based behavioral treatment for young children with autism. J Autism Dev Disord 28:15–23

Stone WL, Yoder PJ 2001 Predicting spoken language level in children with autism spectrum disorders. Autism: 5:341–361

Venter A, Lord C, Schopler E 1992 A follow-up study of high functioning autistic children. J Child Psychol Psychiatry 33:489–507

DISCUSSION

Sigman: I agree with you: I think we need to have focused interventions that are essentially experiments to see whether or not we can identify causality in terms of our interventions. But even before that, there are two other areas where we need more research before we can begin to plan our interventions. One is that we have focused so much on identifying core deficits in autism that we have neglected to investigate individual differences among autistic children to see what predicts later development. We need a lot more research on these individual differences and their consequences if we are going to plan our interventions. The second area is the effect of environmental influences on autistic children. I suddenly realised about three years ago that we have practically no studies of environmental effects on the development of children with autism. Our neglect of this area is probably due to the history of theories about autism. After Bruno Bettleheim, we were all hesitant to think about the environment: we didn't want to make the same mistake he did by suggesting that there was anything the environment could do in terms of causing autism. Aside from biological environmental causes, we were right to be wary. But certainly autistic children should be able to be influenced by the environment. We know that they are, and yet we have had almost no studies of ways in which caregivers or teachers can influence the development of children with autism. This is not the case if you look at the normal developmental literature or the literature on mentally retarded children. So we need many more studies to look at what is going on with autistic children growing up that may account for the differences within the groups. These are two areas where we need more research before we start doing intervention studies. We will never know causality from these correlational studies, but they at least give us the beginnings, and then we can go ahead and try to bring about changes.

Howlin: One of the problems is that there is a desperate need for help here and now and while we are debating the best ways of evaluating treatment effectiveness parents are being seduced by these claims for all sorts of completely wacky therapies. We are biding out time working out proper designs, and they are being sold unproven interventions often at large cost.

Sigman: Coming from the land of miracle cures, I have to deal with this issue all the time. I think we need to warn parents about potentially risky interventions. I follow children long-term, and I worry about the side-effects of treatments, for example the possible cardiac problems in some of the children who were treated with fenfluramine.

Bishop: You mentioned in passing about how you have been involved in training theory of mind, but that it was probably too late. Was this totally ineffective? If so, why? To what extent has there been any input from speech and language therapy approaches to autism? I know that Merzenich and Tallal have argued that their 'FastForword' program, originally designed for children with specific language impairment (SLI), can be applied to children with autism (Merzenich et al 1999), though there is a lot of scepticism about this, and I know of no controlled trials.

Howlin: On the whole, the results of programmes designed to enhance theory of mind are similar to those of social skills training programmes. The evidence is that they can work, but on a fairly circumscribed range of skills. There is some generalization, but not a huge amount. There is very little evidence that such programmes really change children's spontaneous social interactions or fundamental social understanding.

Bishop: Would you therefore not do it?

Howlin: Theory of mind training can enhance skills in certain areas, and there may be knock-on effects of a positive kind. As a component of general teaching programmes it seems a potentially useful thing to incorporate. For most of our educational programmes, whether in mainstream or special schools, there is no evidence for most aspects of them working anyway. However, as the children have specific deficits in these areas, why not try to help them? In terms of language therapy programmes with children with autism, these should, in principle, be, helpful, but the evidence is limited.

Bishop: Many years ago when I trained, I read a bit about the early behaviour interventions. The sorts of things that they were doing with language were so different from what a speech therapist would do. It was like behaviourism, training a child to say something in a given context. Now, of course, the emphasis is much more on communication and trying to develop skills that will generalize.

Howlin: On an individual level, we see children who seem to have responded well to programmes of that kind, but I don't know of any long-term outcome studies.

Rogers: We have hardly begun to explore teaching theory of mind to people with autism. There are only a couple of studies and they have been done in a very traditional way in terms of false beliefs. This ability is also constructed out of other elements of understanding other people. We don't have data on the effectiveness of social skills groups with autistic adults. There are reasons to think that there are many effective ways to start to help both adults and children become more sensitive to other people's behaviours. We have to help people with autism learn to use their cognitive abilities to develop concepts of behaviour. In terms of language intervention, in terms of speech therapy, I only know of one study. Wendy Stone published this recently showing that for outcome measures in her longitudinal studies, the number of hours of speech and language therapy was one of the predictors of better outcomes (Stone & Yoder 2001). But more of the work has been done inside how the language intervention has been developed.

While there are many studies that demonstrate that the discrete trial ABA (applied behaviour analysis) approach does produce speech in children, and you can teach a number of different kinds of language skills in that way, there has also been a nice set of comparisons. Lynn Koegel has done particularly nice work looking at different ways of using applied behavioural approaches to teaching language (Koegel et al 1987). There are pretty consistent findings that the more naturalistic approaches, such as those used in incidental teaching or PRT (pivotal response training), which are still using behavioural principles but are working on child's motivation and natural reinforcers, and following children's own communicated requests, result in better generalization and better maintenance of skills over the long term. In general there does seem to be evidence that one approach is more useful than another in autism for getting to better language (see Delprato 2001). Also, early intervention increases the rate of language. In the published studies that I reviewed looking at the mean rate of useful speech by the age of five, 75–95% of children coming from a range of intensive intervention programs had useful multiword speech by the age of five. McGee has demonstrated that 90–95% of her children talk (McGee et al 1999).

Bishop: This is with a very basic behavioural approach that might be frowned upon by a lot of conventional speech therapists.

Rogers: Yes, it is behavioural, but it is sophisticated. It is quite intensive and carefully delivered. It's built from the communicative functions of language rather than viewing speech as one of many motor behaviours.

Charman: I think that some individual speech and language therapists interested in delivering treatment programs have learned and borrowed a lot from this more pragmatic approach to developing communication understanding, rather than word production. A number of groups are actively conducting research evaluations with those sorts of approaches. Here in the UK there are a number of groups doing that. The problem at the moment is that we don't have any evidence yet. I think we would, if we were to do good enough research studies. Pat Howlin correctly points out that one of the difficulties is that often groups responsible for developing and delivering treatments have been the people reporting on their effectiveness. This raises concerns about objectivity and how applicable these sorts of findings might be. There is a shared responsibility: scientists who understand methodology of treatment trials and autism actually work alongside people who are delivering programmes to help establish whether these programmes are having any effect.

Bailey: It is helpful to be overt about whether what one is trying to do is to reverse a primary deficit, or to teach a compensatory strategy. Very often we are encouraging alternative ways of getting by in the world, particularly in adult life. Pat's emphasis on adult life is terribly important. What strikes me is that able people with autism are acquiring useful skills in their 20s and 30s: their quality of life can be radically different at the end of that period. The problem with the focus on early intervention is not that we don't need to be doing it, but that it is as though we can stop thinking about intervention after that point. This is clearly not the case.

Bauman: I would like someone to address this business of five years of age being a critical point. Many times families will come into the office and say that their child is not doing such and such, and they are almost five years old. It is almost like they think everything is all over if nothing happens by the age of five. Second, the adult functioning population is coming down. The idea that we really need to make these children as functional as they can be, despite their autism: that we are not really looking for a 'cure'. You can be a functioning adult and still be autistic.

Lord: Part of the reason that people treat age 5 as significant is that the original Rutter & Locker study used this age. People assume that there was a systematic assessment of the age, out of all possible ages, at which later verbal fluency could be predicted. However, in fact, in the original study, the authors made a good guess of the age where it was possible to make a distinction without comparing it to other possible ages. This then became a political issue in getting services for children before the age of 5 in the USA. There isn't any evidence that there is a precipitous drop-off of acquisition of language at 5. Where we see the most progress in language is between 2 and 3, and next between 3 and 4. The children who will be really fluent speakers are probably making progress even earlier than 5. But we have to be so careful that if we use arbitrary cut-offs in research, it will be interpreted to mean more than we intended.

Bishop: Is it common to see children who were not talking at 5 and then did at 6?

Howlin: Yes. There are children who were not scoring at all on verbal tests when they were 5, 6 and 7. In our own follow-up (Howlin et al 2003) a good proportion of these, around a third, then went on to speak.

Bishop: I was curious about those. I was wondering whether these are the children who are talking but who are producing rather stereotyped language and are not very responsive in a test situation, or whether they were children who really were not talking.

Howlin: These are data we need to look at more closely.

Bailey: We should do a bidding about what is the oldest that someone has acquired language! It would be into the early teens.

Howlin: Yes, it would be 12–13.

Rutter: I'd emphasize that this was not acquisition of normal speech.

Bishop: Often this is what parents are asking about: they want to know whether their child will ever speak.

Sigman: The problem is that we don't have any longitudinal studies that have looked finely enough to say exactly when it is that language skills have emerged. In my longitudinal studies, not many children acquire speech late.

Dawson: If you are talking about what to say to parents, I can tell you what I tell parents who ask that question. A lot of what we know from the literature is based on correlations. These correlations are pretty small. Predicting what an individual will do on the basis of correlational data is risky. There are a lot of surprises on the individual level. I wouldn't want to base my predictions about a given child on a study that looked just at correlational data. The other thing is that, in the developmentally delayed brain, we don't know whether the windows might be different. It is possible that the same period of sensitivity in a normal child could be extended if brain development were slowed. We know very little about plasticity in individuals with developmental disorders, and whether they are on the same kind of timetable. The reason I say this is because of some of the surprises I have seen, where one wouldn't expect things to come in late and they do. I have seen children develop attachment when they were 18! Therefore, you have to look very carefully at that individual and what the issues are with respect to that child's communication skills, try to assess why they are not making progress, and target those issues. For some children I would suggest using an augmentative approach to language intervention, while for others I might not.

C. Frith: My impression from reading a bit about the changes in the brain in early life is that the evidence for sensitive periods in normal human development is not that strong.

Rutter: Can I put one double question to you? I like and agree with the way that you bring together the evidence, but let's approach it with scepticism from two different points of view. First, if we were talking about licensing a drug that cost as much as 20 hours of intensive work, I have no doubt that NICE (the UK National Institute for Clinical Excellence, which arbitrates as to whether drugs are prescribed or not on the NHS) would not license it on the basis of the evidence you presented. Second, you, Cathy Lord and Sally Rogers have all said that the more naturalistic use of behavioural approaches is the way to go. But if Lovaas were sitting here, he would disagree, pointing out that he used electric prods and all sorts of artificial means, and his results of years ago were as good as anything you are claiming. What is the evidence that naturalistic behavioural approaches are better?

Howlin: With regard to the 20 h a week treatment, this is a period that many children are in nursery school, for example. I am not thinking of specialist programmes outside this, but rather strategies that can be used within the school situation in particular. In the study we are just about completing, comparing children in nursery schools with children going through an ABA program for 2 years, it looks like the Lovaas children probably have the edge on the other children, but it is only an edge. The results seem marginally significant on a lot of variables, but only marginally. If one had only been able to introduce perhaps a bit more structure or specialized training into nursery schools, then there would have been just as good results. I wouldn't see getting 20 h week of appropriate intervention as something that is out of the question. With regard to the naturalistic approaches, a lot of these are based on behavioural studies but with greater emphasis on spontaneity and individuality.

Rogers: There is experimental evidence, but it doesn't come out of the comprehensive two-year long studies. The evidence comes from teaching a specific cognitive or linguistic skills. There are multiple reports that the incidental neo-behavioural approaches ('naturalistic' isn't quite the right word) can be more effective, but it is about teaching a single skill, not years of intervention.

Fombonne: There is a randomized clinical trial which has assessed Lovaas treatment in a rigorously controlled fashion. This study by Smith et al (2000) showed less spectacular results than Lovaas' initial studies, but there were still IQ gains of 20–30 points in the experimental group as compared to the control although, to a large extent, the gains were accounted for by children with an initial diagnosis of PDDNOS as opposed to autistic disorder. Thus, those children with less severe impairments benefited most from the intervention.

References

Delprato DJ 2001 Comparisons of discrete-trial and normalized behavioral language intervention for young children with autism. J Autism Dev Disord 31:315–325

Howlin P, Goode S, Hutton J, Rutter M 2003 Adult outcomes for children with autism. J Child Psychol Psychiatry, in press

Koegel RL, O'Dell MC, Koegel LK 1987 A natural language paradigm for teaching nonverbal autistic children. J Autism Dev Disord 17:187–200

McGee GG, Morrier MJ, Daly T 1999 An incidental teaching approach to early intervention for toddlers with autism. J Assoc Persons Severe Handicaps 24:133–146

Merzenich MM, Saunders G, Jenkins WM, Miller S, Peterson B, Tallal P 1999 Pervasive developmental disorders: listening training and language abilities. In: Broman SH, Fletcher JM (eds) The changing nervous system: neurobehavioral consequences of early brain disorders. Oxford University Press, New York, p 365–385

Smith T, Groen AD, Wynn JW 2000 Randomized trial of intensive early intervention for children with pervasive developmental disorder. Am J Ment Retard 105:269–285

Stone WL, Yoder PJ 2001 Predicting spoken language level in children with autism spectrum disorders. Autism 5:341–361

Early intervention and brain plasticity in autism

Geraldine Dawson and Kathleen Zanolli

Autism Center, Center on Human Development and Disability, University of Washington, Seattle, WA 98195, USA

Abstract. Autism is associated with impairments in brain systems that come on line very early in life. One such system supports the development of face processing. Dawson and colleagues found that 3 year old children with autism failed to show differential event-related potentials (ERPs) to photographs of their mother's versus a stranger's face. Since differential ERP activity to familiar and unfamiliar faces is typically present by 6 months, this represents early brain dysfunction. McPartland and colleagues found that the face-specific ERP component ('N170') is atypical in older individuals with autism. N170 is typically larger to faces than non-faces, and prominent over the right hemisphere. In individuals with autism, N170 was larger for furniture than faces and bilaterally distributed. Biology and experience contribute to the development of face-processing systems. Newborns are capable of recognizing faces. Early face recognition abilities are thought to be served by a subcortical system, which is replaced by an experience-dependent cortical system. Development of a neural system specialized for faces may depend on experience with faces during an early sensitive period. Because children with autism fail to attend to faces, they might not acquire the expertise needed for a specialized face processing system to develop normally. Early interventions that enhance social attention should result in changes in brain activity, as reflected in ERPs to face stimuli, with those children showing the greatest social improvement exhibiting more normal brain activity.

2003 Autism: neural basis and treatment possibilities. Wiley, Chichester (Novartis Foundation Symposium 251) p 266–280

Studies have shown that behavioural intervention during the preschool period can be effective for many children with autism, presumably because of the plasticity of neural systems during that time. Such studies raise several important questions: What brain systems might be affected by early behavioural intervention? How might such effects be measured in young children? Are there sensitive periods in development during which behavioural intervention is likely to have its greatest impact on brain development? This paper briefly considers these questions.

Early behavioural intervention

Several studies suggest that early intensive behavioural intervention (EIBI) can result in dramatic improvements for some children with autism (e.g. Anderson et al 1987, Fenske et al 1985, Lovaas 1987, Smith et al 2000). As reviewed by Dawson & Osterling (1997) and Rogers (1998), although intervention models have varied across studies, most of these intervention models had several common features including (1) a curriculum that is comprehensive and includes core domains of attention, imitation, language, toy play and social interaction, (2) sensitivity to normal developmental sequence, (3) highly supportive teaching strategies, most often based on applied behaviour analysis, (4) behavioural strategies for reducing interfering behaviours, (5) involvement of parents in the intervention process, (6) gradual and careful transition from a highly-supportive environment to more complex, naturalistic environments, (7) intensive intervention consisting of about 25 hours a week of structured intervention lasting for at least 2 years, and (8) onset of intervention by 2–4 years. When these features are present, results have been impressive for a subgroup of children including robust gains in IQ, communication and educational placement.

Although previous studies have examined the efficacy of EIBI for improving behavioural outcome, no published studies have examined the effects of EIBI on brain development. Because autism involves core impairments in language and social relatedness, of particular interest is how EIBI affects the early development of social and language brain circuitry and function. This paper focuses specifically on how EIBI might affect the development of brain systems specialized for social processing, and in particular, face processing.

Autism involves a basic impairment in face processing

Evidence suggests that autism involves a fundamental impairment in processing information from faces. Indeed, one of the first recognizable symptoms of autism involves a failure to attend to faces. In a study of home videotapes of 1st birthday parties of infants with autism, a failure to attend to others' faces was the single best discriminator between one-year-olds with autism vs. typical development (Osterling & Dawson 1994). Face recognition impairments have been found in several studies of older individuals with autism (e.g. Boucher et al 1998, Klin et al 1999). Older individuals with autism fail to show the 'face-inversion effect' that has been demonstrated in normal individuals (i.e. superior ability to recall upright as compared to inverted faces; Hobson et al 1988). In fact, individuals with autism recognized inverted faces *better* than normal individuals, suggesting that they may be using a different information processing approach. In fMRI studies, the fusiform gyrus is activated during face processing, typically more on the right than left (McCarthy et al 1997). Schultz and colleagues found that high-functioning individuals with autism spectrum disorder failed to activate the fusiform face area during face processing (Schultz et al 2000). A study by Dawson et al (2002) using EEG recordings found that children with autism as young as 3–4 years of age exhibit atypical brain activity to faces, whereas they showed normal brain activity in response to objects. In this study, 64-channel event-related potential (ERP) recordings to digitized photos of mother's versus a stranger's face and photos of a favorite versus unfamiliar object were collected from 3–4 year old children with autistic spectrum disorder, CA-matched children with typical development, and MA- and CA-matched children with developmental delay without autism. Typically developing children showed significant ERP amplitude differences in all three components measured (P400, Nc, Positive Slow Wave) to familiar versus unfamiliar faces, and differences in P400 and Nc components to familiar versus unfamiliar objects. Developmentally delayed children showed significant ERP amplitude differences in the positive slow wave for both faces and objects. In contrast, children with ASD did not show differential ERPs to familiar versus unfamiliar faces, but like the typically developing children, they did show P400 and Nc amplitude differences to familiar versus unfamiliar objects. Increased P400 latency to faces was associated with greater joint attention impairment in the children with autism ($r=0.63$, $P<0.0001$). Because face recognition ability emerges very early in infancy, this impairment likely reflects a very early brain abnormality.

In an ERP study of adolescents and adults with autism, Dawson and colleagues (McPartland et al 2001) found that the ERP face-specific component ('N170') differed between participants with autism and normal participants. This ERP component is found in response to face stimuli in normal individuals as early as 4 years of age, and shows reliable increases in amplitude and decreases in latency throughout childhood (Taylor et al 1999), making it a sensitive neural marker of the developmental course of face processing. Consistent with previous research (e.g. Bentin et al 1996), typical subjects demonstrated a pattern of negative electrical activity over the posterior scalp at $\sim$170 ms that was right-lateralized and larger in amplitude in response to faces than furniture. In contrast, participants with autism failed to show hemispheric lateralization, showing equal activity in right and left hemispheres. Whereas typical participants showed the expected shorter right hemisphere latency to upright faces than furniture, individuals with autism exhibited the opposite pattern, i.e. longer latency to faces. Typical participants also exhibited longer N170 latency to inverted than upright faces, whereas participants with autism did not.

The role of experience in the development of face processing

The neural systems that mediate face processing appear to exist very early in life. A visual preference for faces (Goren et al 1975) and the capacity for rapid face recognition (Walton & Bower 1993) are present at birth. By 4 months, infants recognize upright better than upside down faces (Fagan 1972). By 6 months, infants show differential ERPs to familiar versus unfamiliar faces (de Haan & Nelson 1997).

Studies of face recognition in adult human and non-human primates have been informative in describing the neural systems that mediate face processing. In monkeys, face selective neurons have been found in inferior temporal areas, TEa and TEm, superior temporal sensory area, amygdala, ventral striatum (which receives input from the amygdala), and inferior convexity (Desimone et al 1984, Rolls 1992, Williams et al 1993). Face recognition impairment results from damage to fusiform gyrus and the amygdala (e.g. Damasio et al 1982). Neurons that respond to faces have been found in the amygdala (e.g. Rolls 1984). The anterior inferior temporal cortex and the superior temporal sulcus project to the lateral nucleus of the amygdala (Amaral et al 1992). Parts of the inferior and medial temporal cortex may work together to process faces (Nelson 2001). The amygdala is important for assigning emotional relevance to faces, and the emotional arousal that results from amygdala activation may affect both attention and memory for faces. The amygdala is activated during eye-to-eye gaze and has been suggested to play a role in the emotional responses evoked during eye contact between persons (Kawashima et al 1999).

Morton & Johnson (1991) have hypothesized that early face processing abilities are mediated by subcortical systems which are replaced by cortical systems at about 6 months of age. These neural changes in the face processing system reflect 'experience expectant developments' (Greenough et al 1987, Nelson 2001). In other words, a sensitive period might exist during which there is a readiness of the brain to receive experience with faces. Such input is a reliable experience for most human infants, and Nelson (2001) has argued that this experience is important for the development of a specialized face processing system. Nelson and colleagues (Pascalis et al 2002) found human infants superior to adults in discriminating monkey faces, suggesting that experience with human faces results in a 'perceptual narrowing' similar to what is observed with speech perception (Doupe & Kuhl 1999). They demonstrated that younger infants, 6 months of age, were better at discriminating individuals of both human and monkey species, compared to older infants and adults. They argue that this is due in large part to the cortical specialization that occurs with experience in viewing faces. Based on the similarity in timing of perceptual narrowing for both face and speech perception (by about 6 months), these systems may develop in parallel and mutually influence one another.

Role of experience in abnormal face processing in autism

Experience may also play a role in atypical development of the face processing system in autism (Carver & Dawson 2002). As mentioned above, by 3–4 years of age, there is evidence of atypical brain activity in response to faces in young children with autism. In adolescents and adults with autism, there is evidence of slowed neural speed of face processing, as reflected in N170 latency, and atypical cortical specialization for faces in autism. *We hypothesize that the abnormalities in face processing found in autism may be related to abnormalities in social attention, and more specifically that the neural mechanisms that normally draw an infant's attention to others' faces are dysfunctional in autism.* As mentioned above, the amygdala is important for assigning affective significance to faces, and is activated during eye to eye contact. Such neural mechanisms normally facilitate mutual gaze and the acquisition of knowledge about others' faces, including their familiarity and expressions. Beginning early in life, in autism there may be a deprivation of critical experience-driven input that results from a failure to pay normal attention to faces.

What might be the impact of early intervention on the development of brain systems related to face processing? One goal of early behavioural intervention is to teach children to pay attention to social information, including faces, by rewarding them for doing so. Although prompting strategies and selection of skills to be taught vary among EIBI programs, the interaction between adult and child always includes two features (Anderson & Romanczyk 1999). First, the child emits a behavioural skill that is prompted or facilitated by the adult in some way. In the early stages of EIBI, nearly all these skills include eye contact, either in isolation or in conjunction with joint attention, imitation, language, cognitive and social skills. Second, contingent on the child's use of the skill (eye contact), the therapist immediately provides a reinforcer. Because reinforcers are defined functionally, in terms of their effect on the child's behaviour, the specific reinforcer used varies widely depending upon the child's momentary preferences. In practice, most children with autism receive access to a highly preferred non-social stimulus, typically related to the child's sensory or restricted interests. Often, at the same time, the child receives social feedback, such as praise, touch, smiling or applause.

There are several potential effects of this behaviour–reinforcer interaction. First, the ostensible and intended effect is that the skill (e.g. eye contact) is more likely to be repeated. Increased use of eye contact may improve the facial processing system simply by increasing the child's experience with faces.

This experiential input alone, however, may or may not affect the emotional arousal, or motivational, component of facial processing. Another second potential effect of behaviour–reinforcer interactions, which may influence the motivational significance of faces, is conditioned reinforcement. When a previously neutral stimulus (face) is frequently associated with a reinforcer (e.g. access to toy), the neutral stimulus can acquire reinforcer value, that is, it can function as a reinforcer for that individual's behaviour in the future. In EIBI, neutral social stimuli, including the face, facial gestures and language, frequently co-occur with reinforcing non-social stimuli. This could result in faces and social stimuli acquiring reinforcer value (see Fig. 1). Factors that enhance the acquisition of reinforcer value by a previously neutral stimulus are (1) temporal contiguity of the behaviour, neutral stimulus, and reinforcer; (2) consistent association between the neutral stimulus and reinforcer, such that one is never or rarely presented without the other; (3) similarity or overlap between the neutral and reinforcing stimuli; and (4) the presence of a contingency which requires the child to do more than simply attend to the reinforcer and neutral stimulus (Williams 1996, Mazur 1994, Silverstein & Lipsitt 1974). All of these factors are present in EIBI.

Thus, it is possible that EIBI might facilitate the development of the face processing system in two ways: (1) by providing enhanced early exposure to faces by increasing the child's

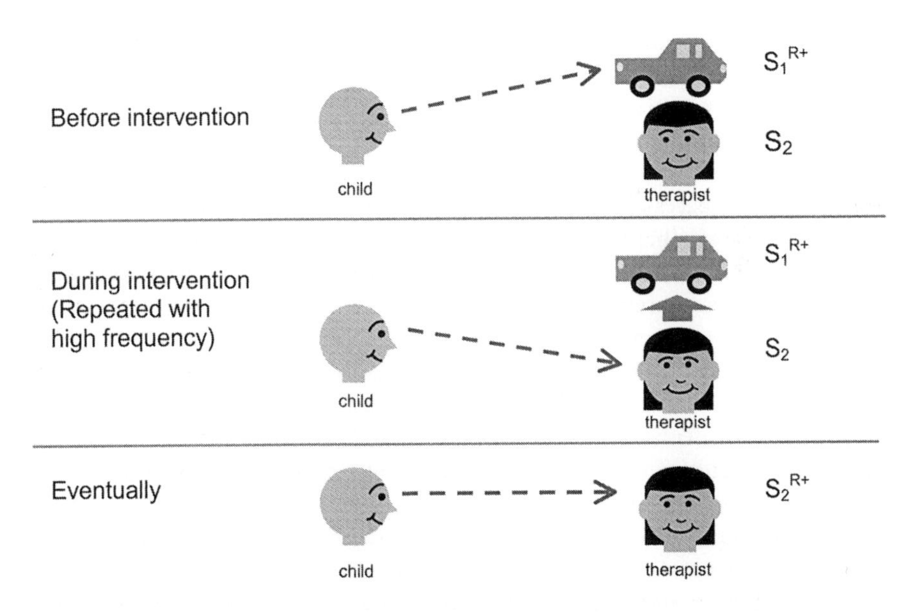

FIG. 1. Acquisition of conditioned social motivation in autism. Before intervention, child does not attend to therapist's face, a neutral stimulus which lacks reinforcer value. During the intervention, the therapist requires child to make eye contact in order to have access to highly preferred object or activity. Through conditioned reinforcement, the previously neutral stimulus, i.e. the face/eyes, eventually begins to function as a reinforcer itself. Through such interventions, brain regions responsible for encoding face stimuli, such as the fusiform gyrus and superior temporal sulcus, receive increased input and stimulation. Furthermore, brain regions responsible for 'affectively tagging' face stimuli, such as the amygdala, are activated through conditioned reinforcement. S, stimulus; R, positive reinforcement.

attention to faces, and (2) by altering the child's motivational preferences, such that engaging in face-to-face interaction becomes more rewarding, and more frequent. The latter mechanism may be important in explaining the robust and durable clinical effects of EIBI in some children, who seem to engage in social interaction and to learn new skills in social contexts after EIBI therapy is discontinued (Smith 1999).

It is possible that EIBI may result in improved *behavioural* performance on tasks related to face processing, but this improved performance may occur via alternative, atypical brain systems. Alternatively, early intervention may facilitate a more normal trajectory of brain development and thus have a fundamental impact on early developing brain systems. In the case of face processing, this might be reflected in more normal patterns of brain activity, as in electrophysiological and fMRI measures, e.g. a normal hemispheric specialization for faces.

The question of whether there might exist a sensitive period during which early intervention is most effective for altering the course of brain systems related to face processing is an important one. It has been shown that adult-level expertise in face processing develops gradually over years and is evidenced by increasing reliance of configural processing strategies (Carey & Diamond 1994). Le Grand et al (2001) found that deprivation of patterned visual input from birth until 2–6 months (due to bilateral congenital cataracts) results in permanent deficits in configural face processing. These results indicated that visual experience during the first few months of life is necessary for normal development of the face processing system. It is unknown whether the early abnormal experiences of young children with autism—i.e. reduced experience actively attending to faces co-occurring with normal or even accentuated experience with nonsocial patterned visual experience—has long-term effects on the development of the face processing system. It is clear, however, that the face processing system comes on line very early during infancy and is sensitive to experiential effects. Thus, very early intervention that enhances attention to faces and social interaction by making faces more rewarding may be optimal for best outcome in autism.

Early behavioural intervention may impact neural systems related to a wide range of domains, such as language. Given that electrophysiological measures of brain activity, such as high-density ERP, can be used with very young children to assess neural processing of a wide range of stimuli, such as speech and faces, this methodology can be useful for assessing whether and how early intervention affects brain development in young children with autism and other disabilities.

Acknowledgements

Supported by a grant from the National Institute of Child Health and Human Development and the National Institute on Deafness and Communication Disorders (PO1HD34565).

References

Amaral DG, Price JL, Pitkanen A, Carmichael T 1992 Anatomical organization of the primate amygdaloid complex In: Aggleton J (ed) The amygdala. Wiley, New York, p 1–66
Anderson S, Romanczyk R 1999 Early intervention for young children with autism: continuum-based behavioral models. J Assoc Pers Sev Handicaps 24:162–173
Anderson SR, Avery DL, DiPietro EK, Edwards GL, Christian WP 1987 Intensive home-based early intervention with autistic children. Educ Treat Children 10:352–366
Bentin S, Allison T, Puce A, Perez E, McCarthy G 1996 Electrophysiological studies of face perception in humans. J Cogn Neurosci 8:551–565
Boucher J, Lewis V, Collis G 1998 Familiar face and voice matching and recognition in children with autism. J Child Psychol Psychiatry 39:171–181
Carey S, Diamond R 1994 Are faces perceived as configurations more by adults than by children? Vis Cogn 1:253–274
Carver L, Dawson G 2002 Development and neural bases of face recognition in autism. Mol Psychiatry 7:S18–S20
Damasio AR, Damasio J, Van Hoesen GW 1982 Prosopagnosia: anatomic basis and behavioral mechanisms. Neurology 32:331–341
Dawson G, Osterling J 1997 Early intervention in autism. In: Guralinick MJ (ed) The effectiveness of early intervention. Brookes Publishing Company, Baltimore, MD, p 307–326
Dawson G, Carver L, Meltzoff AN, Panagiotides H, McPartland J 2002 Neural correlates of face recognition in young children with autism spectrum disorder, developmental delay, and typical development. Child Dev 73:700–717
de Haan M, Nelson CA 1997 Recognition of the mother's face by 6-month-old infants: a neurobehavioral study. Child Dev 68:187–210
Desimone R, Albright TD, Gross CG, Bruce C 1984 Stimulus-selective properties of inferior temporal neurons in the macaque. J Neurosci 4:2051–2062
Doupe AJ, Kuhl PK 1999 Birdsong and human speech: common themes and mechanisms. Annu Rev Neurosci 22:567–631
Fagan J 1972 Infants' recognition memory for face. J Exp Child Psychol 14:453–476
Fenske EC, Zalenski S, Krantz PJ, McClannahand LE 1985 Age at intervention and treatment outcome for autistic children in a comprehensive intervention program. Special issue: early intervention. Anal Interv Dev Disabil 5:49–58
Goren CC, Sarty M, Wu PY 1975 Visual following and pattern discrimination of face-like stimuli by newborn infants. Pediatrics 56:544–549
Greenough WT, Black JE, Wallace CS 1987 Experience and brain development. Child Dev 58:539–559
Hobson RP, Ouston J, Lee A 1988a What's in a face? The case of autism. Br J Psychol 79:441–453
Kawashima R, Sugiura M, Kato T et al 1999 The human amygdala plays an important role in gaze monitoring: a PET study. Brain 122:779–783
Klin A, Sparrow SS, de Bildt A, Cicchetti DV, Cohen DJ, Volkmar FR 1999 A normed study of face recognition in autism and related disorders. J Autism Dev Disord 29:499–508
Le Grand R, Mondloch CJ, Maurer D, Brent HP 2001 Neuroperception. Early visual experience and face processing. Nature 410:890
Lovaas OI 1987 Behavioral treatment and normal educational and intellectual functioning in young autistic children. J Consult Clin Psychol 55:3–9
McCarthy G, Puce A, Gore JC, Allison T 1997 Face-specific processing in the human fusiform gyrus. J Cogn Neurosci 8:605–610

McPartland J, Dawson G, Panagiotides H 2001 Neural correlates of face perception in autism. Poster presented at the meeting of the Society for Research in Child Development, April 2001, Minneapolis, Minnesota

Mazur J 1994 Predicting the strength of a conditioned reinforcer: effects of delay and uncertainty. Curr Dir Psychol Sci 2:70–74

Morton J, Johnson MH 1991 CONSPEC and CONLERN: a two-process theory of infant face recognition. Psychol Rev 98:164–181

Nelson CA 2001 The development and neural bases of face recognition. Infant Child Dev 10:3–18

Osterling J, Dawson G 1994 Early recognition of children with autism: a study of first birthday home videotapes. J Autism Dev Disord 24:247–257

Pascalis O, de Haan M, Nelson CA 2002 Is face processing species-specific during the first year of life? Science 296:1321–1323

Rogers S 1998 Empirically supported treatment for young children with autism. In: Special issue on "Empirically supported psychosocial interventions for children". J Clin Child Psychol 27:168–179

Rolls ET 1984 Neurons in the cortex of the temporal lobe and in the amygdala of the monkey with responses selective for faces. Hum Neurobiol 3:209–222

Rolls ET 1992 Neurophysiology and functions of the primate amygdala. In: Aggleton JP (ed) The amygdala. Wiley, New York, p 143–165

Schultz RT, Gauthier I, Klin A et al 2000 Abnormal ventral temporal cortical activity during face discrimination among individuals with autism and Asperger syndrome. Arch Gen Psychiatry 57:331–340

Silverstein A, Lipsitt L 1974 The role of instrumental responding and contiguity of stimuli in the development of infant secondary reinforcement. J Exp Child Psychol 17:322–331

Smith T 1999 Outcome of early intervention for children with autism. Clin Psychol Sci Pract 6:33–49

Smith T, Groen AD, Wynn JW 2000 Randomized trial of intensive early intervention for children with pervasive developmental disorder. Am J Ment Retard 105:269–285

Taylor MJ, McCarthy G, Saliba E, Degiovanni E 1999 ERP evidence of developmental changes in processing of faces. Clin Neurophysiol 110:910–915

Walton GE, Bower TGR 1993 Newborns form 'prototypes' in less than 1 minute. Psychol Sci 4:203–205

Williams B 1996 Conditioned reinforcement: neglected or outmoded explanatory construct? Psychonomic Bull Rev 1:457–475

Williams GV, Rolls ET, Leonard CM, Stern C 1993 Neuronal responses in the ventral striatum of the behaving macaque. Behav Brain Res 55:243–252

DISCUSSION

Skuse: I would like to come back to something Chris Frith mentioned at the end of the last discussion, which was the lack of evidence for sensitive periods during early human development, and tie this in with the results in LeGrand et al (2001). I thought you developed rather beautifully some of the ideas that might be linked to that. Putting these ideas in a slightly different context, we were talking earlier about why the Romanian orphans might have developed autistic features. I suggested that there could be a sensitive period for eye contact which didn't happen, because they didn't have the opportunity to develop eye contact during that time. What I think is happening with some of these autistic infants is we have the same process — lack of eye contact — but for a different reason, because they find eye contact aversive in that early period. You mentioned the fact that at a year eye contact was the best predictor of later autistic behaviours. We propose that it is over-arousal of the threat detection system that lies behind the failure of these infants to make normal eye contact. Effectively, you have two different circumstances: one avoiding eye contact because it is aversive or threatening, the other lack of experience of eye contact, both of which are leading to maldevelopment of the same system. We can easily imagine how to compensate for the lack of experience, but what do you do in order to encourage these infants who may be avoiding eye contact because it is a stressful experience? How are you going to ensure that they have the appropriate experiences so that these neural circuits develop normally and predispose towards better social cognitive skills in later life?

Dawson: The notion that making eye contact is aversive for children with autism is a very attractive one, and it has been around for a long time. But when we try to investigate it, we don't find much evidence for it. Clinically one does not get the experience from the majority of children that they are finding eye contact aversive. I don't think it is a closed question, but the majority of evidence favours the idea that whatever mechanisms drive attention to faces are not working properly, but not necessarily that children find eye contact aversive. There are conditions, such as fragile X syndrome, where that kind of aversive reaction can be demonstrated with physiological measures. It would be nice to do some systematic experiments comparing children with autism and those with fragile X. Marion Sigman's study did this in young children with autism and Down syndrome (Corona et al 1998). They measured autonomic arousal in response to social stimuli, and found that children with autism were hyporesponsive.

C. Frith: The interesting thing about the Romanian orphans is that they recovered to a large extent. But going back to your point about social stimuli not being rewarding for autistic children, it seems to me that there are two aspects to this. In order for social stimuli to be rewarding, the first stage in this process is that the brain has to have a way of distinguishing between social stimuli and non-social stimuli. In the case of faces this may be easier, but there are all sorts of other social stimuli. What I am particularly interested in is how this distinction is made. Perhaps the problem lies in making the distinction rather than with the reward.

Dawson: It could be that the problem is at the encoding level. This would make things somewhat simpler.

Schultz: Yes, but there is also a fast pathway that gets to the amygdala even before it gets to the perceptual areas of the ventral temporal and occipital cortex that can discriminate between faces and objects. This so-called fast pathway may be intimately tied up with the reward system; certainly there are close anatomical connections between the amygdala and the ventral striatum. My point, I guess, is that I don't think we should conceive of perceptual encoding as a singular phenomenon, one that occurs through information processing along just one pathway: there is more than one feed into the fusiform face area, and some feeds may already be contextualized with reward value or information on emotional salience. For the most part, processing of reward value happens before the type of perceptual recognition processing that we are speaking about, but this may well be a reverberating process.

Bailey: There's also a direct fast pathway to the fusiform, as well.

Rutter: As the Romanian adoptees have come up twice, let me outline a few key findings, because they are intriguing in providing partial support and also partial querying of the programming notion. I will focus on two outcomes, cognitive level and attachment dysfunction (meaning mainly disinhibited unselective social interaction). We found remarkable catch-up in cognition, but the effects of duration of institutional care on outcome at age six were enormous (O'Connor et al 2000). There was a 20-point difference between those who had institutional care for more than two years as against those who had it for less than 6 months. Interestingly, the difference was as great at 6 as it was at 4. Cognitive impairment was significantly associated with a smaller head circumference both at the time of UK entry and at follow-up. This is relevant in terms of what was discussed earlier about whether it was possible to make heads grow: it is possible. But what was also striking was that, in the group who had at least two years of institutional care, the range of outcomes was enormous. The IQ as measured at age 6 ranged from the severely retarded level to above 130. The persistence of deficits suggest some type of programming effect in that there was a massive persisting effect of early experience. But what is the cause of the large individual variation in outcome. The attachment results were both similar and different (O'Connor & Rutter 2000). There was the same massive effect of duration of institutional care. It was different, however, in that there was no association, either initially or at outcome, with head circumference. For me, the findings raise distinctions between what two appear to be different types of programming. There is what Bill Greenough talks about as 'experience expectant'. The best model for this is the Hubel & Wiesel findings on vision. Relevant visual experience is necessary for normal development of the visual cortex of the brain. However, the range of visual experiences that provide what is needed is very wide. Variations in experiences within the normal range are not relevant. Experience-adaptive programming, exemplified by the phonological findings of Pat Kuhl and David Barker's work on nutrition, is different in that it concerns variations within the normal range, as well as outside it. But it is not concerned with whether brain growth is normal or abnormal. Rather, it concerns adaptation of neural functioning to the environment that exists during a particular sensitive period in development (Rutter 2002). In relation to autism, which type of programming might be operative? If it is the experience–expectant kind, can one conceptualize that the defects they are born with provide such a restriction on experiences that it is stopping normal development? If it is experience–adaptive programming, what are the implications for intervention? It has been argued in relation to diet that, if in middle childhood extra nutrition is provided to make up for the early lack, it may make things worse (Rutter 2002).

Dawson: Could you expand on that?

Rutter: The empirical evidence is that early subnutrition (prenatal and first year of life) is associated with a marked increase in the liability to coronary artery disease, late onset diabetes and other conditions. In adult life, it is the opposite way round. David Barker argues, plausibly but without a solid physiological basis yet, that the body metabolism is being programmed to deal with subnutrition. If then you overload it you are making things much worse. Whether there is a psychological equivalent of this is unclear. It is an intriguing notion because it would have radical implications for intervention.

Sigman: When I heard about this I went to speak to nutritionists to see whether they thought the evidence is good. Pretty much everyone was sceptical. In our long-term outcomes we will have to look at these kinds of effects.

Lipkin: This is interesting. I heard Barry Levin of the University of Medicine and Dentistry in New Jersey talk about force feeding rats to obesity during pregnancy. He has found that obese mothers give rise to progeny who subsequently become obese (Levin 2000). This becomes a trait that is carried forward for several generations.

Rutter: It is an intriguing set of notions, but this area requires some experimental physiological evidence and testing of whether the mechanism is correct. The empirical findings are indisputable.

Dawson: It would be interesting to find children very young to see what kinds of changes can occur. But when you think about autism and the kinds of things that are disrupted, and the abnormalities that are present in brain function and behaviour quite young, it makes me think we are looking at very early dysfunction in experience–expectant processes. Intervening early on these processes may be our best hope.

Monaco: I was particularly struck by the non-lateralized, non-specific activation you showed with the face processing. How much does this generalize to other things that could be tested as deficits in autism? If this is true, could this gross activation of the brain increase the susceptibility for epilepsy?

Dawson: Pat Kuhl has looked at brain lateralization using ERP for speech sounds. She found abnormal hemispheric specialization. She didn't look at topography in terms of amount of cortex activated, but it is completely measurable. You also could look at this in an fMRI context. In our ERP study comparing three-year olds with autism to those with typical development and developmental delay, I was struck by just how much the ERPs to objects in children with autism looked just like those of CA-matched typically developing children. The idiopathic developmental delay children showed ERP discrimination for both faces and objects, but their ERP looked different. It was slower.

Bishop: How intellectually impaired were they?

Dawson: They were matched to our autism group. The mean IQ of this group was in the mild to moderately retarded range, with a full range from near normal to severely mentally retarded.

Baird: Just to put what Chris was saying in a slightly different way, rather than the ability of the infant to separate a social versus a non-social response, there is the notion that some of these children are unable to use social information to inform their response. Perhaps certain stimuli, whether facial or sensory, may be given a different meaning because you can use social information about them and it makes them less aversive. If you don't get the social response you develop an aversion to certain types of stimuli. This then has knock-on effects on the rest of their experiences, because there is nothing more aversive to a mother than finding that her social input is not working. Then you get a long period when they are not giving the appropriate input.

Lord: Are there biological mechanisms where it is not the case that if you don't get a stimulus you miss out altogether, but where there are pathways such that if you miss one window, it pulls you in a particular direction there is another opportunity to get on in terms of learning something else? It is not just one path. Are there any analogies to this in brain mechanisms? It seems that so much of social behaviour is redundant and there are ways of getting information from many sources, yet children with autism still miss social meanings. The failure to understand social meanings would in turn affect the environment.

C. Frith: I have a comment about the interesting over-activity in the ERPs of autistic children looking at faces. I think there is something interesting in the comparison of the EEG work with the brain imaging work. In all the brain imaging studies reduced activity is seen in several areas. This is also true in dyslexia; I find this interesting. Why should the activity be reduced? Why isn't the brain working harder if there is a problem? What does this tells us about the relationship between the blood flow and the EEG measures.

I also wanted to ask about this special fast route for face processing to the amygdala. This is extremely interesting, because presumably the resolution of the image being processed must be reduced, because it is not going through V1. How does the amygdala do it? What are the aspects of the face that enable the amygdala to recognize the expression? Does the amygdala distinguish between upside-down and right-way-up faces, for example? This might be very interesting, because if you knew what the differences were, you might be able to develop special face stimuli that you could use to demonstrate behaviourally that the autistics are not using the fast route.

Schultz: I agree, there are many interesting questions, but unfortunately the studies have not yet been done to provide answers to most of these questions. This subcortical pathway passes information from the retina to the superior colliculus, to the thalamus, and then onto the amygdala. Somehow the superior colliculus and thalamus (primarily the pulvinar) is perceiving something about that stimulus to cause it to be prioritized and fed to the amygdala for rapid analysis. I am not sure about the exact role of each of the three nodes in that pathway in perception. We are very interested in the fast pathway with regards to a possible role in the ontogeny of autism, because it is likely to be the one active in the first months of life, before the cortex comes online. In this regard, it may be the first stumbling block for the person with autism.

Bailey: There is some variability in the face-processing findings between groups, and according to the method used. What strikes me is that it is a good model system because we know a lot about face processing and a fair bit about neuroanatomy very early in life. If we give adults a task in which they simply see a range of objects and faces, we do not currently detect this fast response. It is only evident in a paired task where the subject sees one image after another, and the task is to identify whether the second image is the same as the first. It may be that the difference is simply that there are greater attentional demands, or that working memory is being recruited. But we cannot yet see the fast activity in the single image paradigm. When a first face image is seen by normals in a paired paradigm, there is activity that is either parietal or thalamic, followed by anterior temporal/posterior frontal regions and then fusiform gyrus. In individuals with autism or Asperger syndrome the pattern is completely different in that the activity is elicited by the second image and there is hardly any activity in the fusiform gyrus. They seem to be responding fundamentally differently. What we are having problems with is whether this is a difference in innate wiring, or is it to do with social reward mechanisms which generally lead to faces being processed in the fusiform region? When one examines the current dipoles evoked by faces, they orient in the same direction in normals, who are presumably all activating a similar region in which the neurons are oriented in a similar way. In autistic individuals we find that they are activating a whole host of areas around fusiform and the dipoles are randomly oriented. That is, in each individual it as though face associated activity is colonizing a somewhat different bit of cortex.

Amaral: I wanted to comment on Chris Frith's question about these fast pathways to the amygdala. As far as we know, for the auditory system the subcortical fast pathway that goes directly to the amygdala carries information about the volume of the sound, but not very much about the tones. It is very similar in vision: there are nuclei that have some visual information that project to the amygdala: the peduncular nucleus and the medial nucleus of the pulvinar. As far as I know there are no face neurons in these regions. My presumption has always been that you may be able to get some kind of visual information, but it won't be very discriminative. It would only be by going through the V1 infratemporal cortex pathway that you could get discriminative facial information in the amygdala. This subcortical pathway has not been well studied for the visual system. I presume that there is some kind of visual information getting in.

Bailey: The interesting thing is that we do see some activity with other classes of object, but the pattern is different and the response doesn't seem to be as great (Bräutigam et al 2001). What we are wondering is what are the minimum facial features that are needed to elicit this activity? The experiments we have just done with Anneli Kylliäinen, in Riitta Hari's laboratory, are to show affected children faces in pairs in which either the eyes are open or shut, and with gaze either averted or straight ahead. We suspect that eyes may well be the critical features.

C. Frith: In your experiment, it is interesting that you only found this with the pairs of images. I wonder whether there is some sort of priming effect.

References

Bräutigam S, Bailey AJ, Swithenby SJ 2001 Task-dependent early latency (30–60 ms) visual processing of human faces and other objects. Neuroreport 12:1531–1536

Corona R, Dissanayake C, Arbelle S, Wellington P, Sigman M 1998 Is affect aversive to young children with autism? Behavioral and cardiac responses to experimenter distress. Child Dev 69:1494–1502

Le Grand R, Mondloch CJ, Maurer D, Brent HP 2001 Neuroperception. Early visual experience and face processing. Nature 410:890

Levin BE 2000 Metabolic imprinting on genetically predisposed neural circuits perpetuates obesity. Nutrition 16:909–915

O'Connor TG, Rutter M 2000 Attachment disorder behavior following early severe deprivation: extension and longitudinal follow-up. English and Romanian Adoptees Study Team. J Am Acad Child Adolesc Psychiatry 39:703–712

O'Connor TG, Rutter M, Beckett C, Keaveney L, Kreppner JM 2000 The effects of global severe privation on cognitive competence: extension and longitudinal follow-up. English and Romanian Adoptees Study Team. Child Dev 71:376–390

Rutter M 2002 Nature, nurture, and development: from evangelism through science toward policy and practice. Child Dev 73:1–21

Final discussion

Rutter: I would like to use this final discussion to pick up on issues that have been touched on earlier which I think it would be useful to pursue. Chris Frith, what sort of role do you see for magnetic resonance (MR) spectroscopy in the near future?

C. Frith: What struck me is not only have we only mentioned this briefly, but also diffusion tensor imaging hasn't been discussed at all. These are both interesting techniques. My feeling is that there needs to be a very good hypothesis before these techniques are used.

Bishop: Could you explain them for the uninitiated?

C. Frith: In principle, MR spectroscopy enables you to look at any substance you like, with reasonable resolution. Diffusion weighted tensor imaging enables you to trace white matter tracts through the brain, although only at the level of rather large bundles. In principle, this could be used to detect severe deviations in connectivity.

Bishop: How easy is it to do this on children?

C. Frith: It is purely structural, so you could sedate them, if all you are worried about is keeping them still.

Dawson: The data I showed on NAA from Friedman and Dager's study (Friedman et al 2002) were obtained by MR spectroscopy on three-year-old children sedated with propofol.

Rutter: It certainly opens up potential avenues.

C. Frith: I suspect lots of studies will be done that won't be helpful until we have some hypotheses.

Rutter: David Amaral, you focused on the amygdala model, but you didn't talk much about other possible animal models for autism.

Amaral: After listening to the discussion about what autism is and isn't, before an animal model can be designed we need a good definition of what we are trying to model. With Pat Rodier's brain stem model, it has always impressed me that it primarily involves problems in cranial nerve nuclei which don't map very well onto the kinds of pathology that Margaret Bauman and Tom Kemper have found. At the moment one issue is to tell the people who are making the models what we should be modelling. We need to know more about the neuropathology of autism. If it turns out that it is orbital frontal cortex, all the molecular advantages of using the mouse will be wasted because rodents don't have much of an orbital frontal cortex. Tony Wynshaw-Boris has been working with the dishevelled mouse, and this has social deficits. It would be nice to know whether the genetic effects of dishevelled are going to be mapped on to what is coming out of human autism. At this point, however, I think animal models are a little premature. Another problem with mice is that even though they have social behaviours, it is a real stretch to come up with a battery of social tasks. For this reason the monkey is a much more attractive model, although it would be problematic trying to do drug trials in monkeys because of the large numbers that would be required.

Lipkin: I am not sure that there will be a single animal model. There are likely to be multiple animal models in a range of species.

Rutter: As I understand it, the basic point that David Amaral is making is that the model could be based on either the hypothesized biological mechanism, or alternatively some phenotypic feature. Either approach is sound but at the moment many of the models seem to be not well tied to either. Ian Lipkin, can I ask you to help me make sure that I have got the message on immunity mechanisms right. You were pointing to the major methodological problems in relation to some of the autism–MMR work. But you were also saying that the notion of the role of infection and immunity in relation to autism really is potentially interesting and requires much further exploration. Have I got the message right?

Lipkin: I think so, and I was very interested to hear Susan Folstein's observation that there were children with malaria who had social deficits. My feeling, after looking at a number of different animal models — ranging from persistent infection with mumps, to bornavirus, LCMV and unpublished work with non-specific stimuli such as LPS — is that there are some final common pathways which can be triggered by a wide variety of pathogens. If these pathways are introduced at an inappropriate time in a particular context, these can have a wide variety of profound effects. The question then is how do we dissect these? We have to find some way of incorporating the genetic background of the host, we need to know something about the point at which they were introduced and we need to find some way to detect footprints of these effects. This last step will be the most complex. How are we going to know what happened to a mother during day 60 of pregnancy, for example? This is why I think some of these prospective cohorts will be so important, and their implications will be far greater than autism. Typically, when we look at these things in microbiology we are looking after the horse has already bolted. We need to do these studies prospectively, because everyone can come up with an explanation *post hoc*. We need to collect our data in an unbiased way.

Bolton: There is an important implication for what you say there: depending on whether your model is one of prenatal or postnatal exposure, you might be looking at completely the wrong cohort. If you are going for 100 000 new born infants, for example, you would be much better off looking at 100 000 expectant mothers.

Lipkin: This is what is actually being done. The collection begins with the first prenatal visit in the first trimester of gestation.

Rutter: On the first day we touched on obstetric complications and largely put them to one side as a direct cause of autism. What we didn't really touch on are placentation affects. These have been argued to be potentially quite important in relation to schizophrenia. Patrick Bolton, do you see this as something that is important to look at in relation to autism?

Bolton: I'm sure that they are important, but again there is a big problem getting data of this kind to examine. Tony Bailey referred to this earlier: these data are not collected ordinarily. I don't know whether in the Twins Early Development Study (TEDS) sample there is better information about placentation.

Bishop: The only perinatal data come from retrospective report, so no specific information is available on placentation in that study.

There certainly haven't been any other efforts to do anything above what is normally done in hospitals, so there are no data.

Bolton: This is quite a hurdle. People have speculated about the significance.

Bailey: Presumably the way to do this is through prospective cohorts.

Bishop: The base rate of autism is so low there is no way you could do this.

Folstein: You could look at disorder in general.

Lipkin: We are still in the process of putting together a large prospective cohort study. 15 000 people have been entered already. The goal is to recruit 100 000. We can collect new data from the 85 000 remaining subjects; however, it is difficult to go back and collect additional information from those already interviewed.

Rutter: David Skuse, we came back to your notion of an imprinted locus. I wasn't sure that I understood exactly where we are on that. In terms of what you were saying, this does not seem relevant in relation to your imaging findings. Is this right?

Skuse: In so far as the structural brain imaging analysis has gone, we haven't found any differences between the two X-monosomic groups regarding the parental origin of their single X chromosome. This is not something we have examined yet from the perspective of functional MRI, so subtle differences may exist. The neurophysiological and cognitive studies relating to the detection of fear in a face, which appears to be deficient in a large majority of the monosomic subjects, show there is no better performance among the monosomics with a paternal X than those with a maternal X. This was not our prediction, because we have also shown that this ability to detect fear correlates quite nicely with the measurement of theory of mind that we use (Castelli et al 2002).

Rutter: You said that follow-up data showed a similar difference at a later age. However, has the original finding been replicated in another sample?

Skuse: We have shown that the parental origin of the single X chromosome in X-monosomy, which distinguished those with a paternal X from those with a maternal X in childhood, continues to distinguish them in adult life in terms of self-rated social adjustment. However, this finding has not been replicated yet. I don't know of anyone else who has managed to recruit a sample of X-monosomics which is large enough to facilitate this comparison. The key thing is that whoever does try to replicate it should use exactly the same measures that we use. In so far as I have heard anyone talk about an attempt to replicate our original finding, they haven't used the same measures and nothing has been published.

We also have an intriguing finding that there is some X-linked imprinting effect on the relationship between recognition of emotion and face recognition memory. These things are highly correlated in normal females and Turner syndrome females with a single paternal X. They are not at all correlated in normal males (whose X chromosome is maternally derived) or in X-monosomic females with a single maternal X.

I should also say that there are data from our collaborators at the Babraham Institute (Isles & Wilkinson 2000) who have been studying the X monosomic mouse (39,X). For the last few years we have been looking at whether or not there is a behavioural phenotype for this mouse, and whether this is affected by parental origin of the X. We have utterly convincing data that there is an X-linked imprinting effect on behaviour in the mouse, and furthermore that this probably affects behaviour through influencing dopamine in the frontal cortex of the mouse.

Bishop: When you are studying children, whether or not they participate in your studies is largely a function of their parents. When you are studying adults it requires them to agree to come in and have brain scanning, for example. People who are not socially competent are likely to exclude themselves from such studies. Thus, an adult sample is not the ideal one to be looking for these differences, because the recruitment methods will screen out the people with the social problems.

Rutter: Let me raise one last issue of a somewhat different kind. We talked the other day about age of manifestation features, and what needed to be explained. What we haven't said much about is the phenomenon of regression. It seems well documented that in perhaps a quarter of cases there is regression. What does this mean and how should it be tackled? The parallel question is that we have good evidence that there are many children with autism who do show abnormalities quite young—Gerry Dawson's home movie studies would be an example. But equally, there is convincing evidence that there are other children who are perfectly OK for the first 18 months or so. What is the implication of this difference and how might this be tackled?

Lord: On the basis of data from several groups, it is clear that there are two separate issues. That is, when are abnormalities first present, and is there a regression? It is becoming clearer that for many children abnormalities are present fairly early on. Professor Dawson spoke about their first clear documentation, of differences at 8–10 months. Many parents report that their children did not meet milestones early on. Separately, about 20% of children do show a clear loss of words. However, if you go through a list with parents of skills that their children had for a month or so and then lost, you get much higher rates of loss. 60% of parents report that their children had some skills that then disappeared. In the past, we have interpreted regression as meaning that a loss occurred after normal development, which does not seem to occur very often. On the other hand, people have interpreted findings of early abnormalities as meaning that regression does not occur, which is not true either. We have to explain two phenomena: first, why are some children with autism different from the start and some not, and second, why are some children with skills losing them?

Charman: There are two studies relevant here. One is the BABYSIBS study, taking place across a few sites in Canada and the USA, where they are seeing from the first few months of life siblings of children who already had a diagnosis, who are thus at higher genetic risk of developing autism. This is going to provide a way for us to see children prospectively, and will enable us to look at regression systematically, albeit in a relatively small sample.

Sigman: I don't know much about the other sibling studies in the US and Canada. Simon Baron-Cohen, Chris Gilberg, Nurit Yirmiya and I are involved in a small prospective study of infant siblings. Of the 60 infant siblings we followed, we have four who have become autistic. In such a small group it will be hard to make much of a comparison, but an interesting aspect is that the infant siblings of the autistic children are delayed in non-verbal communication at 14 months and 24 months, and they are delayed at language at 24 months and 36 months. In fact (and this is startling to me) 50% of my 20 infant siblings were delayed more than 4 months in language skills when they were 24 months of age, and a third were still delayed more than 6 months when they were 36 months. I expect that these children will grow out of these delays, because we know from the studies of older siblings that they don't look like they are delayed in verbal abilities. Whether or not we will see other problems later on in other areas, I don't know. Michael's studies showing that children who have language delay early have later social difficulties suggest that we will see some kinds of problems if we keep following these children. This sort of research is very important, and several of us in this room would like to do this kind of study with much larger samples. These first findings are surprising: I didn't expect to see this kind of delay in our infant siblings.

Bishop: In language-impaired children in general, later-born children are typically slower to develop (e.g. Bishop 1997). The effect is more apparent in big families. Part of the explanation for this is that later-born children would get less input from the mother and more from other siblings. It occurs to me that this might not be the case if you have an autistic older sibling, but it might be that your mother is preoccupied in this case. It could be an effect of not getting so much good maternal input.

Bailey: Regression is interesting because it has raised in the minds of parents and the public the notion that there might be two sorts of autism: autism that you are 'destined' to have from birth through genetic effects, and a separate sort that is environmentally determined. One way that we have looked at this is in the multiplex IMGSAC sample where the rate of regression (defined as having lost at least language) is 27% (Parr et al 2002). This is a virtually identical rate to all the published studies in singletons. Moreover, there is no concordance in pairs for regression. By this I mean that if you regress you are as likely to have a sib (who presumably has the same genes) who didn't. Our interpretation of this is that regression may be an emergent property of a genetic process, and that this is not an environmentally determined event. But like everything else in autism, we don't seem to be able to predict which particular pathway individuals will follow. It is also a potentially rather handicapping pathway. The group who regressed interestingly acquired language significantly earlier than the children who didn't regress, but their final outcome was significantly worse. They have lower performance and verbal IQ and lower Vineland scores. This raises the possibility that the early language acquisition may be by an abnormal mechanism, as opposed to loss of normal function. We have little idea about the underlying biology, but one possibility is that this reflects the turning on or off of pathways.

U. Frith: We do need some hypotheses about what is happening in regression. Such hypotheses should be guided by what we know about normal brain development. But detailed tracing of normal brain development has hardly been done. So far, we know that there are some dramatic changes in the first two years when one brain system is superseded by another. For example, there are subcortical brain systems that respond to faces very early on which are later superseded by cortical brain systems coming on line. We can observe a walking reflex in newborn infants, but proper walking much later, presumably when the motor system has sufficiently matured. I suspect that the same will hold for language. Infants respond to language from right after birth just as they do to faces before the language areas themselves have matured. One type of 'word' learning, by association of sound and sight, might be available very early. This might be responsible for the acquisition of a few early 'words'. This might be an early talking reflex, so to speak, if we make an analogy to the early walking reflex. It is well accepted that association learning is not enough to sustain language learning. Language learning proper begins from around 12 months, presumably when some critical learning mechanism has matured. Paul Bloom suggests that this mechanism is to do with the 'theory of mind mechanism' which we believe is faulty in autism. Through this mechanism infants automatically understand that human beings have intentions and learn words by tracking these intentions (Bloom 2000).

Children may lose any early reflexive 'words' if they had them, as they only functioned as specific stimulus-response pairs. One explanation of regression in autism would be that it signals the failure of the new system to come online, but where the previous system has been working rather well.

Happé: How much language do these children ever have before they lose it?

Lord: Not much. Fewer than 10 single words, typically.

Bolton: Moreover, the first words that they use are sometimes rather unusual.

Happé: This is important, because it gives clues as to what sort of system you are dealing with.

Lord: Then the question is why do only some children do it?

Happé: I suspect that is true in the normal population also: some children start on an associative route and then come in with the more intentional language learning.

Dawson: I was going to mention some data that we have in the sample of children that we have been referring to. We have been studying 75 young children with autism, and we have looked to see whether they show regression or not. We have looked at regression in three different ways. First, using the Early Development Interview, a parent interview aimed at early development that incorporates mnemonic devices for helping the parent remember events early on, to try to get more reliable information about early development. Second, we used the ADI regression items, and third, we used home videos. We looked to see whether at age 3 there were any differences on several neurocognitive and standardized language and social measures between children with and without a history of regression. Surprisingly, there were not. We could find no differences at age 3. It may of course be that eventually their course will be different. I thought that the regressed children would be more severely impaired.

Bailey: The average age of our sample when they were assessed was about 10–12, but they were significantly different from non-regressors on both performance and verbal IQ and on Vineland social and daily living scores.

Lord: One of the things that we have found longitudinally was that the children who had the best outcome were those who had a regression that was relatively short, at least as described by their parents at age 2 years. When we gave the same parents the ADI with a blind investigator at five years, a significant minority of these parents did not report the regression. Some of this is recall bias.

Folstein: I want to follow up on what Uta Frith just said about different systems coming in. In the children whose parents I interview in some detail about language, it is almost always the case that the words that are lost are imitative. They are not really what I call words. They are always in response to a cue. This regression happens right at the time when you usually see language that the child initiates. This doesn't kick in in these children. I never think of this as regression. What we hear of once in a while is a child who not only does that, but had been socially pretty responsive until a certain point and then regresses socially and starts to do repetitive behaviours. To me, that is regression.

Lord: The two are associated. In children who lose words, the parents also almost always report loss of social skills as well. I think you are correct that word loss is the easiest event to get parents to consistently report, but it is probably wrong to focus on it alone. The social losses are much more pervasive and more common, it is just much harder to get that information.

Folstein: What I was also getting at is that this language change happens more often than this other more pervasive loss of social skills. If we focus mainly on the language we might be over-calling regression.

References

Bishop DVM 1997 Pre- and perinatal hazards and family background in children with specific language impairments: a study of twins. Brain Lang 56:1–26

Bloom P 2000 How children learn the meanings of words. MIT Press, Cambridge, MA

Castelli F, Frith C, Happe F, Frith U 2002 Autism, Asperger syndrome and brain mechanisms for the attributions of mental states to animated shapes. Brain 125:1839–1849

Friedman SD, Shaw DWW, Artru AA et al 2002 Regional brain chemical alterations in young children with autism spectrum disorder. Neurology 60:100–107

Isles AR, Wilkinson LS 2000 Imprinted genes, cognition and behaviour. Trends Cogn Sci 4:309–318

Parr J, Baird G, Le Couteur A, Rutter M, Bailey A and The International Molecular Genetics of Autism Consortium (IMGSAC) 2002 Phenotypic characteristics of autistic regression in an international multiplex sample. Poster presented at The 10th World Congress on Psychiatric Genetics, October 2001, Brussels

Summing-up

Michael Rutter

Social, Genetic and Developmental Psychiatry Research Centre, Institute of Psychiatry, De Crespigny Park, Denmark Hill, London SE5 8AF, UK

Rise in the rate of diagnosed autism

Let me bring the meeting to a conclusion by looking back at some of the questions I raised in my introduction, in order to seek to draw out some inferences and implications. It was agreed that, in large part, the rise in the rate of diagnosed autism is a consequence of a combination of better ascertainment and a broadening of the concept. Nevertheless, it is not possible to rule out the possibility that there has been a real rise in the incidence of autism as well. No-one thought that this could be resolved by any kind of any further going back into records or indeed even by new epidemiological studies. It would be worthwhile to have further epidemiological studies in order to investigate the possible role of specific postulated environmental risk factors but there is a paucity of good leads at the moment and it may well be that genetic evidence will be needed in order to identify environmental risks more clearly. The hypothesis that the combined mumps/measles/rubella vaccine was responsible for the rise in autism is an interesting example in that it postulated, albeit on a totally inadequate basis, a specific cause. It is clear that the epidemiological evidence that has become available over the last few years provides no support for the suggestion that measles/mumps/rubella vaccine (MMR) has caused a large overall rise in the rate of autism. On the other hand, a role for MMR in a very small number of individual cases who have an unusually susceptibility, cannot as yet be firmly ruled out, although most people were very sceptical in view of the lack of any convincing pointers that this might be the case.

Breadth of phenotype

There was universal agreement that the clinical manifestations of autism were indeed very broad, and that both empirical research and theory would have deal with that breadth. We noted the observation that the broader phenotype (meaning the milder manifestations outside the range of the traditional diagnostic criteria for a handicapping disorder) was unassociated with either mental retardation or epilepsy, despite the strength of both associations with the traditional diagnosis. So far, research has been uninformative on the mechanisms involved in the transition from this broader phenotype to the more severe handicapping disorder. It is possible that some form of two-hit mechanism is implied but that did not automatically follow and it could be a consequence simply of variations in the severity of the underlying genetic liability. There was only brief discussion of the phenomenon of regression in relation to the onset of manifestations of autism but it was agreed that this was another phenomenon that still awaited explanation. It was appreciated that the concept of regression remains rather ill-defined and, almost certainly, it covers a range of somewhat different features. Thus, at one end, there is the frequent story that the children had gained a small amount of spoken language (perhaps half a dozen words), which were subsequently lost. It remains uncertain whether this represents any real loss of language. However, there certainly are cases of children whose early development seemed unambiguously normal who, usually about mid-way through the second year of life, do have a more marked loss of language that is associated with a parallel loss of previously acquired skills in play, together with a parallel loss of social responsivity. This appears to constitute a more valid concept of regression but, again, its meaning remains obscure. In addition, there is the even less common phenomenon in which the children lose not only language, play and social skills but also suffer a more general setback in development. This has been termed a disintegrative disorder and it is clear that it has a poor prognosis, although it remains unknown whether or not it is aetiologically distinct from autism.

Originally, autism had been conceptualized as a distinctive disorder that involved several different features that were intrinsically associated as part of the overall syndrome. In recent years, questions have been raised as to whether or not there might be several different traits with different origins, with the syndrome being simply the coincidental coming together of these separate traits. Such traits could involve either behavioural features or different cognitive deficits or both. We had some discussion of how common each of these traits would need to be in order for this to give rise to a syndrome with the frequency of autism as found in epidemiological studies. On the face of it, each of the traits would have to have a frequency that seemed implausibly high. On the other hand, it is possible that synergistic interaction amongst traits could provide an explanation and we considered how that might be tested. The main emphasis was on the great need for general population epidemiological studies that included systematic assessment of these different components and not just the syndrome of autistic spectrum disorders as a whole. For obvious reasons, the assessment of the components would need to be dimensional as well as categorical. We noted the potential for this purpose of the Utrecht study, the TEDS project and the very large scale Norwegian study. Although none were explicitly designed for this purpose, relatively modest modifications would enable them to be used to examine the concept of separable components.

Several papers considered different models for the conjunction of features, asking whether it was likely that autism could be reduced to just one single causal deficit at either the cognitive or the neural level. Clearly, such a parsimonious solution was desirable in principle but most people doubted that it would prove to be the valid model. The use of a probabilistic pleiotropic model (meaning one in which genetic liabilities gave rise to a varied range of outcomes and, in which, the effects of the liabilities were probabilistic, rather than deterministic — substantially increasing the risks of autism but not leading directly to it as such) seemed plausible but numerous questions remain to be addressed.

The discussion of the possible coming together of separate traits also played a substantial part in the collection of papers on genetics. It was agreed that it would be useful to split up samples of individuals with autism spectrum disorders according to different patterns. This might be either with respect to biological features (such as head size or serotonin levels or the presence/absence of congenital anomalies) or clinical features (such as level of language or IQ or behaviour patterns). So far, such splitting up of samples has proved of rather limited value and there are undoubted problems of statistical power. Nevertheless, there was agreement that it was important that genetic studies used dimensional, as well as categorical, approaches and that they should focus on different components as well as on the syndrome as a whole.

Age of first clinical manifestation

There was agreement that the finding that many children with autism spectrum disorders showed abnormalities and social responsiveness by the age of 12 months, and certainly by 18 months, meant that the cognitive hypotheses had to accommodate to this. The proposed mechanisms would have to involve some form of processing that was present at an early age in normal children. There was general doubt as to how far it was possible to separate socio-emotional responsivity and cognitive processing at such an early age. The general population epidemiological/longitudinal studies could be informative but only in so far as the measures included cognitive assessment as well as the reporting or observing of behaviour. It was necessary also to accommodate to the finding that some autistic children showed no evidence of abnormalities until the age of about 18 months. It is not at all clear if they are in any way different in any basic sense from those showing abnormalities in the first year of life. Also, there was the issue of how to accommodate the fact that Asperger's syndrome tended not to give rise to manifestations that were observed and recognized by parents and professionals until even later than that. It was agreed that, although this was a very important group to study, it would be a mistake to regard Asperger's syndrome as the 'pure' condition that should be studied to the exclusion of others. Also, if the cognitive deficits were indeed basic to the social abnormalities, it was not clear why all children with severe mental retardation did not show autism. It may be that this is because they lacked the postulated relevant skills, with autism resulting only when there was the bringing together of both cognitive deficits and cognitive assets. However, the notion that there needed to be cognitive peaks as well as troughs has not, as yet, been subjected to systematic study.

There was general acceptance of the potential value of investigations of atypical populations of various kinds who showed autistic manifestations of one kind or another. It is noteworthy that it appears that individuals with specific developmental disorders of language tend to appear more socially impaired and, in some respects, more like autistic individuals as they grow older despite the apparent lack of autistic features when young. Given that their language impairment markedly diminishes as they grow older, it is curious that the social abnormalities become more evident. By contrast, the autistic-like features seen in children who suffered profound institutional deprivation in infancy seem to become less marked as they grow older. It might be thought that this is what might be expected given their general improvement but it is not at all obvious why the developmental trend appears to be in the opposite direction to that found with children showing specific language impairment.

In discussing cognitive processing, attention was drawn to the much increased leverage that is potentially obtainable through functional imaging (despite the fact that many methodological issues need dealing with). Imaging enables a focus on the possibility of individuals with autistic spectrum disorders performing tasks by cognitive processes that are different from those normally employed. It is possible that compensatory mechanisms may prove to be very important. We noted the considerable potential value of experimental approaches to functional imaging studies using cognitive tasks, in order to differentiate cause from consequence. Clearly this does constitute an important way ahead.

Genetics

With the exception of Alzheimer's disease, progress in the molecular genetic study of psychiatric disorders has been disappointingly slow. However, although no susceptibility gene for autism has been identified as yet, there was general optimism that derives from the fact that there are several partially implicated loci. On the whole, molecular genetic research is in a rather better state than in some areas of psychiatry. Moreover, there was general agreement on the range of molecular genetic research strategies that are going to be needed in order to identify susceptibility genes. There was a recognition that this must take into account the phenotypic research developments discussed above, and that a range of different molecular genetic methods will be needed. Currently, there is an enthusiasm in behaviour genetics for association studies but attention was drawn to the important implications of whether susceptibility genes for autism constituted common or rare variants, with the implication that association studies are less satisfactory for picking up rare variants. On the other hand, association studies have to be part of the overall research strategy. Similarly, there will have to be a combination of categorical and dimensional approaches, with continuing uncertainty as to the relative merits and demerits of each. The quantitative genetic findings suggest the likelihood of synergistic interactions among susceptibility genes and it will be important to search more systematically for epistatic effects once one or more susceptibility genes have been identified. Several participants emphasized the potential value of looking at familial clustering as part of molecular genetic, as well as quantitative genetic, strategies and also the value of examining discordant monozygotic pairs. It is curious that the sib-sib correlation for social communicative deficits was near zero, although there were modest positive correlations for other features and perhaps especially for epilepsy. Again, that constitutes a research strategy well worth further exploitation.

In considering the marked male preponderance for autism spectrum disorders, we noted the hypothesis of an imprinted gene on the X chromosome that could be responsible for an increased male liability to autism spectrum disorders. This has not been replicated as yet in an independent sample but it seems likely that, if relevant, it operates in all males rather than through variable allelic transmission. It cannot explain father to son transmission, which is evident in autism as in other neurodevelopmental disorders, and the pattern of deficits in girls with a missing X chromosome suggests that it is unlikely to explain the male preponderance across the entire range of neurodevelopmental disorders. Although it would be premature to rule out the possibility of risk or protective genes on the X chromosome, the findings to date do not suggest that it is likely that these will provide an answer to the question of why autism spectrum disorders are so much more frequent in males. Epigenetic effects remain a possibility and it is possible that prenatal hormones interact with susceptibility genes, but this possibility can be investigated more satisfactorily once susceptibility genes have been identified.

Structural brain abnormalities

Despite the development of greatly improved technologies, and despite the existence of good research, considerable uncertainties remain on what might constitute the neural basis for autism. There is a general recognition of the importance of considering possible developmental differences in findings. Uncertainty remains on the extent to which the evidence on the existence of such differences is solid and replicated but clearly it is necessary to examine the issue both because a failure to do so will introduce unwanted methodological noise, and because it may throw light on the processes involved in postnatal neural development in autism. The monkey findings in relation to the somewhat different effects of amygdala lesions in the neonate and in the older monkey are illuminating even though amygdalectomy does not seem to be an appropriate animal model for autism. The studies raise important queries as to what processes the amygdala deals with and how this tallies with the somewhat different imaging and post-mortem studies. The plans to combine lesion studies with imaging may well be very helpful. However, probably what came through most clearly from our discussions was the value, indeed the essential need, to combine research strategies — both conceptually, of course, but also practically as part of the basic research design. There is going to be the need for cross-disciplinary barriers if we are to succeed in our study of the structural and functional abnormalities of the brain associated with autism.

Psychological deficits

Functional brain imaging studies in autism are still in their infancy but they provide a considerable potential for understanding better the nature of psychological deficits. In particular, they provide evidence of when individuals with autism are using different strategies to deal with the same tasks tackled by normal individuals in other ways. Studies of face processing have proved very promising in this connection in suggesting that individuals with autism, unlike normals, use a feature-based strategy.

Ordinarily, the brain functions as a predictive system in which the processing of new information is much influenced by the apparent meaning of the stimuli, by the overall gestalt, by past experiences, and by expectations. It seems that individuals with autism do this to a lesser extent or do it in a somewhat different way. Theory of mind concepts have been enormously helpful in providing possible leverage on the question as to how cognitive deficits might underlie social abnormalities in autism. However, such deficits may not be universal in autism, may not be primary (both in the sense that the theory-of-mind deficit may derive from other cognitive features and in the sense that the early social abnormalities must be based on precursors of theory of mind rather than theory-of-mind deficits as such). Also, at least as presently understood, theory-of-mind deficits seem unable to explain the non-social features of autism. Deficits in executive functioning may not be specific to autism, similarly may not be primary, and appear not to explain the superior performance of individuals with autism on some tasks. The concept of central coherence has added a useful alternative approach but currently it is underspecified. The evidence suggests that individuals with autism can process at a global level but they appear not to do so in ordinary circumstances. It seems that compensatory strategies, as well as cognitive deficits may be important in the development of both the social and non-social aspects of autism spectrum disorders.

Infections and immunity

There is a powerful case for the value of considering infection and immunity mechanisms in autism as well as in other conditions and we were reminded of the several well-documented examples in which findings have completely changed concepts of particular diseases. Peptic ulcer constitutes a particularly striking example but there are quite a few others. At the same time, we need to be aware of the many methodological hazards that have to be dealt with and of the ever present need for independent replication of findings. Because of doubts on both points, it is evident that the claims that the MMR vaccine may have induced autism in some susceptible individuals are premature, and public debate in relation to MMR and its supposed association with a type of enterocolitis accompanied by autism has tended to focus on whether or not the virus from the vaccine can be identified in the gut tissues. Doubts remain on the validity of the claimed findings but, even if confirmed, they would leave in considerable doubt the further claim that the connection is causal. The vaccine-induced lifelong immunity to measles implies persistence of a virus, or of its products, in the body. It would be expected that these might well settle in malfunctioning tissues. Other research strategies would be needed to test the causal hypothesis. Nevertheless, the notion of possible involvement of the immune system in autism remains one that should be kept firmly on the research agenda, irrespective of the numerous doubts that surround the MMR hypothesis.

Pharmacological interventions

There is a striking difference in the demonstrated effects of drug treatments on core and non-core symptoms in autism. There is no consistent evidence of marked benefits from any drug with respect to socio-communicative deficits. It is perhaps too early to rule out completely the involvement of serotonin or dopamine systems in autism but the generally negative findings do suggest the need to look in other directions. It is to be hoped that molecular genetic findings may provide the much-needed leads in this connection. By contrast, however, drug treatments have had worthwhile beneficial effects in relation to the range of handicapping, but non-specific, symptoms associated with autism. It is possible that the beneficial effects on these associated features may bring some longer term benefits for socio-communicative functioning, although that has yet to be demonstrated. It was concluded that drug treatments constitute a worthwhile part of the overall therapeutic armamentarium to be employed in the treatment of autism but, so far, drug treatments have not been at all helpful in casting light on causal mechanisms.

Psychological interventions

There is much consistent evidence, albeit mostly from research using imperfect designs, that there are both short-term and long-term benefits from appropriate developmentally oriented behavioural interventions. Their value is now well accepted and there is every reason to expect that they ought to be generally available as part of any adequate service provision. On the other hand, doubts remain on several crucial issues. It has been argued that early interventions can enable a substantial minority of children with autism to attain normal, or near normal, functioning. Clearly, if that were true, it would have very considerable theoretical and practical implications, but the study needed to provide a rigorous test of the claim has yet to be undertaken. Secondly, common sense suggests that starting treatment early ought to be more effective than starting treatment later but, although there are pointers suggesting that this may well be the case, the evidence is less solid than some proponents of early treatment suggest. Third, strong claims have been made that, if early treatment is to be effective, it must be intense—extending over some 30 to 40 hours per week. Again, the evidence that this is so is rather weak. Of course, much depends on what is included in this number of hours. Does it include, for example, high quality preschool education or does it mean just very specialized individual treatment? Fourth, are the treatment approaches that are optimal the same at all ages for children with all degrees and patterns of handicap? It seems unlikely that this would be the case. But, so far, research has not identified any solid set of features that might determine individual differences in need. Fifth, we have yet to determine what are the crucial elements in effective psychological interventions. Equally, much has still to be learned on the factors associated with the very wide range of outcomes in autism. It is obvious that, to a very important extent, these are a consequence of differences in the extent and pattern of the child's initial handicaps but it is possible that treatment features or features of parent–child interaction may play an important contributory role.

It is obvious that there are considerable difficulties in undertaking the necessary evaluations of psychological treatments in autism because of the difficulties in accumulating sufficiently large samples, the problems in using randomized controlled trials, the need for long term follow-up, and the importance of individual differences. Nevertheless, it would be possible to put the key therapeutic claims to test in a much more rigorous fashion than has been done so far. That remains a high priority.

Finally, we considered the possible effects of both psychological and pharmacological interventions on neural development. Our attention was drawn to the likely importance of biological programming for some aspects of neural development and these might prove to be more important in autism than has been appreciated so far. At present, there has to be reliance on speculation but what is obvious is that there needs to be a coming together of biological studies and intervention research.

Implications for the future

Although it is all too apparent how much has still to be learned about autism, there is considerably more agreement on the basic concepts than was the case a few decades ago. Moreover, there is a considerable level of agreement on what needs to be done to provide the answers to the key questions. Discussions during this symposium have had only a modest success in resolving the paradoxes and challenges with which the meeting started but we may reasonably claim that there has been more success in identifying the key features of the research agenda that lies ahead of us.

Index of contributors

Non-participating co-authors are indicated by asterisks. Entries in bold indicate papers; other entries refer to discussion contributions.

L

M

P

R

S

T

Z

Subject index

D

T